Inspection Manual for Highway Structures

Volume 1: Reference Manual

London: TSO   May 2007

TJO

Published by TSO (The Stationery Office) and available from:

Online
www.tsoshop.co.uk

Mail, Telephone, Fax & E-mail
TSO
PO Box 29, Norwich, NR3 1GN
Telephone orders/General enquiries: 0870 600 5522
Fax orders: 0870 600 5533
E-mail: customer.services@tso.co.uk
Textphone 0870 240 3701

TSO Shops
123 Kingsway, London, WC2B 6PQ
020 7242 6393 Fax 020 7242 6394
16 Arthur Street, Belfast BT1 4GD
028 9023 8451 Fax 028 9023 5401
71 Lothian Road, Edinburgh EH3 9AZ
0870 606 5566 Fax 0870 606 5588

TSO@Blackwell and other Accredited Agents

ISBN 978 0 11 552797 5

Printed in Great Britain on material containing a minimum of 75% post-consumer waste and the remainder ECF or TCF pulp.

N8830951 C15 05/07

May 2007

This Manual is supported, endorsed and recommended by

HIGHWAYS AGENCY

County Surveyors' Society

LoBEG

BRIDGES

UK BRIDGES BOARD

TRANSPORT **SCOTLAND**

ROADS Service

An Agency within the Department for
Regional Development
www.drdni.gov.uk

Llywodraeth Cynulliad Cymru
Welsh Assembly Government

Full details of the Project Sponsor, the Technical Project Board and the Project Team are provided in the Acknowledgements section on page F-71.

Foreword

by Ginny Clarke, Chief Highway Engineer – Highways Agency

Inspections play a vital role in the effective management of highway structures ensuring that our highway assets are safe for use and fit for purpose. This document, comprising two complementary Volumes, has been produced to assist engineers and inspectors with this task and draws together a wealth of experience and knowledge reflecting current best practice in the United Kingdom, superseding the Bridge Inspection Guide 'Yellow Book', which has served us well for many years.

The manual reflects a very successful collaboration with other highway authorities, owners, interested bodies and the supply chain and I would like to thank all those who have contributed to its drafting, review and publication.

Ginny Clarke
Chief Highway Engineer – Highways Agency

Contents

Part D: Defects Descriptions and Causes D-1

List of Figures

List of Tables

Summary

It is widely recognised that a well managed transport infrastructure is vital to the economic stability, growth, and social well being of a country. Bridges and other highway structures are fundamental to the transport infrastructure because they form essential links in the highway network. They are relied upon to remain in service year after year, carrying ever-increasing traffic flows, while being managed in a manner that ensures they are Safe for Use and Fit for Purpose.

The majority of highway structures are managed and maintained at public expense. It is therefore important that their management minimises disruption, risk and consequential costs to road users, and makes economic and efficient use of resources. Fundamental to effective management is an inspection regime that provides timely, accurate and appropriately detailed information on asset condition and performance. The overall purpose of inspection, testing and monitoring is to check that highway structures are Safe for Use and Fit for Purpose and to provide the data required to support effective maintenance management and planning.

The purpose of this Inspection Manual is to provide guidance on the inspection process for all staff involved in the management of highway structures. It is also considered that this Manual provides a sound basis for the development of formal inspector training courses.

This Manual aims to ensure that inspections are carried out efficiently, uniformly and to a high standard. The Manual is intended for use as guidance, outlining typical procedures and defining the normal requirements for the various categories of inspection. It is not intended to provide the definitive solution in all situations, as the party best able to decide on the appropriate course of action is the inspector or engineer undertaking the work.

The Manual is divided into two separate but complementary Volumes. Each of the two Volumes is sub-divided into several Parts that provide guidance on a wide range of issues as listed below.

- **Volume 1: Reference Manual** – covers all aspects of highway structures inspection that both inspectors and engineers should be aware of.

    o *Part A. Introduction and Implementation*: Provides an overview of the Manual; includes an introduction to the role and need for the Manual, describes the purpose, objectives and scope of the Manual and how it should be implemented. It also offers an overview of the inspection regime and summarises the general competence and training requirements for inspection staff.

    o *Part B. Behaviour of Structures*: Provides inspectors with an overview of the engineering topics that they should be aware of and offers a consistent basis for inspector knowledge development and training. It contains a summary of topics such as structural mechanics, structural materials and their properties, common types of highway structures and structural elements. It aims to provide organisations with a check list of criteria they should include on training courses but does not provide detailed coverage of the topics.

- *Part C. The Inspection Process*: Provides general information on scheduling inspections, planning and preparing for inspections, access considerations, health and safety aspects, and advice on performing inspections on differing types of structures.

- *Part D. Defect Descriptions and Causes*: Provides background information and guidance on describing and categorising defects. It describes the principal defects that are likely to be encountered in concrete structures, metal and metal/concrete composite structures, masonry structures, and structures built from other materials, with the emphasis placed on identification and likely causes.

- *Part E. Investigation and Testing*: Summarises a wide range of testing methods available for highway structures, including tests for material properties, defects and causes of defects. This aims to make inspection staff aware of the tests that may be used to inform management.

- *Part F. Appendices*: Contains a selection of information which collectively support the guidance provided in Parts A to E.

- **Volume 2: Inspector's Handbook** – acts as a quick reference for inspectors on site.

  - *Part A. Inspector's Guide*: Highlights the key points from Volume 1 that inspectors should be aware of before, during and after undertaking iinspections. Cross-references to Volume 1 are provided, enabling inspectors to quickly look up more detailed advice/guidance as and when required.

  - *Part B. Defect Photographs*: Provides a library of photographs illustrating some of the different types of defects that are likely to be encountered on highway structures.

Acronyms and Abbreviations

AAR	Alkali-Aggregate Reaction
ACR	Alkali-Carbonate Reaction
AQMA	Air Quality Management Areas
ASR	Alkali-Silica Reaction
BA	Breathing Apparatus
BRE	Building Research Establishment Ltd
BS	British Standard
CAWR	The Control of Asbestos at Work Regulations
CDM	The Construction (Design and Management) Regulations
CSBS	Corrugated Steel Buried Structure
CSS	County Surveyors' Society
DMRB	Design Manual for Roads and Bridges
EHD	Environmental Health Department
FRP	Fibre Reinforced Polymer
GPR	Ground Penetrating Radar
GPS	Geographical Positioning System
HAZ	Heat Affected Zone
HSE	Health and Safety Executive
M&E	Mechanical and Electrical
MCDHW	Manual of Contract Documents for Highway Works
MEWP	Mobile Elevating Work Platform
PPE	Personal Protective Equipment
PRoW	Public Right of Way
SNCO	Statutory Nature Conservation Organisation
TSA	Thaumasite Sulphate Attack

Part A
Introduction and Implementation

This Part introduces the Manual, provides an overview on the purpose of inspections of highway structures and establishes the need for an overarching consistent approach. The overall objectives and scope of the Manual are presented as is an overview of the recommended inspection regime. The general competence and training requirements for inspectors are outlined and the different parts of the Manual summarised.

1   Background

1.1   THE ROLE OF HIGHWAY STRUCTURES

1.1.1   Highway structures represent a significant national investment, with most being publicly owned and many being prominent features in the local environment [1]. In the UK the inspection and maintenance of highway structures is undertaken by a variety of owners or agencies, e.g. local authorities, Trunk Road Agencies, Network Rail, BRB (Residuary) Ltd, Environment Agency, British Waterways, London Underground, Transport for London and many private owners. In this Manual they are collectively referred to as 'owner' or 'authority' as appropriate.

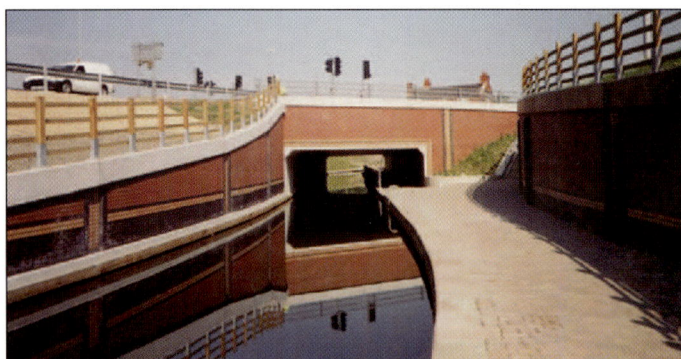

Highways Agency

1.1.2   Bridges and other highway structures are fundamental to the transport infrastructure because they form essential links in the highway network [1]. They are relied upon to remain in service year after year, and are carrying ever-increasing traffic flows. The *Highways Act* [2], *The Roads (Northern Ireland) Order 1993* [3] and *The Roads (Scotland) Act 1984* [4] place a statutory obligation on highway authorities to maintain the public highway. This has been interpreted in the *Code of Practice* [1] as embracing the two essential functions of *Safe for Use* and *Fit for Purpose*, where these are defined as:

- *Safe for Use* requires a highway structure to be managed in such a way that it does not pose an unacceptable risk to public safety.

- *Fit for Purpose* requires a highway structure to be managed in such a way that it remains available for use by traffic permitted for the route.

1.1.3   The majority of highway structures are managed and maintained at public expense. It is therefore also important that their management minimises disruption, risk and consequential costs to road users and makes economic and efficient use of resources. Fundamental to effective management is an inspection regime that provides timely, accurate and appropriately detailed information on asset condition and performance.

1.1.4   A key aspect of any inspection regime is the personnel involved. Guidance on the training and qualifications required by the inspectors and the Supervising Engineer are given in Section 4. In this Inspection Manual (hereafter referred to as the Manual) the term 'Inspector' is used to describe the personnel who actually carry out inspections irrespective of whether they are engineers, technicians or staff with other qualifications/experience. The term 'Supervising Engineer' is used to describe the engineer who is in charge of the inspection regime.

1.2 THE PURPOSE OF INSPECTIONS

1.2.1   As set down by the *Code of Practice* [1], the overall purpose of inspection, testing and monitoring is to check that highway structures are safe for use and fit for purpose and to provide the data required to support effective maintenance management and planning. Inspections, and where required testing and monitoring, should:

- Observe and provide information on the current condition, performance and environment of a structure, e.g. severity and extent of defects, material strength and loading. This enables the safety, functionality and durability of structures to be assessed, and provides sufficient information for actions to be planned where structures do not meet these requirements.

- Inform analyses, assessments and processes, e.g. change in condition, cause of deterioration, rate of deterioration, identification and quantification of maintenance needs, effectiveness of maintenance and structural capacity. This informs management planning and enables cost-effective plans, which deliver the required performance, to be developed.

- Compile, verify and maintain inventory information, e.g. structure type, dimensions and location, for all the highway structures the authority is responsible for.

1.2.2   Although the scope, procedures and work undertaken varies considerably between different inspection types (and testing and monitoring methods), these core objectives remain. As such, the inspector should be able to identify structural defects and clearly document these deficiencies; recognise structural elements that need repair in order to maintain safety and avoid the need for costly replacement; and be on guard for minor problems that could lead to the need for costly repairs. By providing this information, inspectors alert the Supervising Engineer to any defects which might impact the safety of the road user or the integrity of the structure and enable timely corrective action to be taken.

2 Purpose, Objectives and Scope of the Manual

2.1 PURPOSE OF THE MANUAL

2.1.1 The purpose of this Inspection Manual is to provide guidance on the inspection process for all staff involved in the management of highway structures. It is also considered that this Manual provides a sound basis for the development of formal inspector training courses.

2.1.2 This Manual aims to ensure that inspections are carried out efficiently, uniformly and to a high standard. The Manual is intended for use as guidance, outlining typical procedures and defining the normal requirements for the various categories of inspection. It is not intended to provide the definitive solution in all situations, as the party best able to decide on the appropriate course of action is the inspector or engineer undertaking the work.

2.1.3 This Manual supersedes the *Bridge Inspection Guide* [5] and is endorsed by the UK Bridges Board, the Highways Agency, Transport Wales, Transport Scotland, Department for Regional Development Northern Ireland (DRDNI) Roads Service, CSS Bridges Group and the London Bridges Engineering Group (LoBEG).

2.2 OBJECTIVES OF THE MANUAL

2.2.1 The objectives of this Manual are to encourage and assist inspection staff acting on behalf of an authority to:

- Develop and implement an inspection regime that ensures structures are safe for use and fit for purpose.

- Develop and undertake inspection regimes that facilitate the collection of appropriate inventory data to enable the development of robust and sustainable maintenance programmes.

- Harmonise procedures and practices in order to provide a more consistent approach to highway structures inspection throughout the United Kingdom.

- Provide an authoritative account of inspection procedures and practices for highway structures encompassing current national and international best practice.

- Be easy to understand and implement by a wide range of users, primarily inspectors but also Supervising Engineers, designers and planners.

- Provide standard definitions for the different highway structures and their elements and give a glossary of terms relevant to the inspection of highway structures.

2.3   SCOPE OF THE MANUAL

2.3.1   This Manual covers structural (e.g. beams, columns) and non-structural (waterproofing, expansion joints) aspects of highway structures; fittings such as signs, lighting or electrical equipment are not normally included. This Manual does not cover mechanical and electrical (M&E) equipment that may form part of or are attached to a highway structure, for example M&E in tunnels and moveable bridges. Advice on the inspection of M&E equipment is provided in *BD53: Inspection and Records for Road Tunnels* [6].

2.3.2   Inspection includes all visual inspections, testing and investigations carried out to ascertain the condition of a highway structure or for other purposes associated with the management of the structure. It includes work associated with inspection or testing operations, such as traffic management and the provision of access equipment. However, it does not include carrying out repairs or refurbishment, other than making good any drill holes, excavations or other minor damage caused by the inspection.

2.3.3   The principles and procedures described in this Manual are applicable to the inspection of all highway structures and are equally applicable to structures on other transport networks, e.g. heavy rail, light rail and waterways. Highway structures are typically considered to include those structures within the footprint of the highway that align with the following definitions:

- Bridges, buried structures, subways, culverts and other similar structures with a clear span or internal diameter greater than 0.9m.

- Earth retaining structures – structures associated with the highway where the dominant function is to retain earth with an effective retained height, i.e. the level of fill at the back of the structure above the finished ground level at the front of the structure, of 1.5m or greater.

- Reinforced/strengthened soil/fill structures with hard facings – structures associated with the highway where the dominant function is to stabilise the slope and/or retain earth with an effective retained height of 1.5m or greater.

- Structural aspects of all sign/signal gantries.

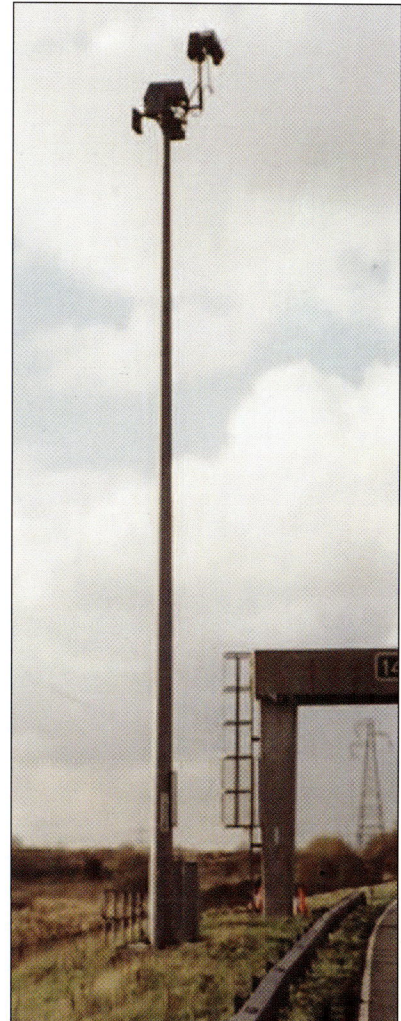

Highways Agency

- Structural aspects of masts, e.g. cantilever masts for traffic signals, lighting masts of 20m or greater, masts for camera, radio, speed camera and telecommunication transmission equipment, catenary lighting support systems, highway signs on posts.

- Access gantries – defined as moveable structures providing access to a highway asset, typically for bridge inspection and maintenance. All access gantries should be subject to inspections in accordance with the Institution of Structural Engineers publication *The Operation and Maintenance of Bridge Access Gantries and Runways* [7].

- Structural aspects of tunnels – where a tunnel is defined as an enclosed length of road of 150m or more.

- Other structures that are within the footprint of the highway, e.g. service/utility crossings that provide crossings either above or below the carriageway.

2.4 DUTY OF CARE

2.4.1 The scope described above does not negate the inspector's duty of care under Health and Safety legislation to report any safety hazards they encounter that are outside the scope of their inspection, e.g. defects which may have safety implications on fittings, or in particular at their attachment to the structure. Inspectors should be careful to note any deficiencies at or near the structure that, in their opinion, may constitute a significant safety hazard. They should report these to the engineer in charge at the earliest possible opportunity. There may be instances where inspection of fittings is required, for example to ascertain their effects on the loading or capacity of the structure.

3 Overview of the Inspection Regime

3.1.1   Cost-effective management of the maintenance of a structure relies on detailed, accurate and up-to-date information about its current condition and rate of deterioration. This objective can best be achieved through the development and implementation of an inspection regime tailored to meet the specific requirements of each structure [1].

Highways Agency

3.1.2   The inspection regime should include a combination of Safety, General, Principal and Acceptance Inspections of the whole structure; and more detailed Special Inspections or Inspections for Assessment concentrating on known or suspected areas of deterioration or inadequacy. The inspection schedule for each structure may be unique to that structure but should be designed to provide the appropriate frequency and detail of information.

3.1.3   The principal features of each type of inspection are summarised in Table A.1, with further guidance provided in Volume1: Part C: Section 2.2. Guidance on the inspection regime is also provided in the *Management of Highway Structures: A Code of Practice* [1] and *BD 63* [8].

3.1.4   Safety Inspections are undertaken at frequencies which ensure the timely identification of safety related defects but are not specific to highway structures; they generally cover all fixed assets on the highway network. Safety inspections may also be undertaken following notification of a defect by a third party e.g. the public or the police. Safety Inspections are normally carried out from a slow moving vehicle and provide a cursory check of those parts of a highway structure that are visible from the highway; in certain instances staff may need to proceed on foot.

3.1.5   General and Principal Inspections have set requirements, but differ in scope and intensity. General Inspections comprise a visual inspection (undertaken from ground level) of all parts of the structure that can be inspected without the need for special access equipment or traffic management arrangements. Principal Inspections, on the other hand, are more comprehensive than General Inspections and comprise a close examination, within a touching distance, of all inspectable parts of a structure. A Principal Inspection should utilise as necessary suitable inspection techniques, access equipment and/or traffic management works. Suitable inspection techniques that may be considered for a Principal Inspection include hammer tapping to detect loose concrete cover and paint and steel thickness measurements. All highway structures should be subjected to a General Inspection not more than two years following the previous General or Principal Inspection; and to a Principal Inspection not more than six years following the previous Principal Inspection. Formal guidance on increasing or decreasing the aforementioned inspection intervals is provided in the *Code of Practice* [1] and *BD 63* [8] (also see Volume1: Part C: Section2: paragraphs 2.2.20 and 2.4.18).

3.1.6　Special Inspections are undertaken for a wide variety of reasons but mainly to provide detailed information on a particular part, area or defect that is causing concern, which is beyond the requirements of the General/Principal Inspection regime. They may comprise a close visual inspection, testing and/or monitoring and may involve a one-off inspection, a series of inspections or an on-going programme of inspections. As such, Special Inspections are tailored to specific needs and are carried out when a need is identified or for some structures are programmed in advance.

3.1.7　Acceptance inspections are undertaken when necessary for exchanging information and documentation and agreeing the current status of, and outstanding work on, a structure prior to change over of responsibility for operation, maintenance and safety [1]. Acceptance inspections usually take the form of a Principal or General Inspection. However, the format content and timing depends on the circumstances (e.g. handover of a new structure, transfer of an existing structure, hand-back of a structure after a concession period) and on the specific purpose (i.e. Pre Opening Inspection, Defects Liability Inspection, or Transfer Inspection). The different types of Acceptance Inspections are described in more detail in *BD 63* [8].

3.1.8　Inspections for Assessment are undertaken when necessary to provide the information required to undertake a structural assessment. *BD 21* [9] provides guidance on undertaking an Inspection for Assessment.

Table A.1 – Summary of Inspection Types		
Inspection Type	**Nominal Interval**	**Description**
Safety Inspection (or Routine Surveillance)	At frequencies which ensure timely identification of safety defects and reflect the importance of a particular route or asset.	Cursory inspection carried out from a slow moving vehicle; in certain instances staff may need to proceed on foot.
General Inspection	2 years.	Visual inspection from the ground level. Report on the physical condition of all structural elements visible from the ground level.
Principal Inspection	6 years.	Close visual examination, within touching distance; utilising, as necessary, suitable inspection techniques. Report on the physical condition of all inspectable structural parts.
Special Inspection	Programmed or when needed.	Detailed investigation (including as required inspection, testing and/or monitoring) of particular areas of concern or following certain events.
Acceptance inspections	When needed.	A formal mechanism for exchanging information prior to changeover of responsibility.
Inspection for Assessment	When needed.	Inspection undertaken to provide information required to undertake a structural assessment.

4 Training and Qualifications

4.1 SUPERVISING ENGINEER

4.1.1 Inspections of highway structures should be carried out under the direction of the Supervising Engineer. The Supervising Engineer should be a Chartered Civil or Structural Engineer with appropriate experience in design, construction and maintenance of highway structures. The responsibilities of the Supervising Engineer normally include checking and countersigning all Principal and Special Inspection reports, including those prepared by other parties, to indicate agreement with their content.

4.1.2 The Supervising Engineer should give due consideration to the inspection requirements set down by the highways authority and ensure that all inspections are undertaken by personnel that satisfy the minimum health, experience and, where appropriate, qualification requirements for the particular inspection types.

4.2 INSPECTOR

4.2.1 The most important part of any inspection regime is the inspector, who is relied upon to perform their duties accurately, consistently, thoroughly and safely. At least one experienced inspector, who should give due consideration to the inspection requirements set down by the authority, should always be present on site during an inspection. The qualities of this experienced inspector should include, but should not be limited to the following:

- knowledge of the safe working practices and methods of access required for inspection;

- ability to recognise and evaluate defects on highway structures;

- an understanding of the behaviour of highway structures;

- knowledge of the construction methods and materials used in the construction of highway structures;

- knowledge of the causes of defects and suitable testing methods to identify, confirm or investigate these; and

- ability to record defects accurately, clearly and consistently.

4.2.2 All the inspectors in a team should be in sound health and have a realistic appreciation of their own limits of experience and ability. Inspectors with limited experience should work under the supervision of experienced staff.

Highways Agency

4.2.3   Particular training in safe working practices is essential for all those required to work in hazardous situations such as in confined spaces, at height or near railways (see Volume1: Part C: Section 3.5). All members of the inspection team must be made aware of the particular risks associated with an inspection before starting work. This will normally take the form of a site-specific briefing by the team leader before starting work. Such a briefing is in addition to training in safe working practices and is intended to highlight particular features of the site.

4.2.4   In addition to any engineering qualifications, inspectors should receive appropriate training in inspection procedures and techniques, including any necessary formal training and accreditation. Inspectors should be encouraged to obtain other qualifications which could be useful during inspection work, e.g. training and qualifications in first aid, qualifications and experience in specialised forms of access, notably diving and abseiling.

5 Implementation of the Manual

5.1.1 The Manual contains a large body of information on the inspection of highway structures. Authorities should encourage all relevant staff to study, understand and use this information. Supervising Engineers and inspectors alike should seek to develop a sound understanding of the Manual, and undertake the activities listed below and outlined in Figure A.1:

1. **Examine the Manual** – authorities should encourage all personnel responsible for the inspection of highway structures to review the content of all parts of the Manual with the aim of using the advice and information contained in the Manual.

2. **Disseminate Key Advice** – the Supervising Engineer should maintain an overview of the whole Manual and should plan and hold internal workshops with relevant personnel to disseminate the guidance and key advice outlined in the Manual.

3. **Identify or Develop a Formal Training Programme** – authorities should encourage Supervising Engineers to identify and/or develop a formal training programme for inspectors based on the topics covered by the Manual. The training programme may be a combination of external and internal lectures/courses or on-the-job training; it may be prudent to develop training courses in collaboration with other authorities in order to share resources. The Supervising Engineer should seek continual improvement for the formal training programme and this should be subjected to a process of continual monitoring and review.

4. **Knowledge Gap Analysis** – with the support of the Supervising Engineer, inspectors should undertake a knowledge gap analysis using the contents of the Manual as a benchmark against which their current knowledge should be compared. The aim of the analysis would be to identify both current competencies as well as knowledge gaps requiring further development and learning. This activity should be undertaken periodically, for example, during an annual appraisal.

5. **Produce a Development Action Plan** – the outcome of the gap analysis should be a prioritised list of training activities aimed at closing the inspectors' knowledge gaps. Having prioritised their training needs, inspectors should be encouraged to prepare, in conjunction with the Supervising Engineer, a Development Action Plan detailing how their training needs will be met, i.e. the plan should include proposed training activities, necessary resources and appropriate timescales. This plan should take a balanced account of different aspects of the inspector's training needs, which should give due consideration to the requirements of the authority. Examples include short and long term development goals, and both technical and legislative components of performance.

6. **Attend Training Activities** – inspectors should be encouraged to implement their Development Action Plan by completing prioritised actions and by attending appropriate training activities. Where necessary inspectors should be encouraged to seek engineering qualifications or attend relevant NVQ courses.

7. **Record and Assess Training Activities** – once inspectors have attended/completed training activities, it is important that they not only record their attendance, but they also identify what they have learned, and with the support of the Supervising Engineer evaluate the benefits gained.

8. **Review the Development Action Plan** – with the support of the Supervising Engineer, inspectors should regularly review their development action plans to ensure that completed actions are signed off and new actions are added to reflect local and national developments or requirements and career development. The outcome of this review should be an updated prioritised list of training needs and activities.

5.1.2 It is an authority's best interest to adopt a formal approach to implementing the Manual, preferably in collaboration with other authorities. The benefits provided by adopting this Manual are likely to include improved quality and consistency in inspection information and more formal recognition of the important role of inspection staff.

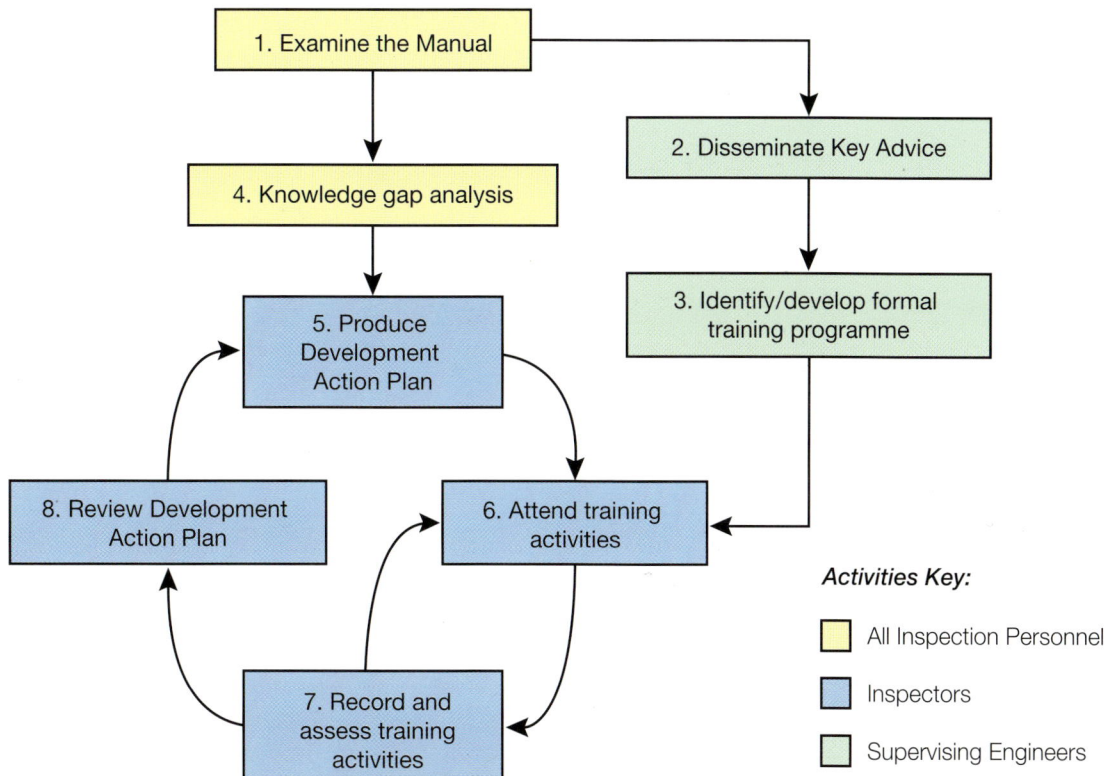

Figure A.1 – Process for the implementation of the Inspection Manual

6 Layout of the Manual

6.1.1 The Manual is divided into two separate but complementary Volumes:

- **Volume 1: Reference Manual** – covers all aspects of highway structures inspection that both inspectors and engineers should be aware of.

- **Volume 2: Inspector's Handbook** – acts as a quick reference for inspectors on site.

6.1.2 Each of the two Volumes is sub-divided into several Parts that provide guidance on a wide range of issues. Table A.2 provides an overview of the purpose and content of each Part of the Manual.

Table A.2 – Layout of the Manual	
Part	**Summary of purpose and content of each part**
Volume 1: Inspection Manual for Highway Structures: Reference Manual (A4 format)	
A. Introduction and Implementation	This part provides an overview of the Manual; includes an introduction to the role and need for the Manual, describes the purpose, objectives and scope of the Manual and how it should be implemented. It also offers an overview of the inspection regime and summarises the general competence and training requirements for inspection staff.
B. Behaviour of Structures	This part provides inspectors with an overview of the engineering topics that they should be aware of and offers a consistent basis for inspector knowledge development and training. It contains a summary of topics such as structural mechanics, structural materials and their properties, common types of highway structures and structural elements. It aims to provide organisations with a check list of criteria they should include on training courses but does not provide detailed coverage of the topics.
C. The Inspection Process	This part provides general information on scheduling inspections, planning and preparing for inspections, access considerations, health and safety aspects, and advice on performing inspections on different types of structures.
D. Defect Descriptions and Causes	This part provides background information and guidance on describing and categorising defects. It describes the principal defects that are likely to be encountered in concrete structures, metal and metal/concrete composite structures, masonry structures and structures built from other materials, with the emphasis placed on identification and likely causes.
E. Investigation and Testing	This part summarises a wide range of testing methods available for the investigation of particular material properties, defects, causes of defects, etc. This aims to make inspection staff aware of the tests that may be used to inform structures management.
F. Appendices	Contains a selection of information which collectively support the guidance provided in Parts A to E.
Volume 2: Inspection Manual for Highway Structures: Inspector's Handbook (A5 format)	
A. Inspector's Guide	This part highlights the key points from Volume 1 that inspectors should be aware of before, during and after undertaking inspections. Cross-references to Volume 1 are provided, enabling inspectors to quickly look up more detailed advice/guidance as and when required.
B. Defect Photographs	This part provides a library of photographs illustrating some of the different types of defects that are likely to be encountered on highway structures.

7 National and Regional Variations

7.1.1   Some of the guidance contained in this Manual is subject to national and regional variations in legislation (e.g. environmental legislation) and to the requirements placed on each highway authority (e.g. specific variations to the DMRB [10]). As such, the guidance provided in this Manual should always be applied within the context of the authority's relevant (and latest) legislation and requirements.

7.1.2   The glossary of terms and definitions provided in Part B of this Manual relate to terms that are used in England and, as such, it is prudent that the users identify and adopt each authority's equivalent terms, e.g. 'highway' = 'road', etc.

8 References for Part A

1. *Management of Highway Structures: A Code of Practice*, TSO, 2005.

2. *Highways Act 1980*, HMSO.

3. *The Roads (Northern Ireland) Order 1993* (SI 1993, No. 3160), HMSO.

4. *The Roads (Scotland) Act 1984* (SI 1990, No. 2622), HMSO.

5. *Bridge Inspection Guide*, Department of Transport et al., HMSO, London, 1984 (out of print).

6. *BD 53 Inspection and Records for Road Tunnels*, DMRB 3.1.6, TSO.

7. *The Operation and Maintenance of Bridge Access Gantries and Runways*, 2nd Edition, Institution of Structural Engineers, London, 2007.

8. *BD 63 Inspection of Highway Structures*, DMRB 3.1.4, TSO.

9. *BD 21 The Assessment of Highway Bridges and Structures*, DMRB 3.4.3, TSO.

10. *Design Manual for Roads and Bridges* (DMRB), TSO.

Part B
Behaviour of Structures

This Part of the Manual provides an overview of common types of highway structures, structural elements and materials. Background information and guidance on the fundamentals of structural behaviour, the basic principles of structural mechanics and material properties that inspectors should be aware of are also provided.

1 Introduction

1.1.1 Inspection staff should be familiar with the different types of highway structures that they are likely to encounter, i.e. any structure that is within the footprint of the highway or that materially affects the support of the highway or land immediately adjacent to it [1]. The purpose of this Part of the Manual is to provide an overview of the types, form, function and behaviour of highway structures.

1.1.2 Important information is provided on terminology, types of highway structures, component types, material types and the fundamentals of structural mechanics and material properties. Awareness of these issues will enable inspection staff to undertake more informed inspections and to improve the quality of information they provide in inspection reports. Table B.1 provides an overview of the layout and content of this Part of the Manual.

Table B.1 – Layout of Part B	
Section	**Summary of Purpose and Content**
2. Bridges	Bridges are the most common and complex type of highway structure that inspection staff are likely to encounter. This section provides an overview of the typical arrangement (anatomy) of a bridge, the different types of bridges and the common types of bridge elements.
3. Other Highway Structures	Other common types of highway structures include culverts, retaining walls, sign/signal gantries and masts. This section provides an overview of their different structural forms and elements.
4. Structural Mechanics	This section provides a high level introduction to structural mechanics, introducing terminology and topics like force, stress and strain.
5. Properties of Construction Materials	This section provides an overview of the characteristics of the common types of materials used in the construction and maintenance of highway structures, e.g. concrete, steel, masonry, wrought and cast iron, timber and advanced composite materials.

1.1.3 This Part of the Manual presents the basic principles of structural mechanics and material properties; if further information is required the references listed in Section 6 should be consulted. Where appropriate, inspectors should seek to follow a training programme (as set out in Volume 1: Part A: Section 5) that will enhance their understanding of the topics introduced here.

1.1.4 A number of the diagrams used in this Part of the Manual are based on those originally printed in the CSS Guidance Documents for bridge inspection (*Bridge Condition Indicators* Volume 2 [2] and Volume 2 Addendum [3]).

2 Bridges

2.1 OVERVIEW

2.1.1 A bridge is a structure built to provide passage over a physical obstacle, e.g. watercourse, railway, road, valley. Bridges also include subways, footbridges and underpasses. Bridges are the most common and complex type of highway structure that inspection staff are likely to encounter and for the purpose of inspection are normally defined as those having a span of 1.5m or greater [1, 4]. This section:

- Introduces the general anatomy of a bridge, Section 2.2.

- Provides definitions for the typical elements found on a bridge, and in some cases (e.g. bearings and joints) provides common examples, Section 2.3.

- Summarises the typical types of highway bridges that inspection staff are likely to encounter; Section 2.4.

2.2 ANATOMY OF A BRIDGE

2.2.1 The success of a thorough bridge inspection lies with the ability of the inspector to identify and understand the function of the major bridge elements. As such, inspection staff should develop an understanding of the typical composition, or anatomy, of a bridge.

2.2.2 In general, bridge elements can be categorised under a number of broad headings; for example: Superstructure, Substructure, Safety Elements, Durability Elements and Ancillary Elements. Definitions of these categories, and a suggested grouping of typical elements under these headings, are given in Table B.2.

Table B.2 – Grouping of Bridge Elements		
Category	**Description**	**Typical Elements**
Superstructure	The horizontal elements of a bridge, generally above the bearings, that directly support traffic loads and transfer loads to the substructure [1].	Primary, secondary and tertiary load bearing elements (e.g. slabs, beams and arches); parapet beams and cantilevers.
Substructure	The vertical elements of a structure, generally below the bearings, that support the superstructure and transfer the loads to the supporting ground [1].	Foundations; abutments (including arch springing); piers; columns; cross-heads, bearings*.
Durability Elements	Elements that have the primary, or significant, function to protect the structure from or delay/mitigate the effects of deterioration and damage.	Drainage; waterproofing; expansion joints; surface finishes.
Safety Elements	Elements that have the primary, or significant, function to safeguard the user and/or those working on the structure.	Handrails; parapets; safety fences; walkways; access gantries.
Ancillary Elements	Other elements associated with highway bridges.	Invert; aprons; wing walls; embankments; lighting; carriageway; footway/verge.

* Bearings are normally considered as the boundary between Superstructure and Substructure. Above they are grouped with the Substructure because they help transfer the Superstructure loads to the Substructure.

2.2.3   Identifying *Durability, Safety* and *Ancillary* bridge elements is normally straightforward (common types are presented in Section 2.3), however, distinguishing between the *Superstructure* and the *Substructure*, and identifying the associated elements, can be more difficult. Figure B.1 and Figure B.2 show two typical types of bridge construction and distinguish between their Superstructure and Substructure elements; other bridge elements are also shown.

2.2.4   The elements present on a bridge depend on a number of criteria, but as illustrated by Figure B.1 and Figure B.2, it is primarily the structural form and material that dictate which elements are on the bridge.

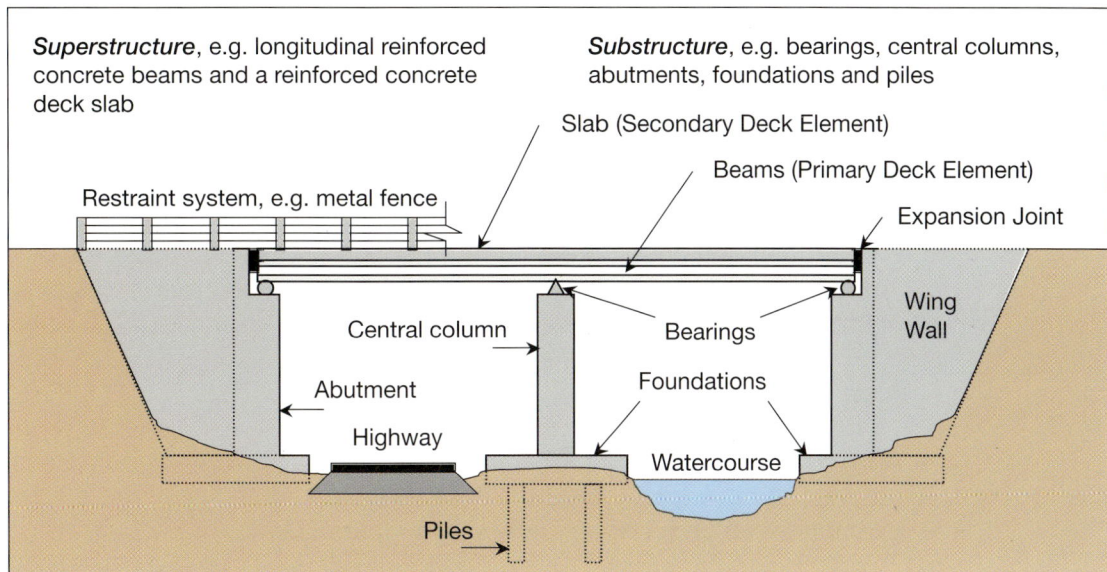

Figure B.1 – Typical modern bridge

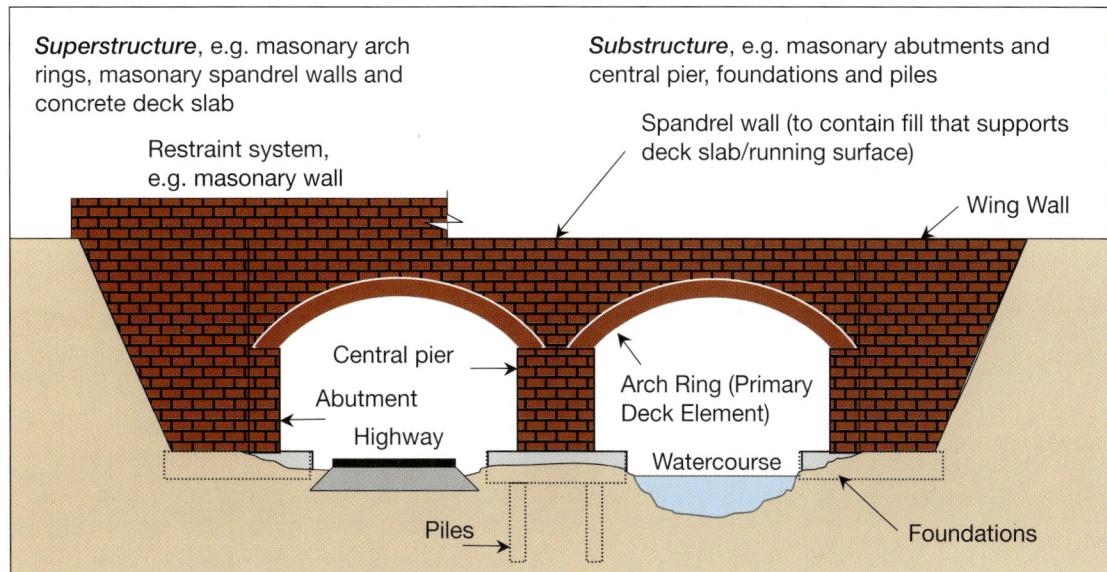

Figure B.2 – Typical older bridge

2.3   BRIDGE ELEMENTS

2.3.1   This section presents definitions of typical bridge elements, grouped under the headings of Superstructure, Substructure, Durability Elements, Safety Elements and Ancillary Elements. In some cases common examples are provided to help expand on the definition. The following definitions align with those provided in the Code of Practice [1].

Superstructure

2.3.2   The bridge superstructure includes one or more of the following: primary, secondary and tertiary load bearing elements. The structure form and material normally dictate what are actually present on a structure, for example:

- In Figure B.1 the primary load carrying element is likely to be some for of longitudinal beam, while the secondary load carrying element is likely to be the deck slab supported by the longitudinal beams

- In Figure B.2 the primary load carrying element is the arch ring, while the secondary load carrying element may be a concrete slab, resting on the arch fill, which supports the running surface.

2.3.3   As such, it is not possible to give a definitive description of the superstructure and its constituent elements, instead it is important to appreciate the different forms this can take and how these influence the form of the primary, secondary and tertiary load carrying elements. Section 2.4 presents a range of superstructure examples, and the following definitions relate to typical elements found in bridge superstructures.

Arch

2.3.4   An arch is a curved beam or slab that functions primarily in compression and produces both vertical and horizontal reactions at its supports, a typical example of an arch bridge is shown in Figure B.2.

Beam

2.3.5   A beam can be defined as a linear structural member that spans from one support to another. Beams are part of the superstructure and common types include:

- **Primary Load Carrying Beam** – a primary, or main, load carrying beam in a bridge that supports the bridge deck and transfers the traffic loads and superstructure weight (including their own self weight) to the substructure (via bearings on some forms of structure). Primary beams normally span parallel to the direction of traffic flow.

- **Girder** – performs the same function as a beam but is normally metal and consists of two flanges with a web (could be a complete rolled section or built up from rolled plates connected together).

- **Parapet Beam/Cantilever** – the main function of a parapet beam/cantilever is to support the parapet, although it may also support the carriageway and/or footway/verge. A cantilever is a structural element, normally a slab or beam, which has one unsupported (free) end and one supported (fixed or built in) end.

- **Transverse Beam** – a secondary load carrying beam that transfers the traffic loads to the main beams. Transverse beams normally span between the primary beams and are perpendicular to the direction of traffic flow.

Bracing

2.3.6   Bracing is elements that provide longitudinal, lateral and/or torsional stiffness to the primary members and/or to the bridge deck.

Bridge Deck

2.3.7   The part of a bridge superstructure that directly supports the running surface and traffic. It is normally defined as a secondary load bearing component because it transfers the traffic loads to the primary load bearing components, e.g. main beams. However, the deck may be the primary load bearing element if it is a slab bridge, i.e. the slab is the bridge deck, or even a tertiary load bearing component if there are transverse beams. A subsidiary function of the deck is to accommodate other elements which include the carriageway (surfacing), kerbs, footways, deck drainage systems, utilities, restraint systems, signs and lighting.

Truss

2.3.8   A truss, when present on a bridge, is normally the primary load carrying element. A truss is built up from individual members, normally arranged and connected in a triangular/rectangular pattern, and consisting of a top chord, bottom chord and internal members.

Substructure

2.3.9   The substructure elements defined in this section are abutments, bearings, piers, columns, and foundations.

Abutments

2.3.10   Abutments support the extreme ends of the superstructure and transfer the loads to the foundations or ground. Abutments generally retain or support the approach embankment and bearings and should provide adequate clearance between the superstructure and obstacle crossed. Common types of abutments (also see Figure B.3) include:

- **Bank-seat abutment** – abutments of small vertical height that are normally situated on top of a natural or man-made bank (e.g. banks of watercourses, embankments) and do not provide a significant retaining function. The combination of the bank and the bank-seat abutment provide the clearance required.

- **Gravity abutment** – an abutment that resists horizontal forces through its own self-weight and normally transfers vertical loads directly to the ground. Similar to a gravity retaining wall.

- **Cantilever abutment** – an abutment wall that is rigidly fixed to the foundation and transfers traffic loads and earth pressures to the foundations principally by bending action. Similar to a cantilever retaining wall.

- **Embedded abutment** – an embedded abutment is similar to a cantilever abutment except there is no horizontal foundation component; instead stability is achieved through the embedded depth.

- **Spill through abutment** – this is effectively a cantilever abutment where the wall is replaced by columns in order to reduce the earth pressure acting on the abutment and the embankment spills through between the columns.

2.3.11   Current practice is to make decks integral with the abutments, i.e. avoid the use of joints over abutments and piers [5, 6].

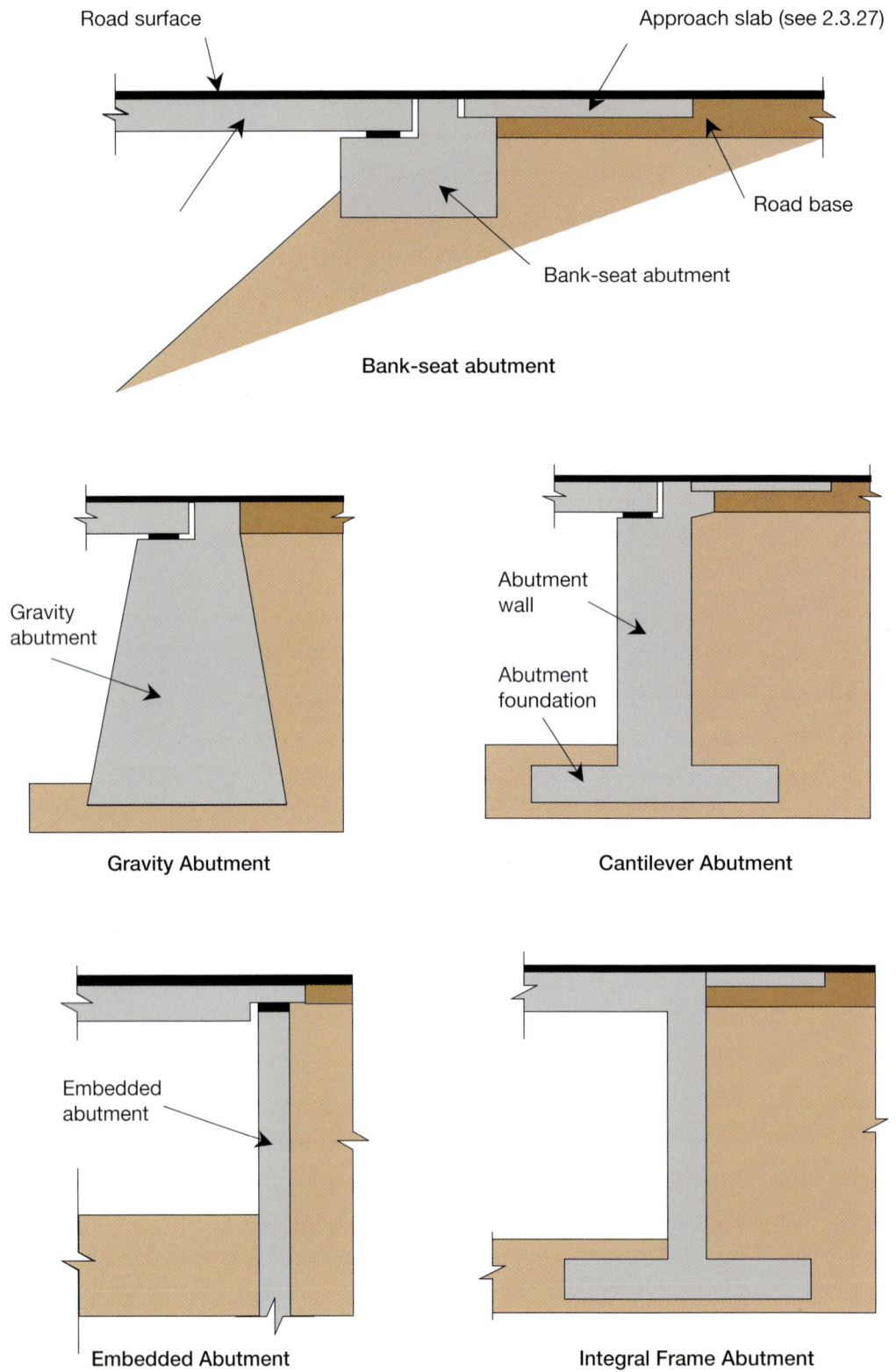

Figure B.3 – Cross-sections of typical abutments

Bearings

2.3.12   Bearings provide connections between the superstructure and substructure, the purpose of which includes all or some of the following:

- To transfer vertical/horizontal loads from the superstructure to the substructure.

- To allow longitudinal/transverse movement of the superstructure.

- To allow rotation of beam/slab ends due to dead and live loading.

2.3.13   Bearings accommodate these movements by deforming (elastomeric), rotating, sliding and/or rolling, see Figure B.4 for typical examples. A large variety of bearings have evolved using various combinations of these mechanisms, including pot, rocker and spherical bearings.

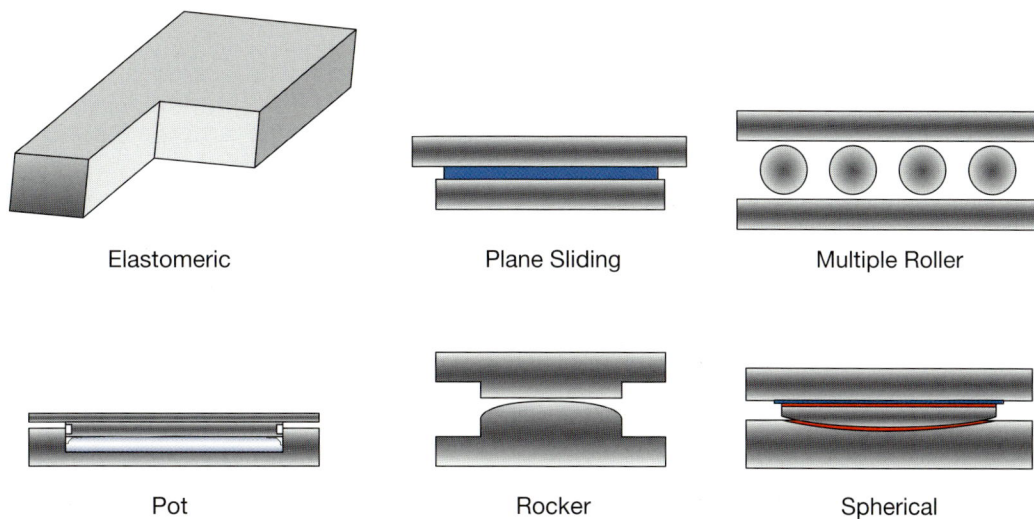

Elastomeric   Plane Sliding   Multiple Roller

Pot   Rocker   Spherical

Figure B.4 – Bearings types [7]

2.3.14   The shelf/surface upon which the bearing sits either directly or via a plinth, is referred to as the bearing shelf. The bearing shelf is normally on the top of the abutment/pier

Piers and Cross heads

2.3.15   Piers, or columns, provide support to the superstructure at intermediate points along multiple bridge spans [8]. Piers transfer loads to the ground/foundation and may be of column, wall or frame construction. A pier, as with the abutment, provides adequate clearance between the superstructure and the obstacle crossed. On some bridges there is a capping beam/cross head on the top of the piers, the purpose of which is to distribute loads to the piers and to provide a base for bearings.

Foundations

2.3.16   Foundations are a construction below ground level that supports piers and/or abutments. The purpose of the foundations is to provide a solid and stable base for the bridge and to distribute loads to the ground. Foundation types depend primarily on the depth and safe bearing pressures of the bearing stratum, and also the restrictions placed on differential settlement due to the type of bridge deck. Foundations for highway structures normally fall into two categories: spread (e.g. slab, strip) or piled. Piled foundations normally have a slab or strip to distribute the abutment loads to the piles.

Durability Elements

2.3.17   The durability elements defined in this section are drainage, waterproofing, joints and surface finishes.

Drainage

2.3.18   The main function of a drainage system is to remove water away from the bridge superstructure and substructure. The superstructure drainage system collects and disposes of water from the deck and deck joints while the substructure drainage system allows the backfill material behind abutments, wing walls or other earth retaining structures to drain accumulated water [6, 7, 9].

Waterproofing

2.3.19   Superstructure waterproofing is a protective coating placed between the road construction and the bridge deck in order to protect the bridge deck from the ingress of harmful agents, e.g. chloride contaminated water. Substructure waterproofing fulfils a similar function as there may be harmful agents present in the ground as well.

Joints

2.3.20   Joints provide a running surface across the expansion gap, i.e. the area between adjacent bridge deck spans or the bridge deck and abutment [10, 11, 12]. Joints allow movement (see Table B.3) and/or are a feature of the construction form (see Table B.4). Joints may be open (allow water/debris to pass through) or closed (do not allow water/debris to pass through).

Table B.3 – Types of Expansion Joints [10, 11, 12]

Joint Cross-Section	Description
Surfacing, Protective layer, Deck Waterproofing, Flexible filler, Flashing, Deck joint gap, Elastometric pad	**Buried Joint** – A joint which is formed in-situ using components, such as an elastomeric pad or a flashing, to support the surfacing which is continuous over the deck joint gap.
Flexible material, Surfacing, Protective layer, Plate, Deck Waterproofing	**Asphaltic Plug Joint** – An in-situ joint comprising a band of specially formulated flexible material, which also forms the surfacing, supported over the deck joint gap by thin metal plates or other suitable components.
Nosing material, Surfacing, Protective layer, Compression seal, Deck Waterproofing	**Nosing Joint** – In-situ material or fabricated component to protect the adjacent edges of the surfacing at the expansion joint, often with a compression seal between the nosings.
Transition strips, Surfacing, Protective layer, Drainage membrane, Bedding, Deck waterproofing, Elastomer reinforced with metal plates	**Reinforced Elastomeric Joint** – A prefabricated joint comprising an elastomer with or without bonded metal plates.
Surfacing, Transition strip, Comb or tooth plates, Securing bolts, Plate, Drainage membrane, Deck joint gap, Bedding, Deck waterproofing	**Cantilever Comb or Tooth Joint** – A prefabricated joint comprising principally mating metal comb or saw-tooth plates which bridge the deck joint gap.
Surfacing, Elastomeric elements, Deck Waterproofing, Securing framework, Support beam, Deck joint gap, Sliding bearing	**Elastomeric in Metal Runners** – A pre-fabricated joint comprising an elastomeric seal fixed between metal runners, either in a single element or multi-element form.

Table B.4 – Types of Structural Joints	
Joint Cross-Section	**Description**
	Half joint – Normally found on three-span beam and slab bridges, located at the points of contraflexure (i.e. at cross-sections of low bending and high shear forces) in the centre span. A half-joint includes bearings that allow rotation and sometimes longitudinal movement [1].
	Hinge joint – Normally found on some three-span beam and slab bridges and located at the points of contraflexure (i.e. at cross-section of low bending and high shear forces) in the centre span. A hinge joint allows rotation but not longitudinal movement. Also found in some bridge piers [1].

* Both half and hinge joints will have one of the preceding expansion joints above them (usually either a buried or asphaltic plug joint).

Surface Finishes

2.3.21   Surface finishes are used to protect bridge elements and/or provide the appropriate (aesthetic) finish to the element. Typical surface finishes include paint, hydrophobic pore lining impregnants [13], mortar rendering, concrete or advanced composite cladding and masonry facing.

Safety Elements

2.3.22   The safety elements defined in this section are vehicle restraint systems and access gantries.

Vehicle Restraint Systems

2.3.23   A vehicle restraint system is installed on the edge of a bridge or on a retaining wall or similar structure where there is a vertical drop and may contain additional protection and restraint for pedestrians and other road users. The vehicle restraint system is normally referred to as a parapet when it runs along the outside edges of the bridge deck parallel to the direction of traffic flow; a parapet typically takes the form of a wall, rail or fence.

Access Gantries

2.3.24   Some bridges have permanent access gantries which provide access for bridge inspection and maintenance; these may be stationary or moveable gantries.

Ancillary Elements

2.3.25   The ancillary bridge elements defined and discussed in this section are the approach embankment and slab, apron, highway, lighting, revetment, signs, utilities and wing walls.

Approach Embankment

2.3.26　A bank formed above the natural ground level to create the approach to a bridge. The purpose of an approach embankment is to raise the road level so that it aligns with the bridge deck level.

Approach Slab

2.3.27　The approach or run-on slab, is a slab positioned below the road surface on the approach to a bridge, the end of which normally rests on the back of the abutment (see Figure B.3, which shows the typical location of the approach slab in relation to the bridge abutment). The purpose of the approach slab is to provide a smooth transition for traffic from the road to the bridge and vice versa. Approach slabs are normally made of reinforced concrete.

Apron

2.3.28　A horizontal slab/mattress built into the bed of the watercourse around piers, abutments, culverts and watercourse banks, possibly in conjunction with revetments (see paragraphs 2.3.31-2.3.33). The purpose of an apron is to provide scour protection to the bed of the watercourse.

Highway

2.3.29　A highway is a collective term used to describe facilities laid out for all types of users, including but not restricted to: carriageways, footways, footpaths and cycleways [14], where these are defined as:

- **Carriageway** – the part of the highway provided for use by motor vehicles. Surfacing is the carriageway or footway wearing course and base course materials applied upon the deck to provide a smooth riding surface and to protect the deck from the effects of traffic and weathering [14].

- **Footway** – the part of the highway alongside a carriageway provided purely for use by pedestrians [14]. Generally there are kerbs alongside footways to reduce the risk of vehicles crossing onto the footway and endangering pedestrians, see Figure B.5.

- **Footpath** – a highway provided purely for use by pedestrians that is not alongside a carriageway.

- **Cycleway** – the part of a highway or a highway provided for use by cyclists. Sometimes they may also be used by pedestrians.

- **Bridleway** – the part of a highway or a highway provided for use by equestrians, cyclists and pedestrians.

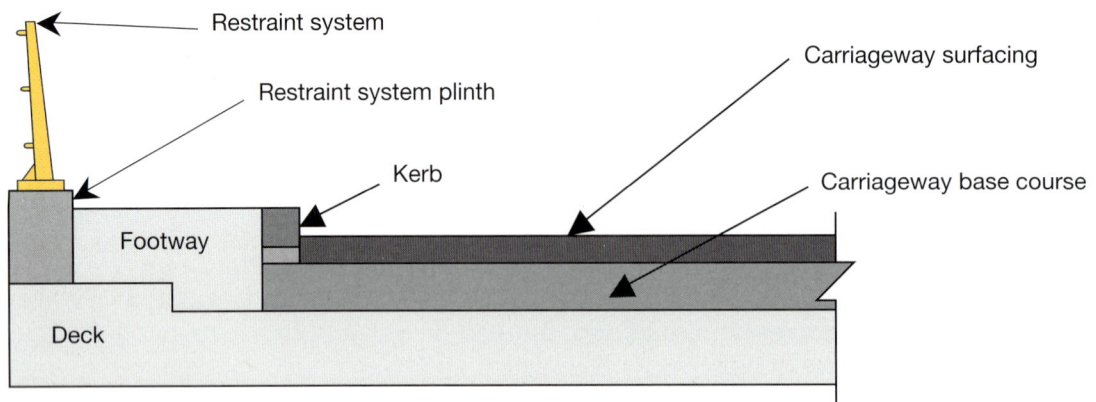

Figure B.5 – Example of concrete footway

Lighting

2.3.30 When present, lighting on highway structures serves several purposes and consists of the following types [15]:

- **Street or Highway Lighting** – lighting for illuminating carriageway, footways, footpaths, cycle tracks and pedestrian subways open to public access.

- **Traffic Control Lights** – light signals used to control traffic.

- **Illuminated Traffic Signs** – internally or externally illuminated traffic signs, e.g., flashing school crossing warning signs, centre island beacons, and pedestrian crossing Belisha beacons.

- **Illuminated Traffic Bollards** – bollards lit by internal or base-mounted lighting units, irrespective of whether they carry a sign or not.

Revetment

2.3.31 A revetment is cladding placed on a soil or rock surface to protect and stabilise it against erosion by water or weathering. A revetment may also be required to accommodate subsidence, surface water drainage, ground water movement, and subsoil drainage. Revetments are not normally classed as structures in their own right, but frequently form ancillary features at bridges, culverts and retaining walls. A revetment may include an apron at the toe of the slope if it is susceptible to scour.

2.3.32 Revetments typically comprise an armour layer, to protect against erosion by weathering, water, currents or wave action, and an underlayer, to restrain subsoil movement and act as a drainage zone. The following are common types of armour layer, although the list is not exhaustive and other materials may be encountered:

- Stone rip-rap;

- Hand placed stone;

- Grouted stone or masonry;

- Gabion mesh mattresses;

- Precast concrete blocks - open jointed or interlocking;

- Cable tied block mattresses;

- Concrete insitu slabs;

- Grassed geotextile mats;

- Grout filled synthetic mattresses;

- Stone asphalt.

2.3.33　The underlayer may be a granular material, a geotextile (typically polypropylene, polyester, polythene or polyamide polymers in filament or fibre form, woven or chemically bound into textiles), or a combination of both. The layer should act as a filter, permitting the flow of water but restraining subsoil. The successful performance of a revetment is dependent on good contact being maintained between the underlayer and the adjacent armour and subsoil.

Signs

2.3.34　Signs are provided to inform the road users about highway conditions and hazards and may include the following types: Regulatory signs, Warning signs, Direction signs, and Information signs. Examples of information that road signs may convey and are of particular importance to the bridge inspector include weight restrictions, speed limits and impaired clearances.

Utilities

2.3.35　Utility companies/bodies use a highway to provide goods and services to the public. As a result some utilities may be attached to or carried by bridge structures in pipes or ducts located in the deck or in the fill/surfacing over the deck and may include one or more of the following: gas, electricity, water, telephone, cable TV, and sewage. There also may be private services.

Wing Walls

2.3.36　Wing walls are essentially retaining walls immediately adjacent to the abutment. The walls may be independent or integral with the abutment wall [6, 8].

2.4　TYPES OF BRIDGES

2.4.1　Depending on the required function and appearance bridges vary widely in construction form and material. The following features are normally used to describe a bridge and by combining these terms it should be possible to give a general description of most bridges:

- Span form, e.g. single, multiple, simply supported, continuous, cantilever

- Construction form, e.g. slab, beam and slab, arch, truss, cable supported

- Construction material, e.g. concrete, steel, masonry

2.4.2　Typical examples of span form and construction form are provided in the following. The construction material type is frequently dependent on the construction form, where this is the case the material associated with a particular construction form is indicated.

Bridge Span Forms

2.4.3   The span of a bridge is the distance between its supports; the superstructure provides passage over this distance. The characteristics of the superstructure can be used to help classify some bridge types, i.e.:

- The number of spans, e.g. single span or multi-span; and

- The relationship between the spans and the supports, e.g. simply supported, continuous and integral, these arrangements are shown in Table B.5.

2.4.4   Table B.6 shows how the number of spans and the span/support relationship can be used to help describe some common bridge types. By combining the span/support descriptions with the material type of the span (superstructure), it is possible to provide a global description for most bridge types, e.g. simply supported, reinforced concrete slab bridge, or three span continuous prestressed concrete beam bridge.

Table B.5 – Span-Support Relationships	
Span-Support Relationship	**Description**
	Simply supported slab or beam – the support allows freedom of rotation of the slab/beam ends. This span-support relationship may take the form of a Pinned (or Hinged) Support which allows freedom of rotation but no movement in any other direction, or a Roller Support which allows rotation and movement in the longitudinal
	Continuous beam/slab construction – the beam/slab is continuous over the support, i.e. there is no 'break' in the beam/slab. This type of arrangement is typically supported by a Roller Support, although it can be a Pinned Support.
	Integral beam/slab and support construction – the beam/slab is monolithically cast with the support creating Fixed (or Rigid) Joints between the beam/slab and the support.

Joint/Support diagrammatic convention:

▲ - Pinned/Hinged Joint   ● - Roller Support   ⫽ - Fixed Joint/Support

Table B.6 – Typical Bridge Span Forms

Bridge Spans	Description
	Simply supported single span bridge – the superstructure rests on the two end supports.
	Single span integral bridge – the superstructure and two end supports are monolithic.
	Simply supported multi-span bridge – a three span arrangement is shown, with two end supports and two intermediate supports. There is a 'break' in the superstructure over the supports.
	Multi-span continuous bridge – a three span arrangement is shown, with two end supports and two intermediate supports. There is no 'break' in the superstructure over the supports.
	Portal Frame (inclined leg) bridge - a three span arrangement is shown, with two end supports and two intermediate supports. There is no 'break' in the superstructure over the supports, and the superstructure is continuous with the substructure.
	Cantilever and suspended span bridge - the cantilever is formed by continuing the side spans over the support to provide a seating for the central simply supported beam or slab, which is referred to as the suspended span. A series of alternating cantilever and suspended spans is sometimes used in multi-span structures

Summary of Span Ranges

2.4.5   Typical span ranges for common types of beam bridges (including beam and slab, girder and box girder) are shown in Table B.7. The span ranges are based on existing structures and values quoted by designers and manufactures, the span ranges take account of functionality and economy and must only be used as a general guide as both economics and technology are constantly changing.

Table B.7 – Typical Span Ranges for Beam and Slab Constructions			
Construction Form		**Typical Span Range**	**Longest Span**
Solid slab	Reinforced concrete	Up to 10m	-
	Prestressed concrete (not common)	Up to 15m	-
Voided slab	Reinforced concrete	Up to 25m	-
	Prestressed concrete	25 to 35m	-
Beam and slab	Reinforced concrete	10 to 25m	-
	Prestressed concrete	25 to 50m	-
	Steel beams encased in a concrete slab	5 to 15m	-
	Steel Universal Beams supporting a concrete slab	Up to 25m	-
	Steel Plate Girders* supporting a concrete slab	20 to 50m	-
Box girder	Prestressed concrete: incrementally launched	45 to 75m	301m, Stolmasundet Bridge, Austevoll Norway, 1974
	Prestressed concrete: span-by-span	45 to 60m	
	Prestressed concrete: balanced cantilever	45 to 75m	
	Steel	45 to 75m	300m, Ponte Costa e Silva Bridge, Rio de Janeiro, Brazil, 1974
Arch	Traditional masonry/iron arch	Up to 25m	-
	Modern concrete/steel arches	40 to 150m	420m, Wanxian Bridge, Wanzhou, China, 1997
Truss	Steel truss	40 to 500m	549m, Pont de Quebec Bridge, Quebec City, Canada, 1917
Cable supported	Cable stayed	150 to 500m	1018m, Stonecutters Bridge, Hong Kong, China, 2008
	Suspension	250 to >1000m	1991m, Akashi-Kaikyo Bridge, Kobe-Naruto, Japan, 1998

Note: The bridges listed in the 'Longest Span' column are special structures that have much longer spans than the typical span range.

Slab Bridges

2.4.6   Slab bridges, typically used for spans of up to 35m, are one of simplest forms of construction where the slab is the primary load carrying component and forms the entire superstructure. They are a descendant of the old clapper bridges (see photo) found in upland areas of the United Kingdom and constructed using large flat slabs of stone supported on stone piers and abutments or river banks.

Wikipedia

2.4.7   Modern slab bridges are normally built from either reinforced concrete or prestressed concrete and may be solid in cross section (Figure B.6) or have voids to reduce their self weight (Figure B.7). Above a span of 10m the dead weight of a solid slab bridge becomes excessive [16]; therefore voided slab bridges are required to achieve spans of up to 35m.

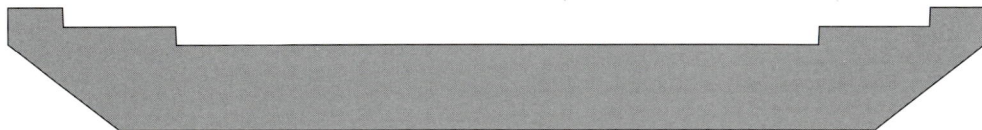

Figure B.6 – Typical cross-section of a solid slab bridge

Figure B.7 – Typical cross-sections of voided slab bridges

Beam Bridges

2.4.8   A typical cross section of a beam and slab construction is shown in Figure B.8. A beam and slab arrangement may be cast as a monolithic concrete element or be cast/constructed separately and structurally connected together (e.g. reinforced concrete slab supported by prestressed beams, or a reinforced concrete slab supported on steel beams). A reinforced concrete slab supported on precast concrete beams or steel beams are known as composite construction.

2.4.9   For spans beyond the range of standard rolled steel sections, girders are built up, e.g. flanges and web rolled separately and connected together (see Figure B.9 for flange and web). Prior to the 1940s the flange and web plates were joined by riveting with angles and cover plates, whereas welding has generally been used since then. In modern steel girder bridges, the most commonly used girders are I-beam girders and box-girders. In bridge construction, I-beams are normally made of steel although other materials like aluminium and timber are also used. A typical cross section of a half through girder bridge is shown in Figure B.9.

2.4.10 Box girders may be constructed of steel, reinforced concrete or prestressed concrete. They may be cast monolithically or constructed of separately fabricated components that are joined together, but in either case they are normally structurally connected with the deck slab that forms the running surface, see Figure B.10. The webs in a box girder add stability and increase resistance to twisting forces. Box girders are therefore more stable than I-beams and are able to span greater distances (see Table B.7). However, the design and fabrication of box girders is more complex than that of I-beams.

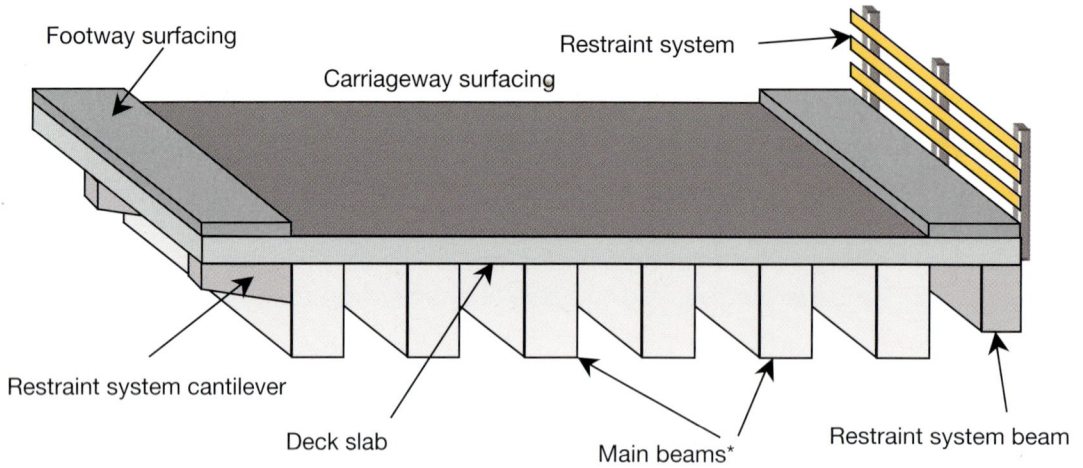

Footway surfacing

Restraint system

Carriageway surfacing

Restraint system cantilever

Deck slab

Main beams*

Restraint system beam

* main beams come in a range of cross section shapes (e.g. T, I, Y, M, rectangular) and material types (e.g. reinforced concrete, prestressed concrete, steel).

Figure B.8 – Typical cross-section of a beam and slab bridge

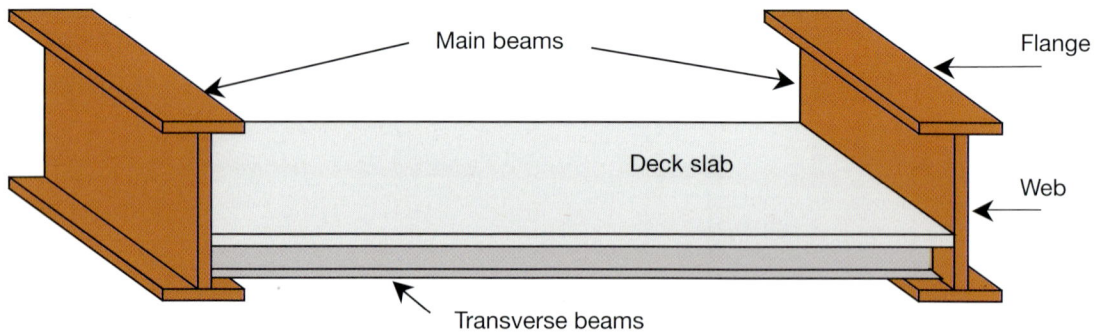

Main beams

Flange

Deck slab

Web

Transverse beams

Figure B.9 – Cross-section of a half through girder bridge

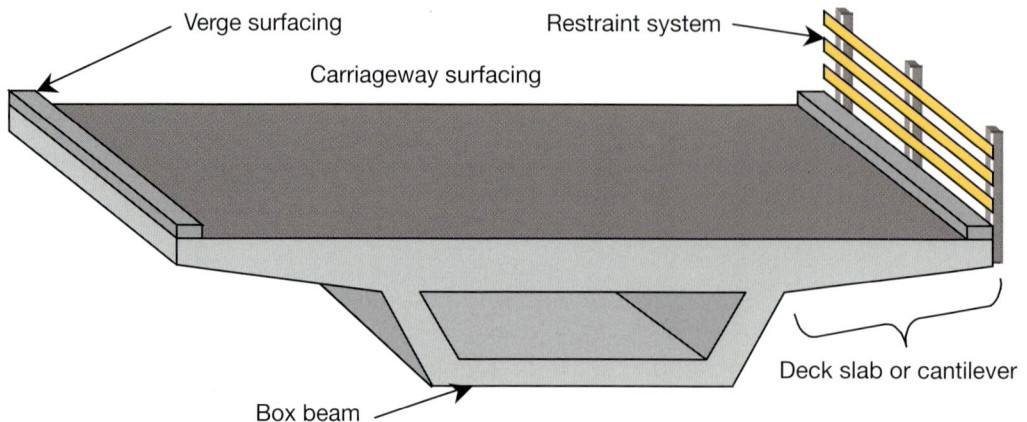

Verge surfacing

Restraint system

Carriageway surfacing

Deck slab or cantilever

Box beam

Figure B.10 – Cross-section of a box beam girder bridge

Arch Bridges

2.4.11 The arch is an ancient form of construction, which has been utilised for thousands of years due to its natural strength. The first arch bridges were built of stone and brick, however modern arch bridges are constructed from materials such as reinforced concrete and steel. Arches may be solid, hollow or formed as a truss, the latter allowing arch bridges to cover longer spans.

2.4.12 Arches use a curved structure with the main carrying element being the arch itself. Both ends of an arch are fixed in the horizontal direction (i.e. no significant horizontal movement is allowed). Thus, when a load is placed on the bridge (e.g. a vehicle passes over it) horizontal forces occur in the springing points of the arch, i.e. the points at which an arch rises from its supports.

2.4.13 Arches can have open or closed/solid spandrels. The bridge deck can be either above, between or underneath the arches. Open spandrel arches are a development of the closed spandrel arch, where the earth fill is, for example, replaced by vertical columns which carry the bridge deck. Different types of arch bridges are shown in Figure B.11a, Figure B.11b and Figure B.11c.

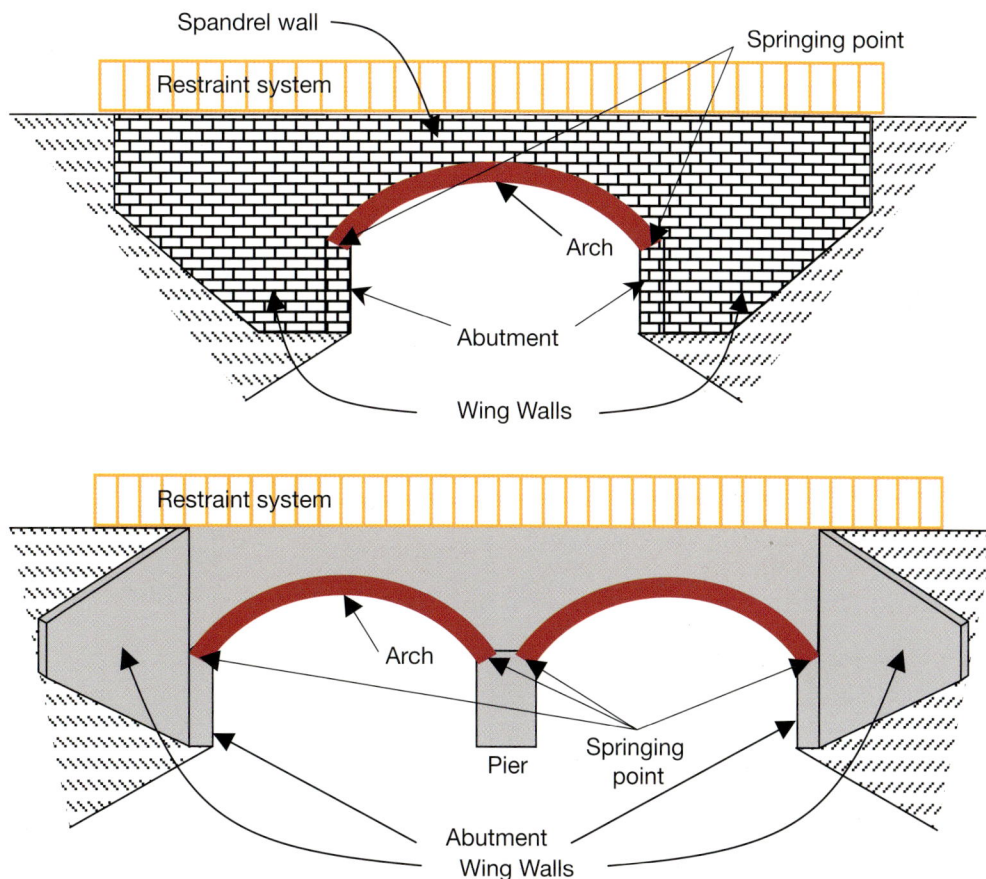

Figure B.11a – Cross-section of closed or solid spandrel arch bridges

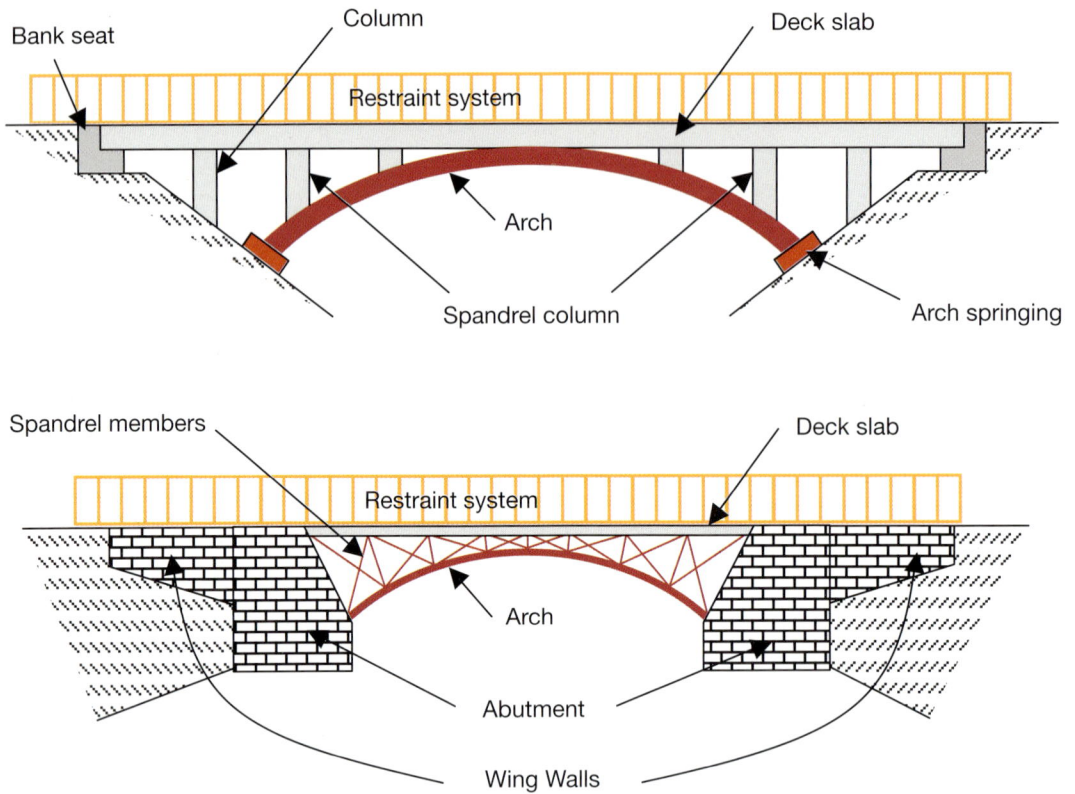

Figure B.11b – Cross-section of open spandrel arch bridges

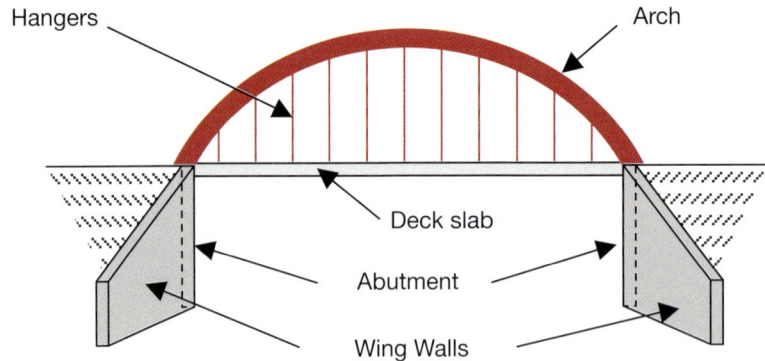

Figure B.11c – Cross-section of a tied arch (Bowstring) bridge

Truss Bridges

2.4.14   Trusses form the primary (main) longitudinal members in a truss bridge. The members that make up a truss are normally straight metal sections (steel in modern constructions), although timber and occasionally reinforced concrete are used.

2.4.15   The position of the bridge deck (and travelling surface) relative to the truss can be used to classify the structure, examples include:

- **Underslung truss-bridge** – the deck is on top of longitudinal trusses, see Figure B.12.

- **Through truss-bridge** – the longitudinal trusses are connected by top and bottom transverse beams and bracing, that forms a 'cage' which the traffic passes through, see Figure B.13.

- *Half through truss-bridges* – the longitudinal trusses are connected by bottom transverse beams and bracing but top beams/bracing is omitted because there would be insufficient headroom for traffic, see Figure B.14.

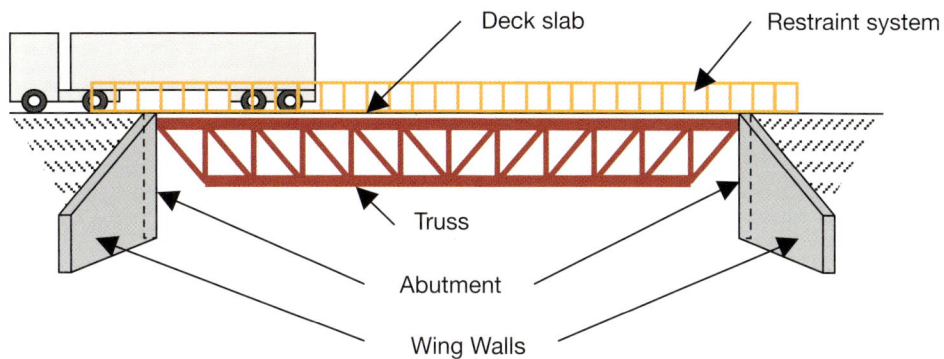

Figure B.12 – Cross-section of an underslung truss bridge

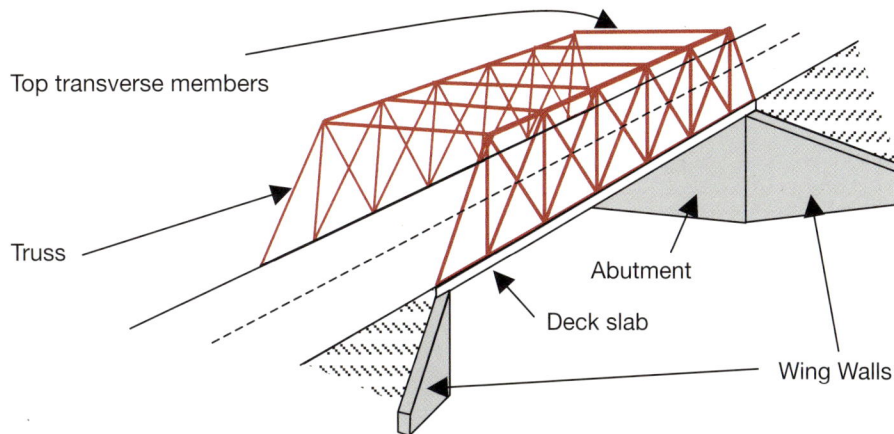

Figure B.13 – Cross-section of a through truss bridge

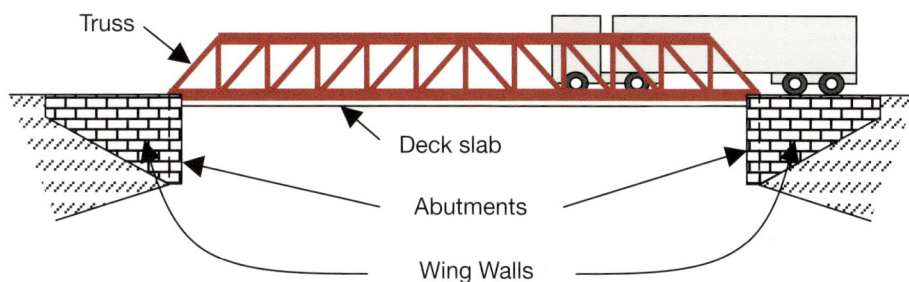

Figure B.14 – Cross-section of a half-through truss bridge

Cable Supported Bridges

2.4.16   A cable supported bridge is where the superstructure is directly or indirectly supported by cables, and where the cables pass over or are attached to one or more towers. There are two types of cable supported bridges: cable stayed bridges and suspension bridges.

2.4.17   Cable stayed bridges and suspension bridges, at first glance, may look similar, i.e. they both have towers and bridge decks that hang from cables, however, the two bridges support the load of the bridge deck in very different ways. The difference lies in how the cables are connected to the towers. In cable-stayed bridges, the cables are attached to the towers, which alone bear the load. In suspension bridges, the cables ride freely across the towers, transmitting loads to the anchorages at either end as well.

Cable Stayed Bridges

2.4.18   Steel cables are strong and flexible as they allow for a slender and light structure which is able to span distances of up to 1000m, although this is continually increasing as technology progresses. The current record is 890mTatara Bridge, in Onomichi-Imabari Japan, but this will be surpassed by Stonecutters Bridge, in Hong Kong, which will have a span of 1018m when completed in 2008. Cable bridges typically have a span of greater than 150m, but shorter span cable stayed bridges have become popular in community and regeneration schemes as they provide a striking landmark. Cable-stayed bridges can be classified by the number of spans, number of towers, girder type and number of cables.

2.4.19   Typically, a cable stayed bridge has a continuous deck (such as a box girder) supported by steel cables stretching down diagonally (usually to both sides) from one or more vertical towers. Types of cable-stayed bridges, differentiated by how the cables are connected to the towers, include:

- **Parallel attachment pattern** – cables are attached at different heights along the tower, running parallel to one other, see Figure B.15.

- **Radial attachment pattern** – the cables extend from a single point at the top of the tower to several points on the bridge deck, see Figure B.16.

- **Hybrid attachment pattern** – the cables have a combination of features common to both the parallel and radial attachment patterns, see Figure B.17.

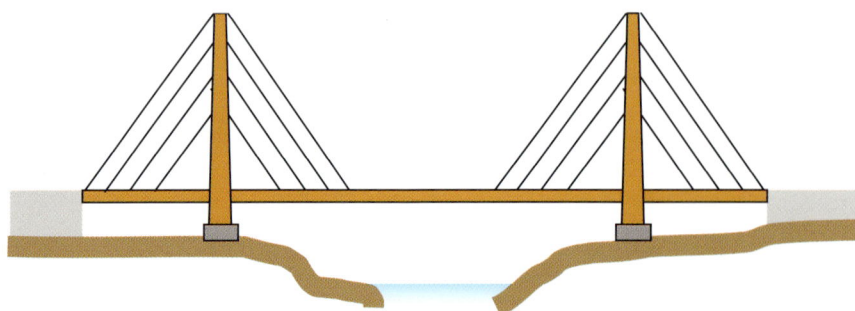

Figure B.15 – Cable stayed bridge: parallel attachment pattern

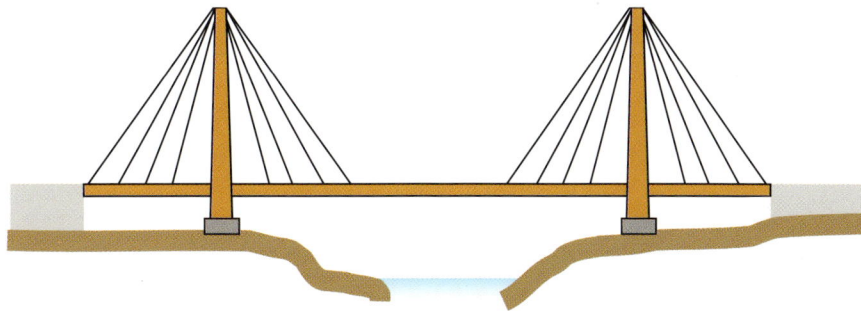

Figure B.16 – Cable stayed bridge: radial attachment pattern

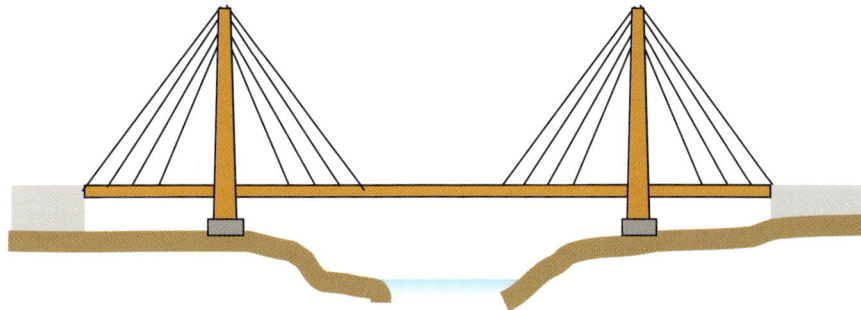

Figure B.17 – Cable stayed bridge: hybrid attachment pattern

2.4.20 There are many variations in the number, type and arrangement of towers, typically including single, double, portal and A-shaped towers; these are shown in Figure B.18.

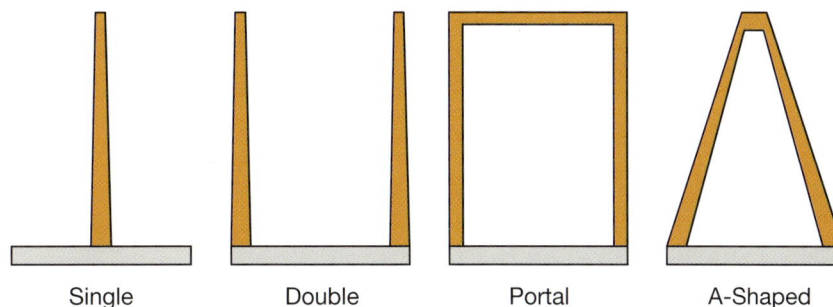

Single   Double   Portal   A-Shaped

Figure B.18 – Typical towers used for cable stayed bridges

Suspension Bridges

2.4.21 In a suspension bridge, shown in Figure B.19, the bridge deck is held by vertical cables, also known as hanger cables, which hang from longitudinal cables. The main longitudinal cables extend from anchorages at either end of the bridge, passing over the towers to provide the required vertical height. Suspension bridges enable long spans to be achieved, i.e. currently spans in excess of 300m are readily achievable, and there are a number of bridges where the span is well in excess of 1000m (the current record for the longest suspension bridge is held by the Akashi Kaikyo Bridge in Japan, also known as the Pearl Bridge, which has a main span of 1,991m).

2.4.22 Though suspension bridges are leading long span technology today, they are in fact a very old form of bridge. In some countries, simple suspension bridges for pedestrians and livestock are still constructed, using techniques and materials similar to the ancient Inca (Latin American) rope bridges, i.e. a shallow downward arc suspended from two high locations using vines and ropes for cables.

2.4.23 The deck of a typical suspension bridge (Figure B.19) is a continuous girder with one or more towers. The girder is usually a truss or box girder, though in shorter spans plate girders are not uncommon. At both ends of the bridge large anchors or counter weights are used to hold the ends of the cables. Some suspension bridges do not use anchors, but instead attach the main cables to the ends of the girder. These self-anchoring suspension bridges rely on the weight of the end spans to balance the centre span and anchor the cable. The main longitudinal cables pass over a special structure known as a saddle (Figure B.19). The saddle allows the cables to slide to accommodate changing loads on the bridge deck, sliding also allows smoothly transfer of the load from the cables to the towers and anchorages.

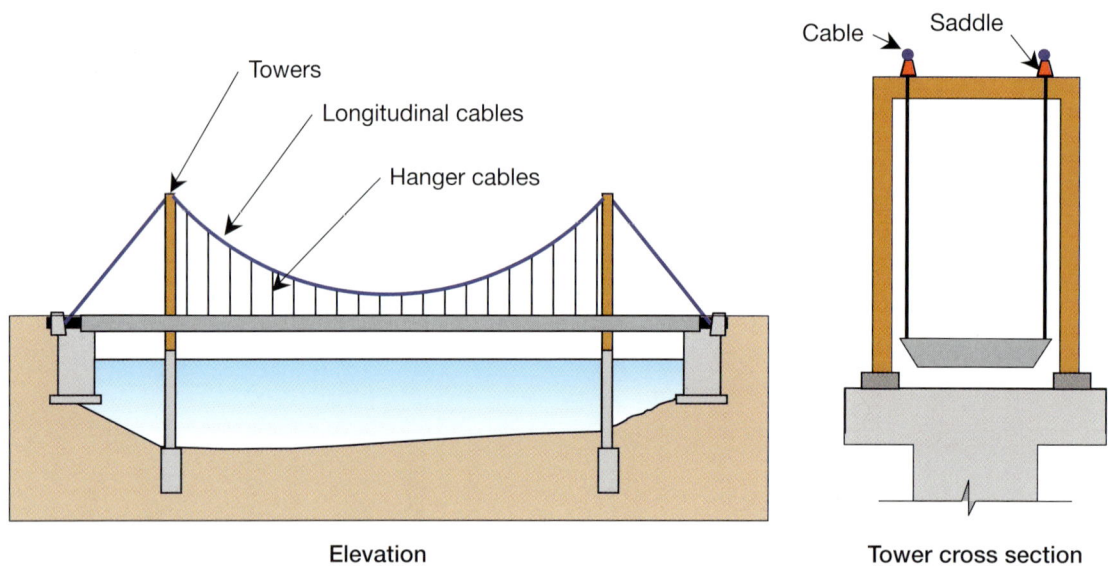

Elevation

Tower cross section

Figure B.19 – Typical suspension bridge

2.4.24 As shown in Figure B.19, the length of the vertical cables changes to accommodate the downward arc of the main load-bearing cables and provide a more comfortable travelling surface. This arrangement allows the deck to be level or to arc slightly upward for additional clearance.

Movable Bridges

2.4.25   A moveable bridge is a bridge having part of all of the superstructure capable of being raised, turned, lifted, or slid from its closed position in order to provide passage to navigable traffic. A moveable bridge is in most cases a beam or girder bridge equipped with machinery which allows the bridge to move in the desired direction.

Wikipedia

3   Other Highway Structures

3.1   OVERVIEW

3.1.1   This section provides details on the construction form and characteristics of common highway structures other than bridges, covering Culverts (Section 3.2), Retaining Walls (Section 3.3), Sign/Signal Gantries (Section 3.4) and Masts (Section 3.5).

3.2   CULVERTS

3.2.1   A culvert is a drainage structure passing beneath a highway embankment that has a proportion of the embankment, rather than a bridge deck, between its uppermost point and the road running courses. Culverts are usually rectangular or circular in cross section [1].

3.2.2   Culverts are made of many different materials, for example, masonry, steel, concrete and polyvinyl chloride (PVC). Steel culverts are usually made from corrugated steel plates and can be shaped like pipes or vaults. The most common type of concrete culvert is the box culvert, but concrete culverts formed as pipes or vaults are also common. They can be cast in situ or prefabricated.

Pipe Culvert

3.2.3   A pipe culvert is circular or elliptical in cross section and made of steel (usually corrugated steel), PVC or concrete. Common types are shown in Figure B.20.

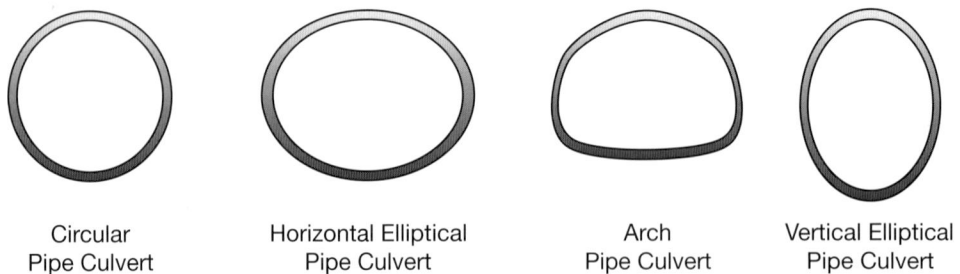

| Circular Pipe Culvert | Horizontal Elliptical Pipe Culvert | Arch Pipe Culvert | Vertical Elliptical Pipe Culvert |

Figure B.20 – Common types and cross-sections of pipe culverts

Box Culvert

3.2.4   A box culvert is rectangular or square in cross-section with rigid connections between the top and bottom slabs and the side walls [1]. It may be single or multi cell and typically made of reinforced concrete. Common types of box culvert are shown in Figure B.21.

Top Slab

Bottom Slab

Interior Wall

Single Cell Box Culvert

Multi Cell Box Culvert

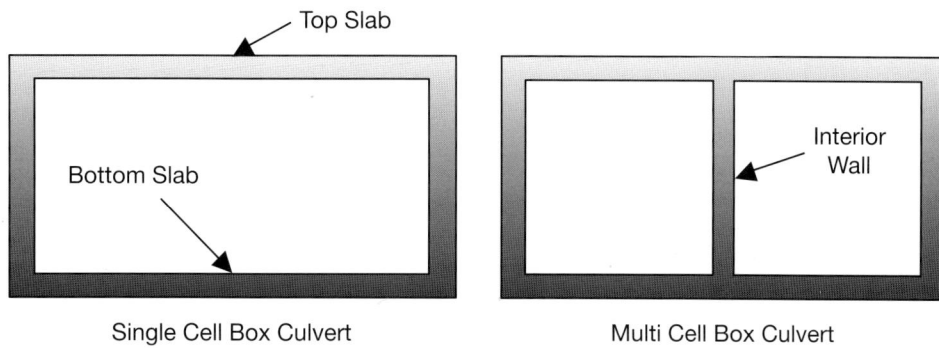

Figure B.21 – Common types and cross-sections of box culverts

Portal or Frame Culvert

3.2.5   A portal frame culvert is rectangular or square in cross-section with only rigid (fixed) connections between the top slab and side walls [1]. These types of structures are usually made of reinforced concrete. An example is shown in Figure B.22.

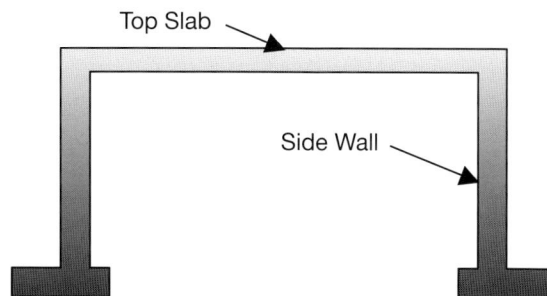

Top Slab

Side Wall

Figure B.22 – Portal or frame culvert

Slab Culverts

3.2.6   A slab culvert is rectangular or square in cross-section without a rigid connection between the top slab and the side walls [1]. Often it does not have a bottom slab. An example is shown in Figure B.23.

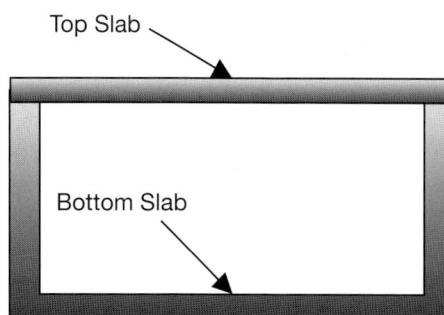

Top Slab

Bottom Slab

Figure B.23 – Slab culvert

3.3 RETAINING WALLS

3.3.1 A retaining wall is any wall, irrespective of height, where the dominant function is to act as a retaining structure for embankments or fill slopes [1]. There are several structural forms for retaining walls including gravity, cantilever on foundation and embedded but other structural forms are also used.

Gravity Retaining Walls

3.3.2 A gravity retaining wall (Figure B.24) resists horizontal earth pressure through its own self-weight [1]. Typical examples of gravity walls, depending on the type of construction material, are:

- **Mass concrete monolithic walls** – Typical profiles of mass concrete walls are shown in Figure B.25. Mass concrete construction is common for walls up to 2 or 3m, but rare above 6m. The surface of the wall may have a cast architectural finish or have been otherwise visually enhanced by bush-hammering or other exposed aggregate treatment; precast concrete facings may have been used as permanent shuttering. Walls may also have independent masonry or concrete cladding attached by mechanical fixings and separated by a cavity.

- **Unreinforced masonry walls** – Stepped or buttressed construction is likely to have been used for masonry walls in excess of 1.5m high. The top of the walls is usually protected with coping stones.

- **Gabions** – Gabions are rectangular boxes (with dimensions that typically range from 0.5m to 4m) normally formed from woven steel wire or welded steel mesh, either galvanised or PVC coated, or from polymer grid fabrications and filled with stone. They may be tied or placed together to form retaining walls. Woven wire and polymer grid boxes are able to deform significantly without loss of strength and are often used where settlement is predicted. Gabion walls are semi-flexible structures, usually constructed at an angle to the vertical to accommodate some settlement after construction and prevent the face of the wall tilting forward. Typical examples of gabion retaining wall layouts are shown in Figure B.26.

- **Crib walls** – Crib walls are composite structures typically constructed of reinforced concrete or timber crib units, backfilled with free draining fill. A variety of types of individual units and facings are available but typical examples are shown in Figure B.27.

- **Reinforced soil walls** – Reinforced soil refers to the use of placed or insitu soil in which tensile reinforcements act through interface friction, bearing or other means to improve stability [17, 18]. It may be used to form a strengthened embankment or, in conjunction with a facing, to form a retaining wall. Techniques used to reinforce ground (virgin soil or existing fill material) are soil nailing and anchoring. Typical systems for reinforced soil walls are shown in Figure B.28.

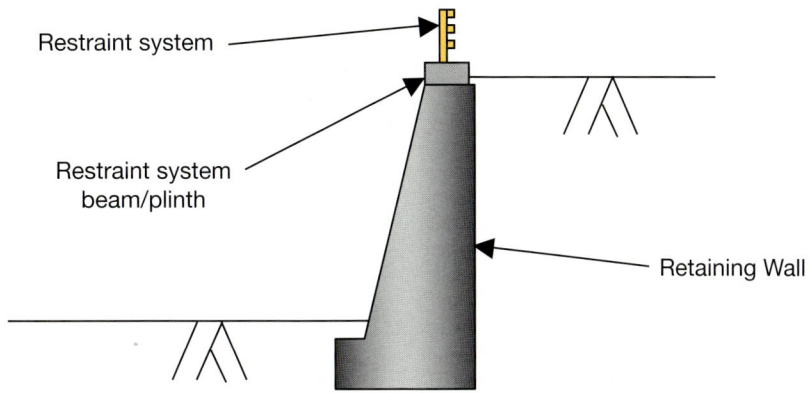

Figure B.24 – Gravity retaining wall

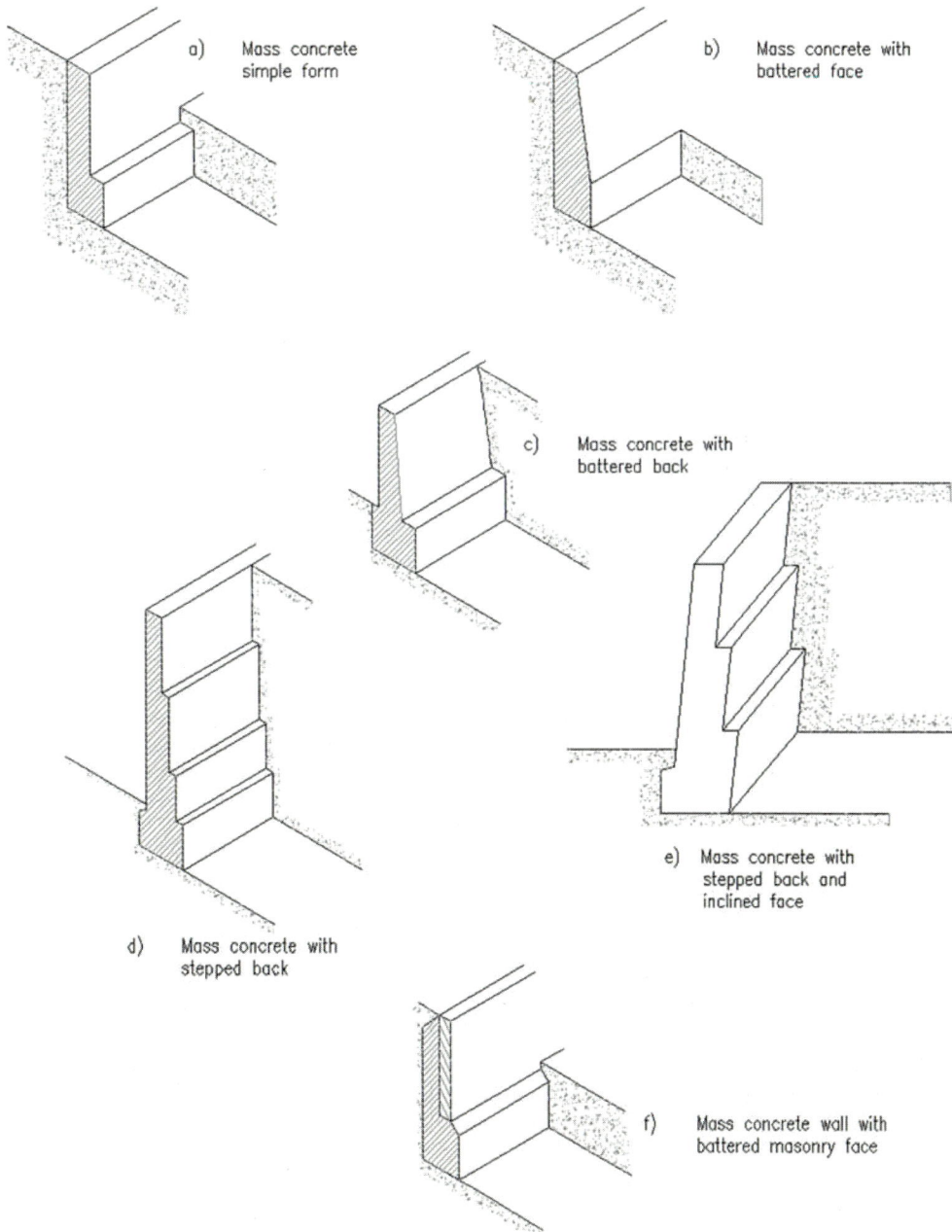

Figure B.25 – Typical forms of mass concrete walls [19]

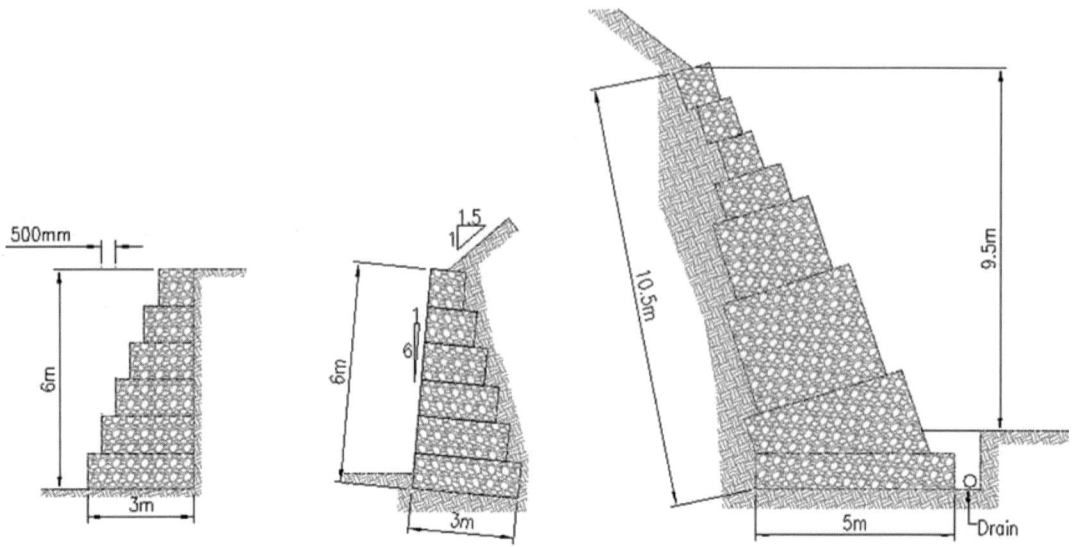

Figure B.26 – Examples of gabion retaining walls [19]

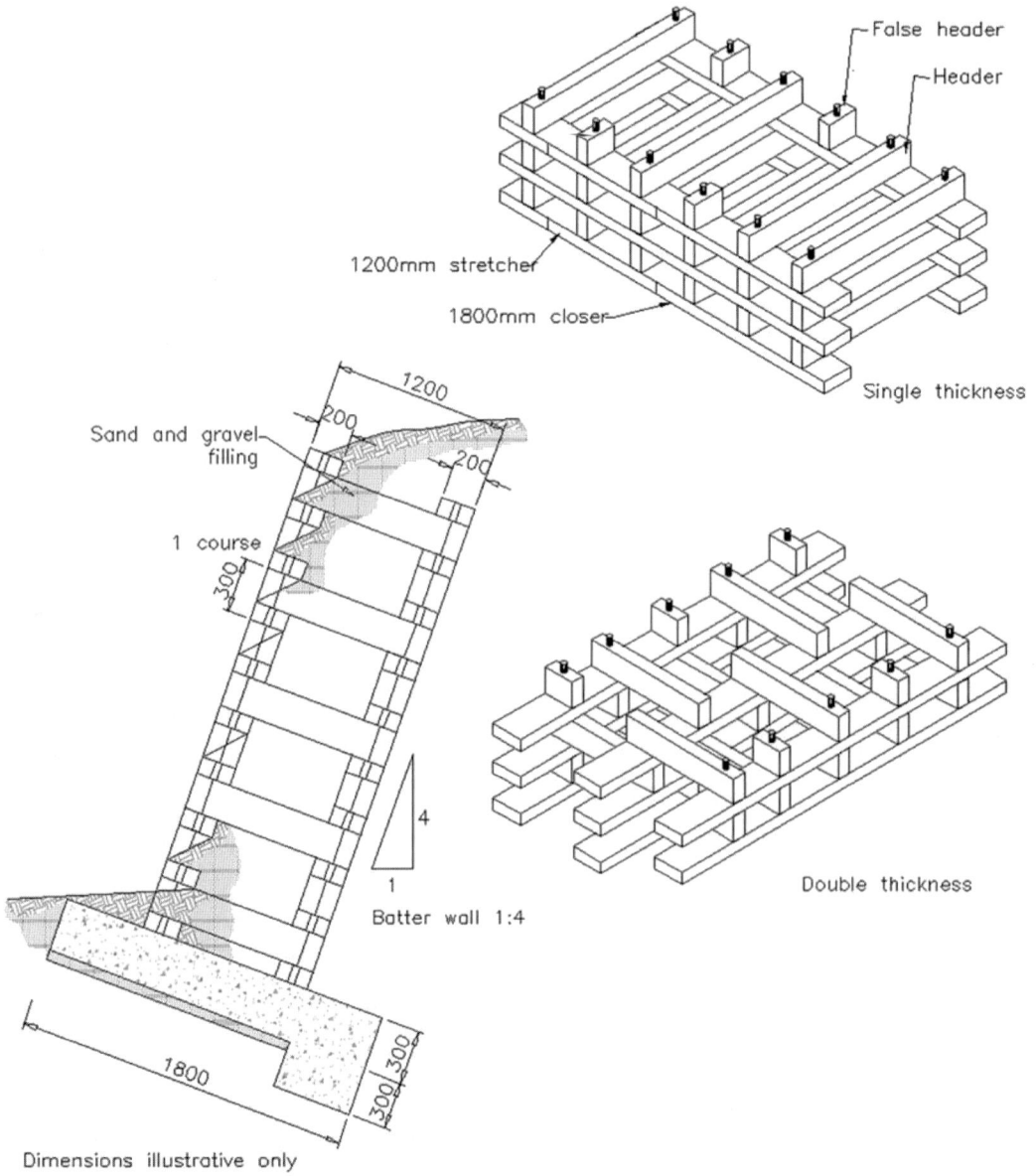

Figure B.27 – Examples of crib walls [19]

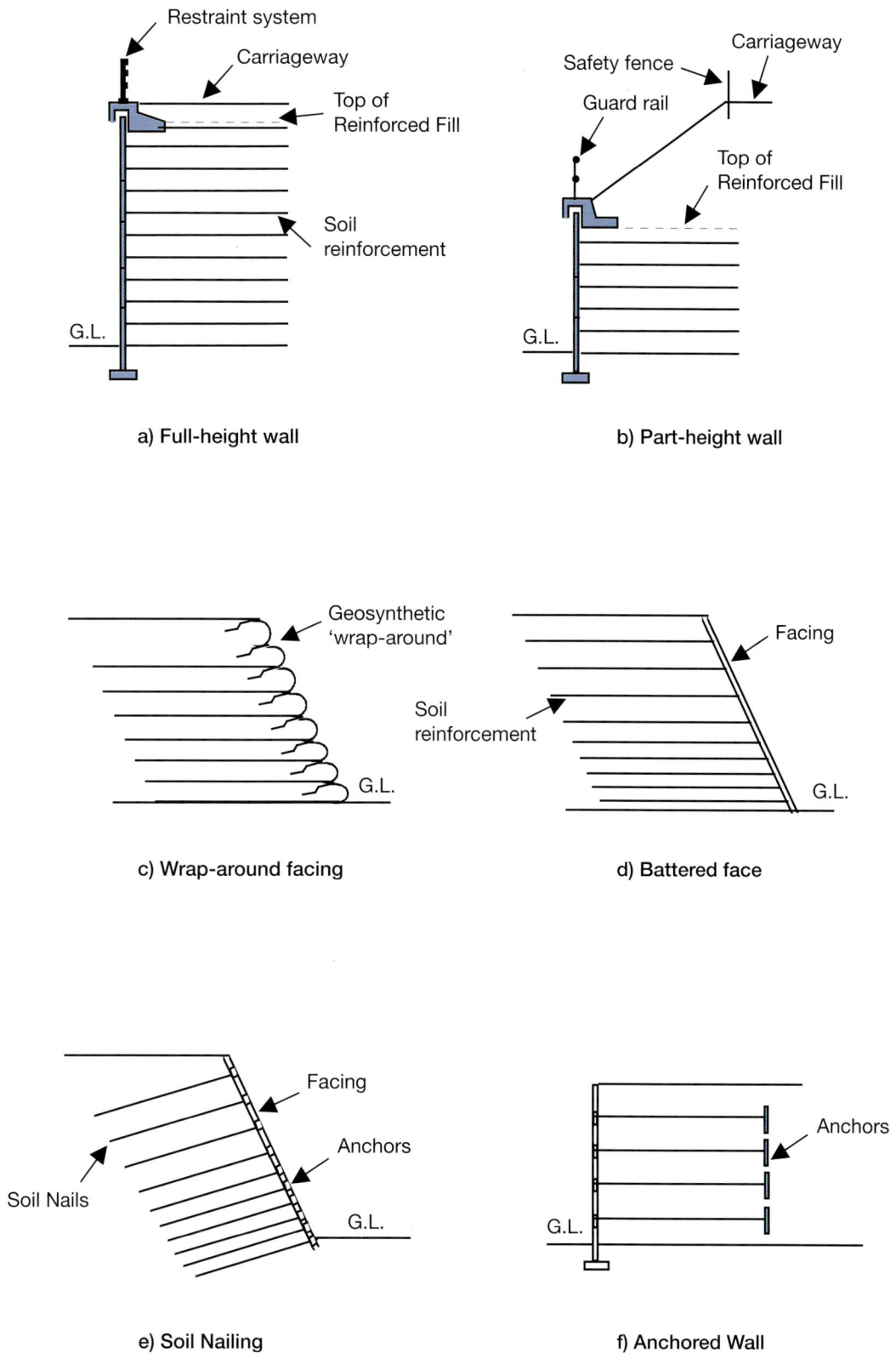

a) Full-height wall

b) Part-height wall

c) Wrap-around facing

d) Battered face

e) Soil Nailing

f) Anchored Wall

Figure B.28 – Typical layouts of reinforced soil walls

Cantilever on Foundation Retaining Wall

3.3.3   A cantilever retaining wall (Figure B.29) is an inverted T or L shaped structure (in cross section) where the vertical section of the wall is rigidly fixed to the horizontal foundation section and transfers horizontal loads to the foundation principally by bending action [1]. They rely on gravity to supply part of the stabilising force, provided by the weight of retained material on the heel of the foundation. Such walls are usually made of reinforced concrete, but may be reinforced masonry.

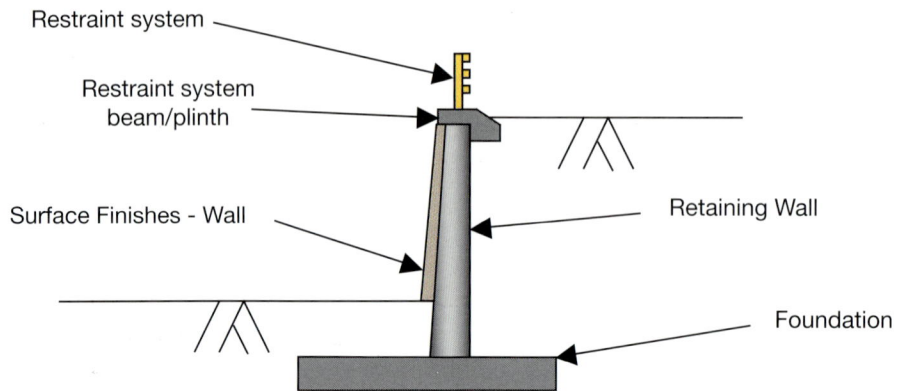

Figure B.29 – Cantilever on foundation retaining wall

Embedded Retaining Walls

3.3.4   An embedded retaining wall (Figure B.30) is similar to a cantilever retaining wall except there is no horizontal foundation component; instead stability is achieved through the embedded depth [1]. Typical examples are:

- **Steel sheet piles** – Steel sheet piling is usually interlocking, producing a continuous, relatively watertight wall. Reinforced concrete capping beams are generally provided to prevent variations in movement at the top of the wall, and also to mask the often irregular line of the piles arising from driving tolerances. Anchored sheet pile walls will have one or more horizontal walings near the top of the wall with ties leading to anchorages located in the retained soil. Anchorages can be ground anchors, mass or reinforced concrete blocks, anchor walls (sheet piles driven singly, in groups, or as continuous walls), and tension piles. The anchors are connected to the wall with steel tie bars or tendons. The only evidence that an anchorage system is present may be the presence of horizontal walings. However, walings may be installed behind the wall, in which case only the projecting tie will be visible on the front face.

- **Concrete sheet piles** – Precast reinforced concrete or prestressed concrete piles may be driven to form embedded walls. Piles are normally 500mm wide, with joints to assist with correct alignment. Watertightness is achieved through the use of interlocks or grouted joints. The heads of the piles, which may have been damaged during piling, are cut back and cast into a concrete capping beam.

- **Timber sheet piles** – Timber piles may have been used for retaining walls, but are more likely to be encountered as river training walls on the approach to structures.

- **Insitu concrete bored pile walls** – Two basic types of bored pile walls may be encountered:

  i.   Contiguous (close bored) piled walls are constructed at centre-to-centre spacings which are equal to or slightly greater than the external diameter of the casing. The piles can be in-line or staggered. The gaps between the piles need to be sealed if a watertight construction is required. Contiguous bored pile walls may have vertical drains incorporated between piles to collect water in front of the finished wall.

ii. Secant piled walls are bored at centres less than the casing diameter. Alternate unreinforced piles are constructed first, which are then partially cut away during the boring of the intermediate piles. The second stage piles are of full circular section and reinforced. This method produces a relatively watertight construction.

- **Diaphragm walls** – Diaphragm wall construction is achieved by excavating a deep narrow trench, which is kept full of a bentonite ('slurry') suspension. The bentonite suspension maintains the stability of the trench during excavation. Walls are formed in panel lengths, either cast insitu or precast. An insitu diaphragm wall is constructed by lowering a reinforcing cage into the bentonite filled trench and placing concrete by tremie pipe, the concrete displaces the bentonite suspension. A precast diaphragm wall is formed by lowering precast units in to the trench.

- **Soldier pile walls** – Soldier piled walls consist of vertical piles driven or cast into the ground at regular centres. Sheeting is then spanned horizontally between the piles or vertically between walings attached to the piles. The piles may be of reinforced concrete or steel sections, whilst the sheeting can be steel, precast concrete or timber. Sheeting planks can be separated by up to 50mm to facilitate drainage of ground water.

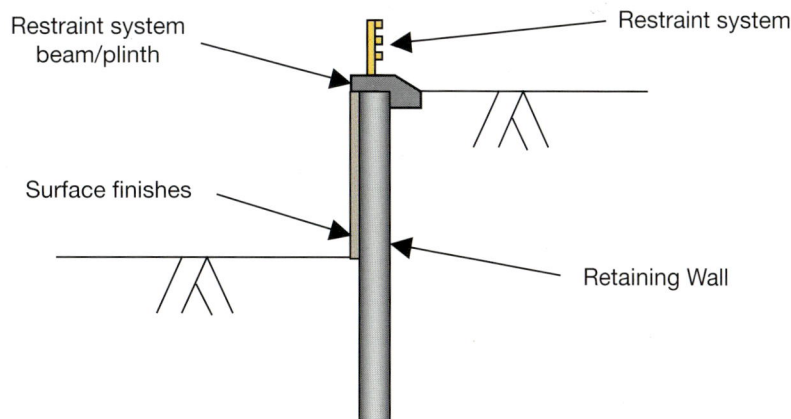

Restraint system beam/plinth

Restraint system

Surface finishes

Retaining Wall

Figure B.30 – Embedded retaining wall

3.4   SIGN/SIGNAL GANTRIES

3.4.1   Sign and signal gantries are structures which either wholly span the carriageway or are partially cantilevered over the carriageway or hard shoulder, and are for supporting large signs, motorway signals or message signs. The main structural elements on a sign/signal gantry are typically made of steel, aluminium, reinforced concrete, prestressed concrete, and more recently, of advanced composites (e.g. FRP and plastic), or some combination of these. Gantries, especially those made of metal and advanced composites, are slender and flexible and may be subject to fatigue stress from wind loading and pressure waves or vibration generated by passing traffic. The typical elements/features found on a sign/signal gantry are:

Highways agency

- **Main load carrying elements** – the main horizontal element/s, perpendicular to the direction of traffic flow, which is normally a truss or beam that either spans the carriageway or cantilevers over part of the carriageway. Roadside cantilever arm sign/signal masts are typically used to position a traffic signal over a single lane of a multiple-lane carriageway.

- **Supports** – the vertical columns (or legs) that support the main load carrying elements.

- **Foundations** – typically reinforced concrete pads that may project above ground level, although piling may have been used on poor load bearing ground.

- **Transverse members** – when there are two or more main load carrying elements in parallel, they are generally connected by transverse members that act as bracing and a platform for an access walkway.

- **Base connection** – the connection between the support and the foundation, which is typically in the form of a metal plate with holding down bolts for steel gantries.

- **Support/main element connection** – on some types of sign/signal gantries (e.g. beam) this connection is a significant, feature, vulnerable to deterioration, that needs to be inspected, e.g. bolted or welded steel plate joints.

- **Access walkway/deck** – this may be on top of transverse beams or cantilevered from the main load carrying element. It enables access to the sign/signal equipment for inspection, cleaning and maintenance.

- **Access ladder** – if present, it is normally found on the supports and provides access to the walkway (some sign/signal gantries do not have access ladders/walkways, instead access for inspection, cleaning and maintenance is achieved using mobile platforms).

- **Sign and signal supports** – used to support the signs/signals and attached them to the main structural elements of the gantry.

- **Traffic signs and signals** – for example, matrix signs.

3.4.2   The following figures present elevations and cross sections of typical highway sign/signal gantries and highlight some of the above elements and features. Recent years have seen a significant increase in the number of new sign/signal gantries being erected because they provide valuable traffic management facilities. This trend is likely to continue and as such the structural form and materials used for sign/signal gantries will continue to evolve.

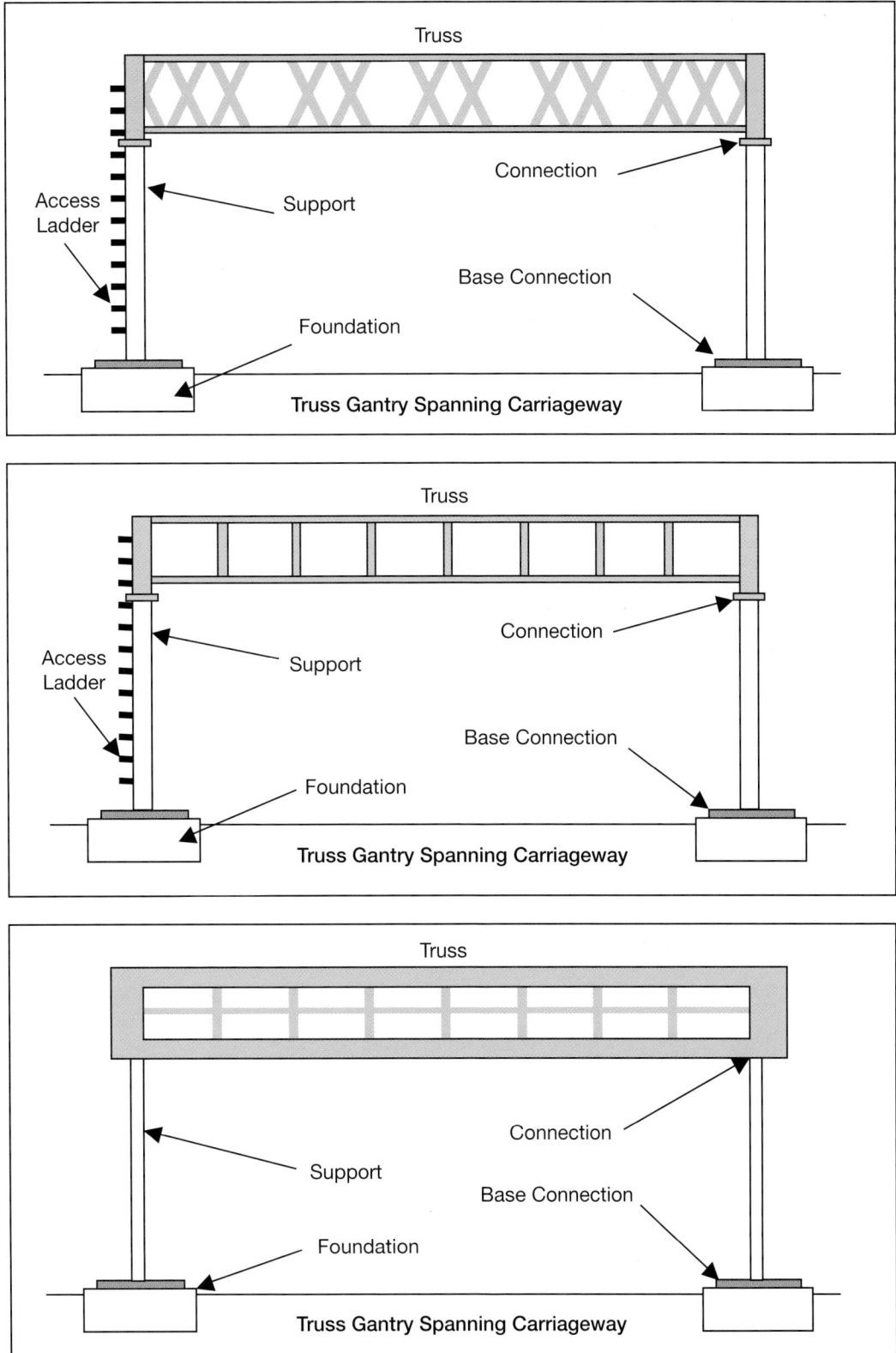

Figure B.31 – Types of sign/signal gantries (1)

Beam

Possible connection position

Access Ladder

Supports

Base Connection

Foundation

Beam Gantry Spanning Carriageway

Cantilever Truss

Connection

Access Ladder

Support

Base Connection

Foundation

Cantilever Truss Gantry

Cantilever Beam

Possible connection positions

Access Ladder

Support

Base Connection

Foundation

Cantilever Beam Gantry

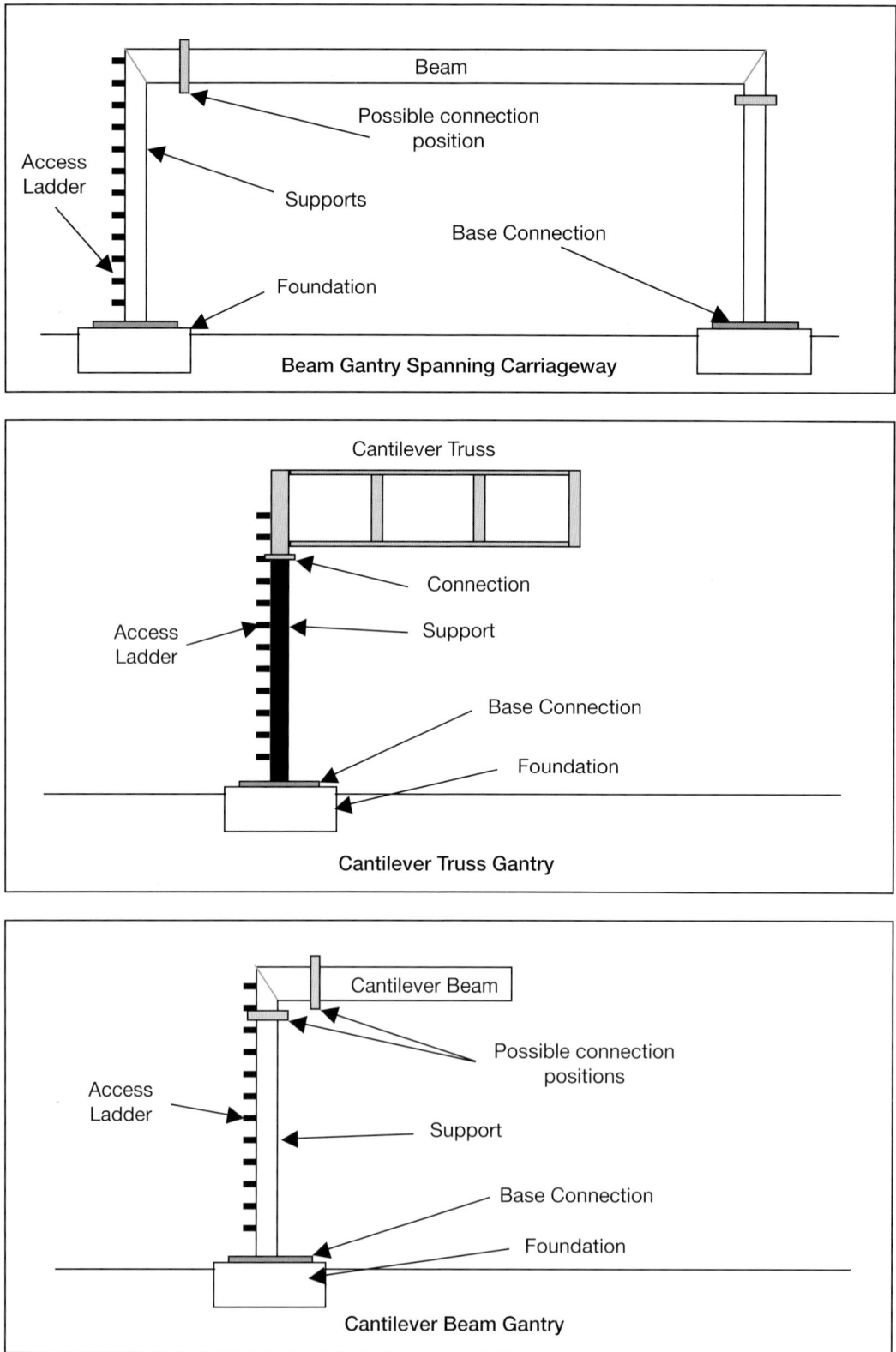

Figure B.32 – Types of sign/signal gantries (2)

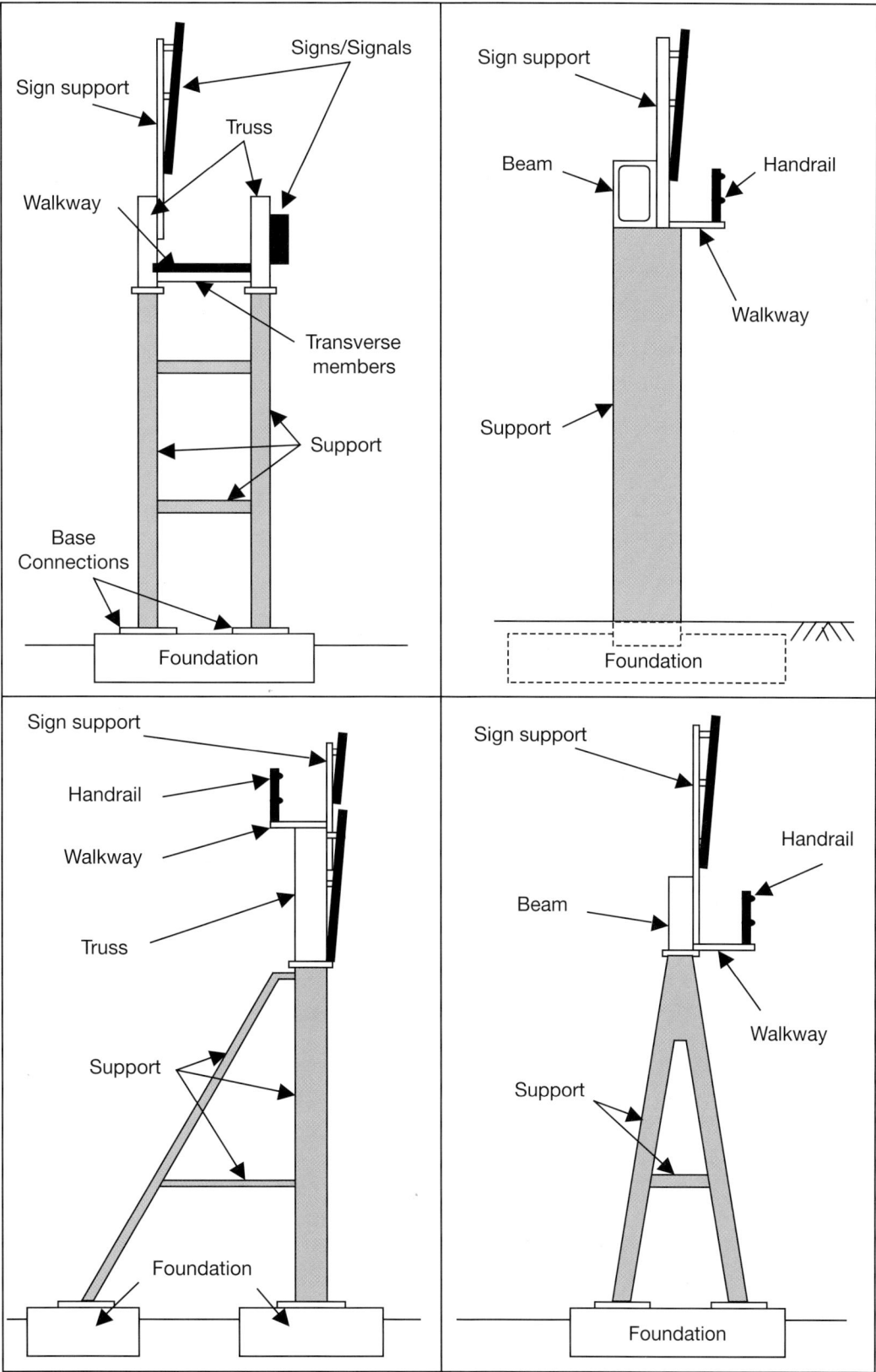

Figure B.33 – Cross-sections of typical sign/signal gantries

3.5   MASTS

3.5.1   Lighting masts (20m or higher) and masts for cameras and telecommunication transmission equipment are usually of steel or reinforced concrete. The main structural element acts as a vertical cantilever, relying on the ground anchorage to provide fixity. Masts are typically bolted onto an anchorage cast into a spread footing or piled foundation. However, some smaller masts may have buried (planted) foundations. When masts are mounted on structures, they are bolted onto anchorages cast into the deck. Some high lighting masts have lifting gear, enabling light fittings to be lowered to the ground for inspection and maintenance.

Highways agency

3.5.2   Catenary lighting systems consist of a row of masts or columns, often in the central reserve, with the lighting units attached to a cable (the catenary) joining the tops of the masts. At each end of a run, the cable drops down to an anchorage at ground level. The cable provides longitudinal restraint to the masts.

4 Structural Mechanics

4.1 OVERVIEW

4.1.1 Mechanics of a solid structure is primarily concerned with the effect of energy and forces acting in and around that structure and their relation to equilibrium, deformation, or motion [9]. Inspectors are be predominantly concerned with physical systems that are in equilibrium, for example, in a bridge the relative positions of components do not vary over time; and both the structure and components are at rest under the action of external forces of equilibrium [20].

4.2 BRIDGE DESIGN LOADINGS

4.2.1 Bridges are designed to carry vehicles, pedestrians and other transient loads (such as those due to temperature and wind effects); all collectively known as 'live loads' (see Figure B.34). In addition, bridges have to withstand their own weight and the weight of any permanent fixtures, such as parapets, surfacing and finishes, which constitute 'dead load' (see Figure B.35). Although dead loads on a structure usually remain constant over time in some cases these may change during the life of a bridge due to work such as installation of thicker surfacing, upgraded restraint systems and safety fencing or additional utilities. Bridge design loadings determine the size and configuration of its members, which are designed to withstand the loads acting on them in a safe and economical manner [21].

4.2.2 In some bridges, such as arch bridges, the dead load greatly exceeds the live load with the result that the bridge is primarily required to sustain its self-weight, i.e. the effects of traffic being small compared to self-weight.

4.2.3 Loads may be concentrated or distributed depending on the way in which they are applied to the structure. A concentrated or point load, is applied at a single location or over a very small area. Vehicle loads are considered to be concentrated loads. A distributed load is applied to all or part of the member, and the amount of load per unit of length is usually constant. The weight of superstructures, bridge decks, carriageway surfacing, and bridge parapets produce distributed loads. Transient loads, such as wind, snow and ice, are also usually distributed loads.

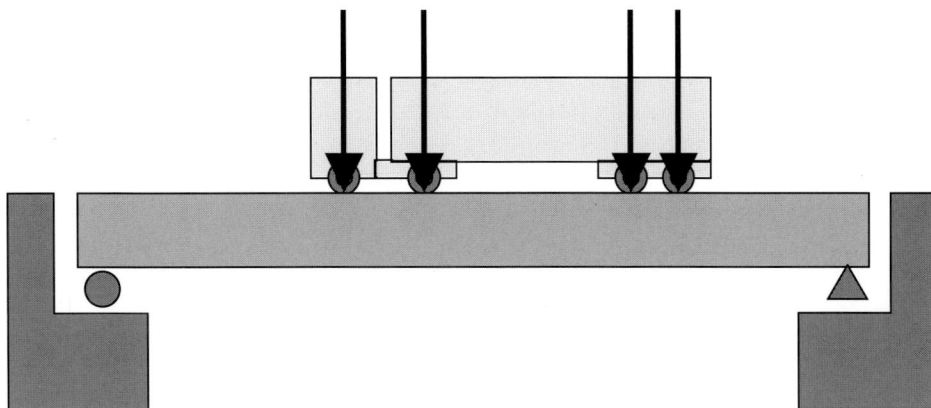

Figure B.34 – Live load on a bridge [9]

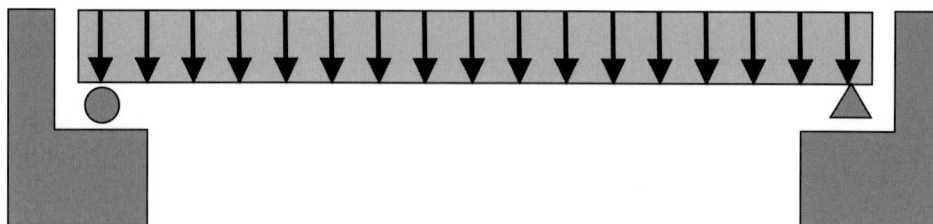

Figure B.35 – Dead load on a bridge [9]

4.2.4   The vehicular loading used in the design and assessment of a bridge falls under two distinct categories; one is the equivalent of a queue of heavy commercial vehicles and the other is the equivalent of one special vehicle carrying exceptionally heavy load. Designing and assessing bridges to comply, as a minimum, with the former meets the UK statutory requirement for bridges to carry vehicles up to 40 tonnes with adequate margins of safety against collapse [21]. However, a bridge may show signs of distress such as cracking, under a succession of much smaller loads, leading to a reduction in durability and/or carrying capacity of the structure or its individual components. Such deterioration affects the 'serviceability' of the structure and current design codes require adequate safeguards to ensure serviceability of all components for specified loading and life [22].

4.2.5   It is important to remember that bridges are long life structures (current design life is 120 years) and therefore the character and intensity of traffic loads are likely to change significantly during the life of a bridge. For instance, older masonry arch bridges may have been originally designed and constructed to cater for loading from horses and carts, but now they are supporting modern traffic loads. Fortunately, many arch bridges have been shown to have a considerable reserve of strength.

4.3   MATERIALS RESPONSE TO LOADINGS

4.3.1   Each member of a bridge has a unique purpose and function, which directly affects the selection of material, shape, and size for that member. Certain terms are used to describe the response of a bridge material to loads and a working knowledge of these terms is essential for the inspector.

Force

4.3.2   A force is the action that a body exerts on another body. A force can occur in any direction, for example, vehicular live load acts vertically on the deck of a bridge, whilst traction forces apply horizontally in the direction of movement. For engineering analysis the resultant force is translated into component forces in the x, y and z coordinate system, i.e. Fx, Fy and Fz as shown in Figure B.36. In most cases, the resultant force is known only by its magnitude, position and direction rather than its individual components.

4.3.3   The unit of measurement of force is normally Newtons (N), or Kilonewtons (kN), where 1 kN = 1000N. The weight of a body with a mass of 1 kg will generate, due to gravity, an equivalent force of 9.81N.

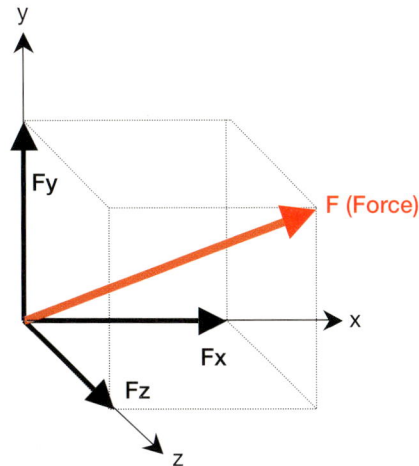

Figure B.36 – Force components [20]

Stress

4.3.4 Stress is defined as the force per unit area within a body that balances and reacts to the loads applied to it. This is expressed in the formula below and the unit of stress is normally N/mm^2 or kN/m^2.

$$Stress\ (S) = \frac{Force\ (F)}{Area\ (A)}$$

Deformation

4.3.5 When a force is applied to a structural member, its shape changes or undergoes local distortion due to stress. This phenomenon is known as deformation. For instance, in Figure B.37 it can be seen that the force (indicated by the arrow) has caused deformation in the cylinder so that the original shape (dashed lines) has changed (deformed) into one with bulging sides. The sides bulge because the material, although strong enough to not crack or otherwise fail, is not strong enough to support the load without change, thus the material is forced out laterally.

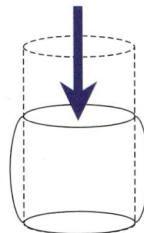

Figure B.37 – Deformation

Strain

4.3.6 Strain is the basic unit of measure used to describe the amount of deformation. For example, strain in a longitudinal direction is computed by dividing the change in length by the original length (as such strain is dimensionless).

$$Stress\ (\varepsilon)\ = \frac{Change\ in\ Length\ (\Delta L)}{Original\ Length\ (L)}$$

Stress-Strain Relationship

4.3.7   For most structural materials, values of stress and strain are directly proportional only up to a particular value called the elastic limit (see Figure B.38). The material property, which defines its stress-strain relationship, is called the modulus of elasticity, or Young's modulus and this is the slope of the elastic portion of the stress-strain curve. The higher the modulus of elasticity of the material, the smaller the strain for a given stress.

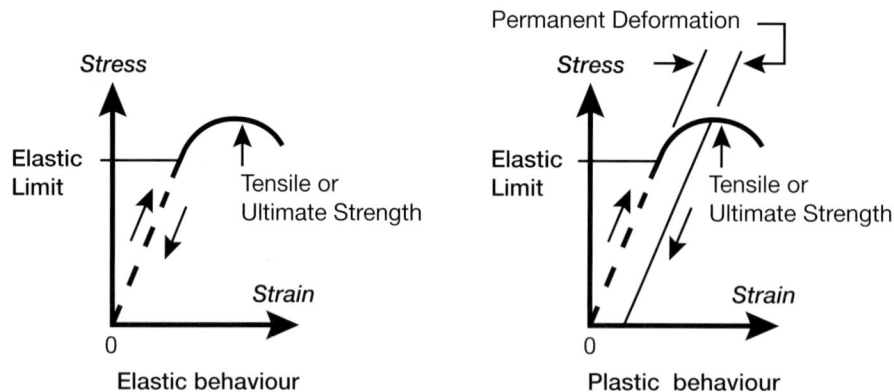

Figure B.38 – Stress-strain relationship [20]

4.3.8   Up to the elastic limit, a material deforms elastically, i.e. the material distortion is reversible and it returns to its original shape upon removal of the stress, this is know as elastic deformation. Bridges are designed to deform elastically and return to their original shape after live loads are removed. Beyond the elastic limit, the deformation is plastic and strain is not directly proportional to a given applied stress, this is known as plastic deformation. Plastic deformation is the irreversible or permanent distortion of a material and the material would retain a deformed shape upon removal of the stress.

4.3.9   Materials respond to loadings in a manner dependent on their mechanical properties. In characterising materials, certain mechanical properties are important, including:

- ***Yield Strength*** – is defined as the stress at which a material begins to plastically deform. Knowledge of the yield point is vital when designing a component since it generally represents an upper limit to the load that can be applied without causing catastrophic failure.

- ***Tensile Strength*** – is the stress level defined by the maximum tensile load that a material can resist without failure. Tensile strength corresponds to the highest point on the stress-strain curve and is sometimes referred to as the ultimate strength.

- ***Compressive Strength*** – is the capacity of a material to withstand axially directed pushing forces (see paragraphs 4.4.2-4.4.4) without failure.

- ***Toughness*** – is the energy required to break a material and this is not necessarily related to strength. A material might have high strength but little toughness. A ductile material (see paragraph 4.3.13) with the same strength as a non-ductile material (see paragraph 4.3.14) will require more energy to break and thus exhibit more toughness.

Creep

4.3.10   Creep is a form of plastic deformation that occurs gradually as a result of long term exposure to levels of stress that are below the yield or ultimate strength of the material. Creep is more severe in materials that are subjected to heat for long periods and near melting point. The rate of this damage is a function of the material properties and the exposure time, exposure temperature and the applied load (stress). Depending on the magnitude of the applied stress and its duration, the deformation may become so large that a component can no longer perform its function. Creep is usually a concern to bridge engineers when evaluating components that operate under high stresses and/or temperatures. Creep is not necessarily a failure mode, but is instead a damage mechanism. Moderate creep in concrete is sometimes welcomed because it relieves tensile stresses that may otherwise have lead to cracking.

Thermal Effects

4.3.11   Materials expand as temperature increases and contract as temperature decreases. The amount of thermal deformation is proportional to:

- the coefficient of thermal expansion of the material

- the temperature change

- the member length

4.3.12   In bridges, thermal effects are most commonly experienced due to the longitudinal expansion and contraction of the superstructure although on wide bridges significant thermal effects can occur in the transverse direction. The forces generated will depend on whether the members are free to expand and contract or not and should be considered by the inspector. Even when expansion and contraction is allowed significant forces can occur due to the resistance of bearings. However, when movement is inhibited or prevented larger forces will be generated.

Ductility and Brittleness

4.3.13   Ductility is the measure of plastic (permanent) strain that a material can endure. A ductile material will undergo a large amount of plastic deformation before breaking (see Figure B.39). It will also have a greatly reduced cross-sectional area before breaking. Structural materials that are generally ductile include steel and aluminium.

4.3.14   Brittle, or non-ductile, materials will not undergo significant plastic deformation before breaking (see Figure B.39). Failure of a brittle material occurs suddenly, with little or no warning. Structural materials that are generally brittle include concrete, cast iron, stone and timber.

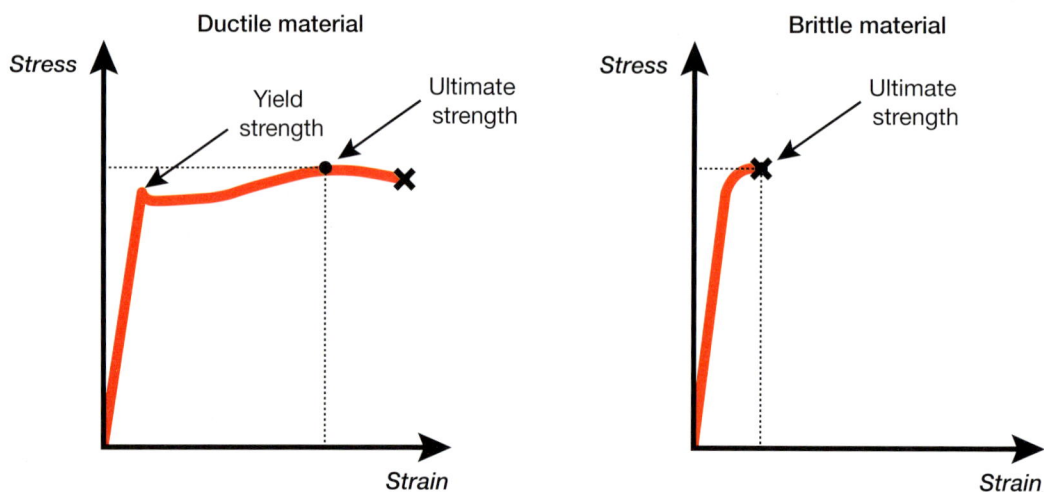

Figure B.39 – Typical stress-strain curves for ductile and brittle materials

Fatigue

4.3.15　Fatigue is a material response that describes the tendency of a material to break when subjected to repeated loading. Fatigue failure occurs within the elastic range of a material after a certain number and magnitude of stress cycles have been applied. Each material has a hypothetical maximum stress value to which it can be loaded and unloaded an infinite number of times. This stress value is referred to as the fatigue limit and is usually lower than the ultimate strength for infrequently applied loads. Ductile materials have high fatigue limits, while brittle materials have low fatigue limits.

4.4　STRUCTURES RESPONSE TO LOADINGS

4.4.1　The structural effects of loads and moments are assessed in terms of axial forces, such as tension and compression, bending forces, shear forces and torsion forces. These forces may act individually or in combination. In calculating these forces, the analysis is governed by equations of equilibrium. Equilibrium equations represent a balanced force system where the sum of all forces and the sum of all moments acting on a body in any direction must be zero [20].

Axial Forces

4.4.2　Axial forces may be tensile or compressive. A member is said to be in tension when forces applied to it tend to pull it apart, as for example in the cables or hangers of a suspension bridge. A member is in compression when the applied force tends to push its ends together as in the columns supporting a bridge deck (Figure B.40).

Axial Compression

Axial Tension

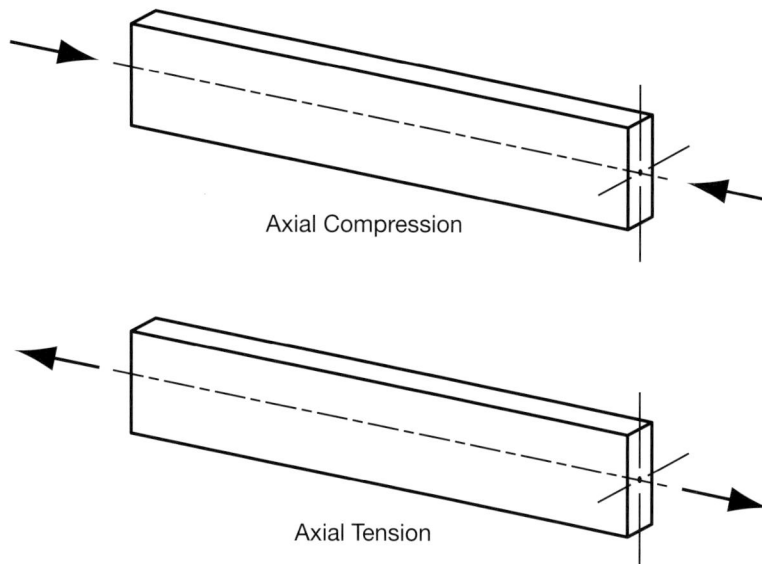

Figure B.40 – Axial forces [9]

4.4.3   Axial forces act uniformly over a cross-sectional area and the axial stress can be calculated by dividing the force by the area on which it acts. Therefore, when structural members are designed to resist pure axial forces, the cross-sectional area will vary depending on the magnitude of the force, whether the force is tensile or compressive, and the type of material used.

4.4.4   The acceptable axial compressive stress is generally lower than that for tension because of a phenomenon called buckling (see Figure B.41). This is a failure mode characterised by a sudden deformation of a structural member at compressive stresses that are lower than the compressive stresses that the material itself is capable of withstanding.

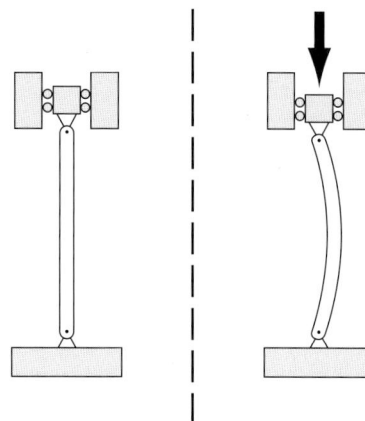

Figure B.41 – A column exhibiting the characteristic deformation of buckling

Bending Forces

4.4.5   Bending (also known as flexure) characterizes the behaviour of a structural element, such as a beam, subjected to loading along its length between the points of support, see Figure B.42. Bending produces reactive forces inside an element as it attempts to accommodate the flexural load and associated deformation. In Figure B.42 the material at the top of the beam is being compressed while the material at the bottom is being stretched.

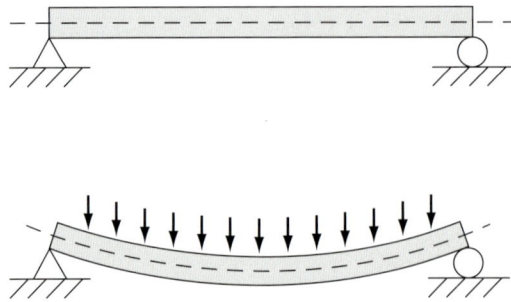

Figure B.42 – A simply supported beam bending under a distributed lateral load

4.4.6   There are three notable internal forces caused by this type of loading, shear parallel to the loading, compression along the top of the beam, and tension along the bottom of the beam; these are shown in Figure B.43. The last two forces form a bending moment as they are equal in magnitude and opposite in direction. This bending moment produces the sagging deformation characteristic of members experiencing bending.

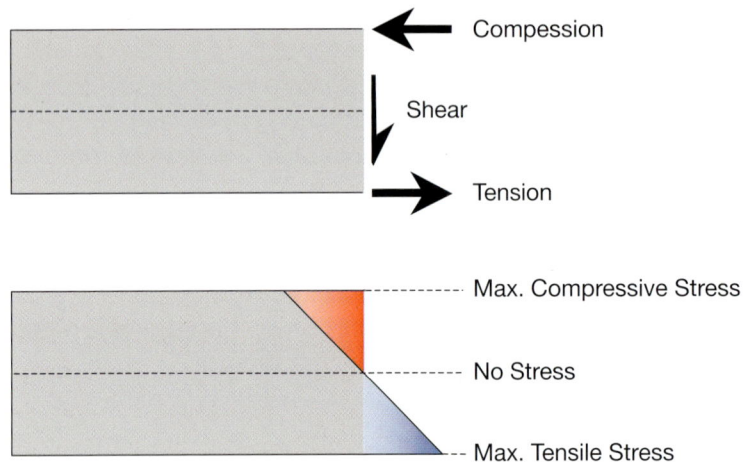

Figure B.43 – Internal stress distribution of a beam in bending under working loads

4.4.7   The compressive and tensile forces shown in Figure B.43 induce stresses on the beam. The maximum compressive stress is found at the uppermost edge of the beam while the maximum tensile stress is located at the lower edge of the beam. Since the stresses between these two opposing maxima vary, there exists a point on the linear path between them where there is no bending stress, i.e. the neutral axis. Because of this area with no stress and the adjacent areas with low stress, using uniform cross section beams in bending is not a particularly efficient means of supporting a load as it does not use the full capacity of the beam until it is on the point of collapse. Wide-flange beams (I-Beams) and truss girders effectively address this inefficiency as they minimize the amount of material in this under-stressed region.

4.4.8   At higher loadings the stress distribution becomes non-linear, and ductile materials will eventually form a plastic hinge where the magnitude of the stress is equal to the yield stress everywhere in the beam, with a discontinuity at the neutral axis where the stress changes from tensile to compressive. This plastic hinge state is typically used as a limit state in the design of steel structures.

Shear Forces

4.4.9   The vertical shear force shown in Figure B.43 produces vertical shear stresses and complimentary equal horizontal shear stresses that cause a slicing or scissors action as shown in Figure B.44. Shear forces normally occur in conjunction with other forces (e.g. bending forces) and may result in the development of tension cracks, e.g. in concrete these appear as diagonal cracks.

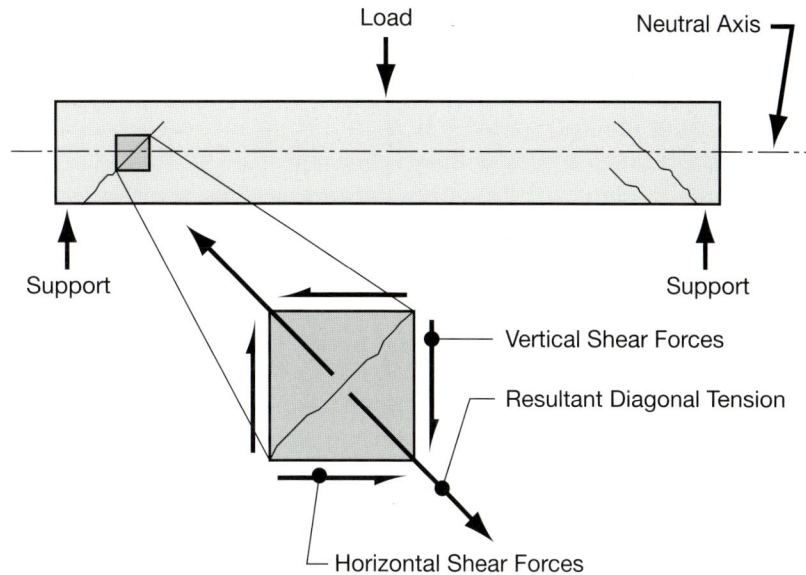

Figure B.44 – Shear forces in an element [9]

4.4.10   Beams and girders are effective shear resisting members. In an I-beam or T-beam, most of the shear is resisted by the web, whilst in a solid rectangular beam shear is resisted by the entire cross-section.

Torsion Forces

4.4.11   Torsion forces are produced when a member is subjected to a twisting motion (see Figure B.45). Beams experience torsion when the load is located away from the longitudinal axis. As with shear, torsion is normally associated with other forces giving rise to complex distributions of stress.

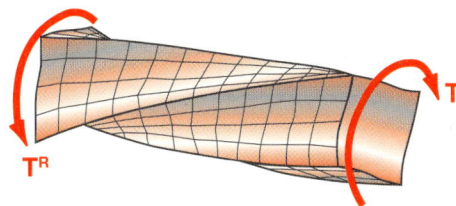

Figure B.45 – Torsion [23]

5 Properties of Construction Materials

5.1 OVERVIEW

5.1.1 Many different materials are utilised in the construction and maintenance of highway structures, the most widely used include concrete, steel, and masonry although wrought iron, cast iron, timber as well as advance composite materials such as fibre reinforced polymer may also be used. The behaviour of a structure under load is significantly influenced by the properties of its materials and as such it is important for the inspector to appreciate these properties and how they influence the safety and functionality of the specific component and the whole structure. The following sections summarise the properties of common structural materials.

5.2 CONCRETE

5.2.1 Concrete is a composite building material made from a combination of a graded range of stone aggregate particles, a cement binder and water. Often cementitious materials such as pulverised fly ash, ground granulated blast furnace slag or silica fume are used as partial cement replacements. Contrary to common belief, concrete does not solidify by drying after mixing and placement. Rather, the water reacts with the cement in a chemical process known as hydration. This water is absorbed by cement, which hardens, gluing the other components together and eventually creating a stone-like material.

5.2.2 During hydration and hardening, concrete develops certain physical and chemical properties including mechanical strength, low permeability to moisture, and chemical stability. Concrete has relatively high compressive strength, but significantly lower tensile strength (about 10% of the compressive strength). The practical implication of this is that concrete elements subjected to tensile stresses must be reinforced or prestressed. For instance a very long but thin slab of un-reinforced concrete, suspended only by its edges may be unable to support its own weight, and therefore crack in two.

5.2.3 Concrete is placed while in a wet (or plastic) state, and therefore can be manipulated and moulded as needed. Hydration and hardening of concrete during the first three days is critical and abnormally fast drying and shrinkage due to factors such as evaporation from wind during placement may lead to increased tensile stresses at a time when it has not yet gained significant strength, resulting in shrinkage cracks. The early strength of the concrete can be increased by keeping it damp for a longer period during the curing process. Minimizing stress prior to curing minimizes cracking. High early-strength concrete is designed to hydrate faster, often by increased use of cement, which increases shrinkage and cracking.

5.2.4 In the curing process concrete shrinks and therefore can crack if shrinkage is restrained. Plastic-shrinkage cracks are immediately apparent, being normally visible within the first 2 days of placement, while drying-shrinkage cracks develop over time. Engineers are familiar with the tendency of concrete to crack, and, where appropriate, special design precautions are taken to ensure crack control. This entails the incorporation of secondary reinforcing placed at the desired spacing to limit the crack width to an acceptable level. The objective is to encourage a large number of very small cracks, rather than a small number of large, randomly-occurring cracks.

5.3 REINFORCED CONCRETE

5.3.1 Reinforced concrete is a composite material using concrete for compressive strength and (usually) steel reinforcing bars to accommodate the tensile forces. The concrete which is cast round the reinforcing bars or mesh shrinks slightly as it hardens thus gripping them and ensuring composite action. The concrete provides good protection to the steel from corrosion provided the cover is adequate and the concrete is dense.

5.4 PRESTRESSED CONCRETE

5.4.1 Prestressing concrete is a method for overcoming concrete's natural weakness in tension. Prestressing may be used to construct beams and bridges with longer spans than is possible with reinforced concrete. Prestressed concrete is put into a state of compression by tension in steel tendons (either cables or wires), after casting. In this way the concrete in a beam is made to resist a bending moment because of the compression included in the concrete due to the prestressing. Prestressing may be classified as pre-tensioned or post tensioned, i.e.:

- **Pre-tensioned** – in pre-tensioned beams or slabs the steel tendons are stressed before the concrete is cast around them and the concrete is allowed to harden and grip the steel before the initial tensile loads are released and transferred into the concrete as compression.

- **Post-tensioned** – in most post-tensioned beams on the other hand, ducts are formed in the concrete during casting and the tendons are placed in these and tensioned after the concrete has hardened. The tendons are protected from corrosion by cement grout, injected into the duct to fill them completely, an objective which is not always achieved. In a few cases the tendons are external to the concrete and are protected by encasing in cementitious or other special materials.

5.5 STEEL

5.5.1 Steel is a metal alloy that can be plastically formed (pounded, rolled, etc.) and comprises of iron as its major component and carbon as the primary alloying material. Varying the amount of carbon and its distribution in the alloy controls qualities such as the hardness, elasticity, ductility, and tensile strength of the resulting steel material. Steel with increased carbon content can be made harder and stronger than iron, but is also more brittle. One classical definition is that steels are iron–carbon alloys with up to 2.1% carbon by weight; alloys with higher carbon content than this are known as cast iron. Steel is also to be distinguished from wrought iron which has little or no carbon. Currently there are several classes of steels in which carbon is partially replaced with other alloying materials.

5.5.2 Steel is primarily used for its tensile characteristics, but it can be used in compression provided care has been taken to ensure that relatively slender plates or sections do not buckle. Steel is used for structural members, box girders, plate girders, rolled sections, tubes and plates and also as reinforcing bars and prestressing tendons in concrete. There are various grades of steel; mild steel is the most commonly used but high yield steels are used for structural members which have to carry higher stresses and for reinforcement and prestressing tendons.

5.5.3   Steel requires protection from the atmosphere if it is not to rust and sound paint systems are very important where steel is exposed [24].

5.5.4   In recent years a type of structural steel has been used on some bridges which, under normal exposure to the atmosphere, is allowed to corrode because it slowly produces an (oxide) patina or rust film that adheres strongly and provides a measure of protection to the steel as it develops. This type of steel is known as weathering steel and under certain conditions it may not require painting.

5.5.5   Stainless steel is as a ferrous alloy with a minimum of 10.5% chromium content, which has higher resistance to oxidation (rust) and corrosion than ordinary steel, because, similarly to weathering steel, a passivation layer is formed when exposed to oxygen protecting the metal beneath. This layer quickly reforms if or when the surface is scratched. Stainless steel is often found in railings, bearings, and fixings, such as cladding ties, etc.; but may also be used as an alternative reinforcing material for new or existing individual structural elements or highway structures that are most vulnerable to corrosion, e.g. restraint system beams, substructures adjacent to carriageway splash zones or structures in marine environments. *BA 84* [25] provides further guidance on situations where the use of stainless steel reinforcement may be beneficial.

5.6   WROUGHT IRON

5.6.1   Wrought iron is commercially pure iron, having a very small carbon content (not exceeding 0.15%), but usually contains many strands of slag mixed into the metal. These slag inclusions give it a 'grain' like wood and distinct look when etched. Also due to the slag, it has a fibrous look when broken or bent past its failure point. It is a material that is tough, malleable, ductile and can be easily welded. Hammering a piece of wrought iron cold causes the material fibres to become packed tighter, which makes the iron both brittle and hard.

5.6.2   Wrought iron is found in some older structures and although its working stress is less than that of modern steels, it is usually a durable material capable of yielding or distorting under excessive loads without fracturing. However, if inadequately protected against the weather, its laminar nature is very susceptible to corrosion.

5.7   CAST IRON

5.7.1   Cast iron usually refers to any of a group of iron-based alloys containing more than 2.1% carbon (alloys with less carbon are carbon steel by definition). It is made by remelting pig iron, often along with substantial quantities of scrap iron and scrap steel, and taking various steps to remove undesirable contaminants such as phosphorus and sulphur, which weaken the material. Carbon and silicon content are reduced to the desired levels, depending on the application. Other elements are then added to the melt mixture before the final form is produced by casting.

5.7.2   Cast iron is a brittle material but suitable for carrying compression loads. Tensile failure in this material happens suddenly because there is little or no yield before fracture. Owing to the difficulties of casting there are likely to be considerable variations in quality and also high possibility of voids being undetected in large castings.

5.8   ALUMINIUM AND ITS ALLOYS

5.8.1   Aluminium is a soft and lightweight metal with a dull silvery appearance, due to a thin layer of oxidation that forms quickly when it is exposed to air. Aluminium is about one-third as dense as steel; is malleable, ductile, and easily machined and cast; and has excellent corrosion resistance and durability due to the protective oxide layer.

5.8.2   Aluminium alloys are usually formed with copper, zinc, manganese, silicon, or magnesium. They are much lighter and more corrosion resistant than plain carbon steel and stronger than but not quite as corrosion resistant as pure aluminium. Bare aluminium alloy surfaces will keep their apparent shine in a dry environment. Galvanic corrosion can be rapid when an aluminium alloy is placed in close proximity to stainless steel in a wet environment.

5.9   MASONRY

5.9.1   Masonry is the building of structures from individual block units laid in and bound together by mortar. The common materials of masonry construction are brick, concrete block and stone such as sandstone, granite or limestone. Basic mortars consist normally of a sand filler and binders such as cement and/or lime. Pure lime mortars were traditionally used and are relatively weak, but able to accommodate movement both from the bricks themselves and the structure as a whole. Pure cement mortars may be stronger than the brick or stone, and hence a proportion of cement was often replaced with lime to reduce strength and increase flexibility. Many modern mortars contain additives, which, for example, may improve plasticity, increase adhesion and tensile strength, reduce permeability, retard or accelerate the initial set or produce a range of colours by the addition of pigments.

5.9.2   Masonry is generally a highly durable form of construction. However, the materials used, the quality of the mortar and workmanship, and the patterns the units are laid in can considerably affect durability. Masonry has good compressive strength but has much lower tensile strength unless reinforced.

5.9.3   Masonry is not commonly used in modern structures except for facework or retaining walls, but will be frequently encountered when inspecting older structures, especially arch bridges/culverts and retaining walls.

5.10   TIMBER

5.10.1   A tree trunk has three layers: the outer bark, which has almost no strength, the sapwood, and the heartwood in the middle of the tree. The heartwood, which may be used as a construction material, is usually harder and darker in colour than the sapwood.

5.10.2   Nowadays, timber is not used in main members of vehicular bridges in the UK but is used for footbridges (sometimes in the form of laminates) and also for the decks of some movable bridges. The main problems in the use of timber are connected with fixings and preservation against rotting.

5.11   ADVANCED COMPOSITES

5.11.1   Advanced composites include materials such as polyvinylchloride, acrylonitrile butadiene styrene, acrylics, polystyrene, polypropylene, polyesters and epoxides, and may be used for the primary load-carrying elements of a structure or as composite ropes and cables.

5.11.2   Advanced composite materials are usually characterised by long continuous filaments contained within a resin. Glass fibre composites are least expensive and therefore most commonly used, but aramid, polyester and carbon fibres are also used. Manufacturing techniques include pultrusion, filament winding, and moulding; these techniques may be used in combination.

5.11.3   Advanced composites have been used in highway structures since the 1970's, mainly within soil-reinforcement systems. Another common application is the use of glass fibre reinforced panels as permanent soffit formwork spanning between deck beams and as deck facia panels. Since 1990 advanced composites have been used as structural elements in bridges, including footbridges, and as bridge enclosure systems. They are also used in repair applications, where thin plates, usually utilising carbon fibre, are bonded onto existing structures to provide extra strength.

5.11.4   The principal reasons driving the development of advanced composites are their anticipated long-term corrosion, fatigue and impact resistance and their very high strength/weight ratio.

6 References for Part B

1. Management of Highway Structures: Code of Practice, TSO, 2005.

2. *Bridge Condition Indicators: Volume 2: Bridge Inspection Reporting: Guidance Note on Evaluation of Bridge Condition Indicators*, CSS Bridges Group, 2002.

3. *Bridge Condition Indicators: Addendum to Volume 2: Bridge Inspection Reporting: Guidance Note on Evaluation of Bridge Condition Indicators*, CSS Bridges Group, 2004.

4. *BD 63 Inspection of Highway Structures*, DMRB 3.1.4, TSO

5. *BA 42 The Design of Integral Bridges*, DMRB 1.3.12, TSO.

6. *BD 30 Backfilled Retaining Walls and Bridge Abutments*, DMRB 2.1.5, TSO.

7. *BS EN 1337, Parts 2-8 Structural Bearings*, British Standards Institution.

8. *BD 37 Loads for Highway Bridges*, DMRB 1.3.14, TSO.

9. *Bridge Inspector's Reference Manual*, Publication No. FHWA NHI 03-001, United States Department of Transportation, Washington, D.C., 2002.

10. *BD 33 Expansion Joints for Use in Highway Bridge Decks*, DMRB 2.3.6, TSO

11. *BA 26 Expansion Joints for Use in Highway Bridge Decks*, DMRB 2.3.7, TSO

12. *Bridge Joints*, Concrete Bridge Development Group, Current Practice Sheet No. 5.

13. *BD 43 The Impregnation of Reinforced and Prestressed Concrete Highway Structures using Hydrophobic Pore-Lining Impregnants*, DMRB 2.4.2, TSO.

14. *Guidance Document for Highway Infrastructure Asset Valuation*, Roads Liaison Group, TSO, 2005.

15. *Well-lit Highways: Code of Practice for Highway Lighting Management*, TSO, 2004.

16. *Manual of Bridge Engineering*, The Institution of Civil Engineers, Thomas Telford Publishing, 2000.

17. *BS 8006 Code of Practice for Strengthened/Reinforced Soils and Other Fills*, British Standards Institution.

18. *HA 68 Design Methods for the Reinforcement of Highway Slopes by Reinforced Soil and Soil Nailing Techniques*, DMRB 4.1.4, TSO.

19. *BS 8002 Code of Practice for Earth Retaining Structures*, British Standards Institution.

20. *Mechanics of Materials*, 2nd Edition, Popov EP, Englewood Cliffs, Prentice-Hall, 1976.

21. *BD 37 Loads for Highway Bridges*, DMRB 1.3.14, TSO.

22. *Highway Structures: General Design*, DMRB Volume 1 Section 3, TSO.

23. *Mechanics of Materials with Tutorial CD*, 3rd Edition, Beer FP, Johnston ER and DeWolf JT, McGraw-Hill, 2001.

24. *BD 87 Maintenance Painting of Steelwork*, DMRB 3.2.2, TSO.

25. *BA 84 Use of Stainless Steel Reinforcement in Highway Structures*, DMRB 1.3.15, TSO.

Part C
The Inspection Process

This Part outlines the fundamentals of the inspection process; providing guidance on scheduling and adequately planning inspections. It also lists the factors that should be considered when preparing for inspections, including consideration of environmental impacts, preparation of risk assessments and selection of the appropriate access equipment and safe methods of working. Special considerations and other details relating to performing inspections on structures constructed of different materials and certain special structures are contained in this Part. A generic process to facilitate recording, rating and reporting defects and inspection results is presented.

1 Introduction

1.1 OVERVIEW

1.1.1 The overall purpose of inspection, testing and monitoring is to check that highway structures are safe for use and fit for purpose and to provide the data required to support effective maintenance management and planning (see Volume 1: Part A: Section 1.2). In order to achieve this, a rational approach is required for the development and delivery of a robust inspection regime. This Part of the Manual presents a rational five step process for developing an inspection regime and this is shown in Figure C.1. It is recommended that owners of highway structures adopt this process.

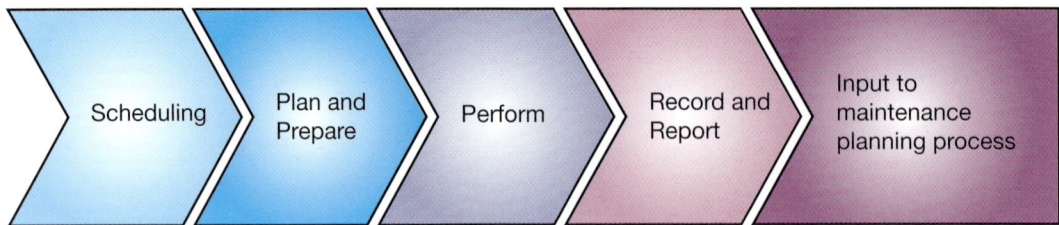

Scheduling > Plan and Prepare > Perform > Record and Report > Input to maintenance planning process

Figure C.1 – The inspection process

1.1.2 An outline of the above inspection process is provided in Table C.1. The layout of this Part of the Manual aligns with this process; with each section providing details on a specific part of the process. The topics covered herein apply to all inspection types, however, the amount of time and effort required should be commensurate with the specific circumstances and inspection type.

Table C.1 – Layout of Part C	
Section	**Summary of purpose and content of each section**
2. Scheduling Inspections	This section provides guidance on scheduling inspections and outlines the factors that should be considered when programming different types of inspections.
3. Planning and Preparing for Inspections	This section provides guidance on adequately planning and preparing for inspections and outlines the factors that should be considered. It provides advice on the preparation of method statements and risk assessments and guidance on the selection of appropriate access and other equipment. This section also outlines Health and Safety considerations and potential environmental impacts.
4. Performing Inspections	This section provides details relating to performing inspections on structures constructed of different materials and certain special structures.
5. Recording Inspection Findings	This section provides recommendations on the adoption of a generic process that would facilitate the recording, rating and reporting of defects and inspection results.
6. Inputs to the Maintenance Planning Process	This section provides brief guidance and appropriate references that should be utilised for the input of the inspection findings into the maintenance planning process.

1.2 A PRAGMATIC APPROACH

1.2.1 It is important that the process described in this Part of the Manual is tailored to needs, i.e. the inspection type (e.g. Safety, General, Principal or Special) and the characteristics of the structure stock. There is a considerable body of information provided in this Part, covering a wide range of areas (e.g. access, safety and environmental considerations). If these considerations are not dealt with in a rational and pragmatic manner then the scheduling, planning and delivering of inspections can quickly become cumbersome and resource intensive. In order to avoid this it is recommended that:

- The Supervising Engineer, or a suitably competent member of the inspection team, develops a sound overall understanding of and stays abreast of the key issues covered by this Part of the Manual, including contacts with staff/organisations that can provide more in-depth guidance on specific issues.

- The structure files/records for each structure are stored and maintained in such a manner that relevant information can be readily accessed, reviewed and where required, updated during the inspection process. This avoids lengthy repetition of data compilation during each inspection cycle. Any information that may need to be checked for relevance (i.e. criteria that are likely to change over time) should be suitably flagged in the files/records.

- The Supervising Engineer, or a suitably competent member of the inspection team, lists those factors that are relevant for the different types of inspections and for different families of structures.

1.2.2 This approach should assist authorities to develop an efficient and effective inspection process, enabling them to avoid unnecessary commitment of resources on issues that are not relevant to a particular inspection or structure type.

2 Scheduling Inspections

2.1 INTRODUCTION

2.1.1 Inspections should be scheduled to make the most efficient use of the resources and to minimise disturbance to the public, e.g. scheduling inspections to take advantage of traffic management planned for other reasons. In all cases, a rational schedule that takes account of the stock characteristics should be developed.

2.1.2 When scheduling inspections, consideration should be given to the issues raised in the following sections, these are summarised in Table C.2.

Table C.2 – Scheduling Inspections			
No	Items to be considered	Comments	Section
1	The type of inspection required	Safety, General, Principal, Special, Acceptance, or Inspection for Assessment, and considerations for decreasing or increasing their inspection intervals	2.2
2	Objectives of the inspection	Why is the inspection being carried out? What information is required? How does this fit in with the management of the structure?	2.3
3	Criteria that influence or constrain the inspection schedule	What criteria may influence the scheduled date of an inspection, e.g. resource availability, co-ordinating with other works, seeking efficiencies in the programme, environmental issues, etc.?	2.4

2.2 INSPECTION TYPES

2.2.1 A summary of the inspection types is provided in Table C.3. Further details on each inspection type are provided in the following paragraphs, this aligns with the guidance provided in the *Code of Practice for the Management of Highway Structures* [1] and *BD 63* [2].

Table C.3 – Summary of Inspection Types		
Inspection Type	**Nominal Interval**	**Description**
Safety Inspection (or Routine Surveillance)	At frequencies which ensure timely identification of safety defects and reflect the importance of a particular route or asset.	Cursory inspection carried out from a slow moving vehicle; in certain instances staff may need to proceed on foot.
General Inspection	2 years	Visual inspection from the ground level. Report on the physical condition of all structural elements visible from the ground level.
Principal Inspection	6 years	Close visual examination, within touching distance; utilising, as necessary, suitable inspection techniques. Report on the physical condition of all inspectable structural parts.
Special Inspection	Programmed or when needed	Detailed investigation (including as required inspection, testing and/or monitoring) of particular areas of concern or following certain events.
Acceptance inspections	When needed	A formal mechanism for exchanging information prior to changeover of responsibility.
Inspection for Assessment	When needed	Inspection undertaken to provide information required to undertake a structural assessment.

Safety Inspection/Routine Surveillance

2.2.2   The purpose of a Safety Inspection (or Routine Surveillance) is to identify obvious deficiencies which represent, or might lead to, a danger to the public and therefore require immediate or urgent attention.

2.2.3   If a Safety Inspection, or other source, reveals a possible defect requiring urgent attention, including defects that may represent a hazard to road, rail and other users, the Supervising Engineer should immediately take such action as is required to safeguard the public and/or sustain structural functionality. In such circumstances the appropriate third parties should also be notified, e.g. the police, the public, other owners.

2.2.4   Safety Inspections are not specific to highway structures and generally cover all fixed assets on the highway network, including carriageways, footways, structures, drainage, verges and lighting. Safety Inspections are normally carried out by trained highway maintenance staff from a slow moving vehicle. In certain circumstances staff may need to proceed on foot either to confirm suspected defects or to complete the inspection. For example, some bridges, such as footbridges or underpasses with high pedestrian usage, may require a weekly or monthly walkover.

2.2.5   Safety Inspections should be scheduled and undertaken at frequencies which ensure the timely identification of safety related defects and reflect the importance of a particular route or asset. Safety inspections may also be a result of a defect notification by a third party e.g. police or public.

2.2.6   As such, a Safety Inspection only provides a cursory check of those parts of a highway structure that are visible from the highway with the aim of identifying any obvious deficiencies or signs of damage and deterioration that may require urgent attention or may lead to accidents or high maintenance costs, e.g. collision damage to superstructure or bridge supports, damage to restraint systems, spalling concrete and insecure expansion joint plates.

General Inspection

2.2.7   The purpose of a General Inspection is to provide information on the physical condition of all visible elements on a highway structure.

2.2.8   A General Inspection comprises a visual inspection of all parts of the structure that can be inspected without the need for special access equipment or traffic management arrangements. This should include adjacent earthworks or waterways where relevant to the behaviour or stability of the structure. Riverbanks, for example, in the vicinity of a bridge should be examined for evidence of scour or flooding or for conditions, such as the deposition of debris or blockages to the waterway, which could lead to scour of bridge supports or flooding.

2.2.9   General Inspections should be scheduled and carried out at two year intervals following the previous General or Principal Inspection. When a General Inspection coincides with a due Principal Inspection, only the Principal Inspection should be undertaken.

Principal Inspection

2.2.10   The purpose of a Principal Inspection is to provide information on the physical condition of all inspectable parts of a highway structure. A Principal Inspection is more comprehensive and provides more detailed information than a General Inspection. Principal Inspections should be scheduled and carried out at six year intervals, as a replacement of a General Inspection.

2.2.11   A Principal Inspection comprises a close examination, within touching distance, of all inspectable parts of a structure. This should include adjacent earthworks and waterways where relevant to the behaviour or stability of the structure. A Principal Inspection should utilise as necessary suitable inspection techniques, access and/or traffic management works. Suitable inspection techniques that should be considered for a Principal Inspection include hammer tapping to detect loose concrete cover and paint thickness measurements. Limited testing may be undertaken during a Principal Inspection; however, this is not a requirement. When appropriate, a Principal Inspection may be combined with a Special Inspection, monitoring activities, detailed testing work or routine/planned maintenance.

Special Inspection

2.2.12   The purpose of a Special Inspection is to provide detailed information on a particular part, area or defect that is causing concern, or an inspection which is beyond the requirements of the General/Principal Inspection regime.

2.2.13 A Special Inspection may comprise a close visual inspection, testing and/or monitoring and may involve a one-off inspection, a series of inspections or an on-going programme of inspections. As such, Special Inspections are tailored to specific needs or special requirements.

2.2.14 Special Inspections are carried out when a need is identified. For example, based on the specific characteristics of the structure, identified by a General, Principal or Safety Inspection, to follow certain events, or to consider parts of the structure more closely or at a more frequent interval that the normal General/Principal Inspection regime. Table C.4 provides guidance on scheduling Special Inspections and a list of situations when special requirements of different types of structures or elements of structures may instigate the need for Special Inspections to be considered. A number of these inspections may be scheduled to coincide and be undertaken in conjunction with Principal Inspections, as appropriate, and this is identified in Table C.4.

Table C.4 – Scheduling Special Inspections		
Reason for Inspection	**Interval**	**Recommendations**
Underwater parts of structures	6 years	A programme of underwater inspections should be prepared and implemented. These inspections may be scheduled to coincide and be undertaken in conjunction with Principal Inspections.
Structures at risk from scour	6 years	An inspection may be required to check for changes to the geometry of the river and for general degradation of the bed. These inspections may be scheduled to coincide and be undertaken in conjunction with Principal Inspections.
Structures at risk from scour located on steep, upland rivers with potential lateral instability	After major floods	An inspection may be required to check for changes in the river channel. Further guidance on assessing the risk of scour is provided in *BA 74* [3].
Settlement, tilting or other movement is observed greater than that allowed for in the design	When needed	An inspection may be required to identify the cause and assess the urgency of remedial measures, and if necessary monitor the rate of movement, by developing a programme of inspections.
Structures in areas of mineral extraction	When needed	An inspection may be required to ascertain the condition of the structure after subsidence occurs and determine the extent of the damage caused.
Structures with weight or other restrictions	6 months	Where the interim measures require monitoring, an inspection should be carried out in accordance with *BD 79* [4]. Where the interim measures consist of temporary propping or width restrictions, a visual check to ensure that these are still functioning correctly is required.

Continued

Table C.4 – Scheduling Special Inspections (continued)		
Reason for Inspection	**Interval**	**Recommendations**
Structures that have to carry an abnormal heavy load	When needed	The structure should be inspected before, during and after the passage of the load if either an assessment has indicated that the margin of safety is below that which would be provided for a design to current standards; or similar loads are not known to have been carried.
Structures subjected to major accident/impact, chemical spillage and/or fire	When needed	The structure should be inspected to investigate possible structural damage. A programme of inspections may be required to monitor the damage until such time as a permanent repair is carried out. The frequency of these inspections will depend on the structural importance of the member affected and the extent and severity of the damage.
Concrete structures at risk from reinforcement corrosion	When needed	An inspection of concrete elements at risk from reinforcement corrosion may include a combination of a visual inspection and any of the tests described in Volume 1: Part E: Section 5. *BA 35* [5] provides guidance on limited site testing that may be employed as part of a Special Inspection, i.e. half-cell potential, chloride level, covermeter and depth of carbonation. The scope of this testing should be compatible with any preventative maintenance strategy for the structure.
Post tensioned concrete bridges	When needed	Post-tensioned concrete bridges with grouted tendon ducts are particularly vulnerable to corrosion and severe deterioration where internal grouting of the ducts is incomplete, allowing moist air, water or de-icing salt to enter the ducting. These types of bridges should be inspected in accordance with *BA 50* [6].
Structures strengthened by the use of externally bonded plates	6 months for the first 2 years	Structures should be inspected to ensure that the externally bonded plates are functioning as intended, e.g. they are not detached. Inspections should be scheduled and carried out at intervals not exceeding six months for the first two years after strengthening and thereafter in accordance with the intervals prescribed in the maintenance records. For concrete structures a manual of the inspection procedures required should be prepared in accordance with *BA 30* [7]. Particular requirements for the inspection of structures strengthened with fibre composites are included in the Concrete Society *Technical Report 57* [8].

Continued

Table C.4 – Scheduling Special Inspections (continued)

Reason for Inspection	Interval	Recommendations
Steel and steel composite structures at risk from corrosion	When needed	Steel is particularly vulnerable to corrosion when exposed to certain environments. Most highway structures steelwork is protected with paint or other protective coatings. Inspections of the protective system may be required in order to identify or monitor corrosion or in order to identify the extent and rate of deterioration.
Weathering steel structures	Various	Special Inspections should be scheduled and carried out throughout the life of these type of structure in accordance with *BD 7* [9], i.e. immediately after construction; every 2 years (visual examination of critical areas to investigate patina irregularities); and every 6 years (to check thickness measurements). These inspections may be scheduled to coincide and be undertaken in conjunction with Principal Inspections.
Cast iron structures	6 months	Cast iron members on structures should be inspected at intervals not exceeding six months to identify any cracked or fractured elements and to determine the cause and extent of deterioration. Monitoring of changes in the cracking may be required; this may be achieved through a programme of follow-on inspections and/or appropriate crack monitoring techniques (see Volume 1: Part E: Section 4).
Bearings.	6 years or as agreed with the authority	Inspections of bearings should be carried out in accordance with *BS EN 1337-10* [10]. For bearings types other than those described in *BS EN 1337-10* [10], engineering judgement should be used to determine the inspections required. These inspections may be scheduled to coincide and be undertaken in conjunction with Principal Inspections.
Hangers in structures with suspended spans	6 years	Hangers are usually critical elements whose failure would be catastrophic. The nature of the inspection, apart from a close visual inspection of all surfaces at touching distance, will depend on the form and materials of construction. Detailed requirements should be agreed with the authority. These inspections may be scheduled to coincide and be undertaken in conjunction with Principal Inspections.
Pre-contract surveys	When needed	When maintenance work is planned, an inspection including relevant tests and measurement will usually be necessary to ascertain the nature and extent of work required.

Continued

Table C.4 – Scheduling Special Inspections (continued)		
Reason for Inspection	**Interval**	**Recommendations**
Geometric data	When needed to supplement structure records	Detailed geometric measurements may be required when as-built or construction drawings do not exist for a structure; or when specific dimensions are subject to doubt. Critical dimensions such as headroom or lateral clearances may need to be checked after maintenance work or modifications are carried out on a structure or adjacent highway. If a detailed check of headroom is required, the procedure set out in Annex A of *BD 65* [11] should be followed. Inspections undertaken to ascertain such measurements, may be scheduled to coincide with a Principal Inspection.
Permanent access gantries	Prior to use and at regular intervals	Permanent access gantries (also see paragraphs 3.6.25-3.6.27) should be inspected prior to use and at regular intervals in accordance the Institution of Structural Engineers report on *The Operation and Maintenance of Bridge Access Gantries and Runways* [12].
Hoists, winches and associated cables	At regular intervals	Hoists, winches and associated cables should be inspected in accordance with the requirements of the relevant orders under the *Factories Act* [13].
Machinery for movable bridges	As prescribed in the maintenance records	Electrical and mechanical equipment for operating movable bridges should be inspected at regular intervals to ensure that the equipment is maintained in a safe and operable condition. A programme of inspections should be prepared for each movable bridge, setting out the specific requirements.
Electrical and mechanical equipment in subways, underpasses and tunnels	As prescribed in maintenance records	Electrical and mechanical equipment (e.g. lighting and ventilation) should be inspected at intervals that comply, as a minimum, with the manufacture's recommendations. In certain circumstances the Supervision Engineer may wish to set down a more onerous regime in the maintenance records, for example, weekly checks of lighting in pedestrian underpasses (Note: inspections of lighting equipment should be in agreement with the authority's lighting engineers and align with the *Lighting Code of Practice* [14]).

Continued

Table C.4 – Scheduling Special Inspections (continued)		
Reason for Inspection	**Interval**	**Recommendations**
Specific problems	When needed	Following the discovery of a specific problem, it may be necessary to arrange a single or series of inspections to confirm the cause and extent of the defects and to enable the appropriate course of action to be determined. If a specific problem is found on a structure, consideration should be given to undertaking inspections on structures of similar construction form and/or material, e.g. the discovery of concrete degradation due to thaumasite sulfate attack or alkali silica reaction.
Other reasons	When needed	There may be a need for Special Inspections at structures due to reasons other than those set out above. Possibilities include, but are not restricted to: structural distress; investigation of a severe defect reported at an inspection; cladding failure; bulging, sliding or tilting of spandrel walls; unexplained cracking; condition of vulnerable details such as half or hinge joints; and water leakage. Except in cases of emergency, the need for the inspection should be agreed with the Supervising Engineer.

Acceptance Inspections

2.2.15   The need for an Acceptance Inspection should be considered when there is a change over of responsibility from one party or highway authority to another, for the operation, maintenance and safety of a structure.

2.2.16   The purpose of an Acceptance Inspection is to provide the party taking over responsibility for the structure with a formal mechanism for documenting and agreeing the current status of, and outstanding work on, a structure prior to handover. The scope of an Acceptance Inspection depends on the circumstances, e.g. handover of a new structure, transfer of an existing structure, handback of a structure after a concession period. Acceptance responsibilities and activities depend upon the form of contract, but the Acceptance Inspection is normally carried out by the party taking over responsibility but who may be accompanied by the other party to facilitate agreement. Further details on the format, content and timing of Acceptance Inspections are included in *BD 63* [2].

Inspection for Assessment

2.2.17   The purpose of an Inspection for Assessment is to provide information required to undertake a structural assessment. Such inspections are necessary to verify the form of construction, the dimensions of the structure and the nature and condition of the structural components. The inspection should cover not only the condition of individual components but also the condition of the structure as an entity, especially noting any signs of distress and its cause.

2.2.18   The inspection should be sufficient to enable the dead and superimposed dead loads on the structure, and the disposition of live loading, to be determined; or to enable the strengths of members to be calculated, including allowance for deficiencies, and to confirm the dimensions and articulation for any structural model.

2.2.19 Samples are often needed to determine material strengths. Where dimensional checks are required in order to determine or confirm the structural details, such checks may require exploratory excavations, probing or boring to determine depth and extent of foundations or extents of internal features. The amount of investigation required will depend on the availability and reliability of the structure records. Site testing may also be required when an Inspection for Assessment is undertaken, in such cases careful planning of these activities is essential. *BD 21* [15] provides guidance and outlines the factors that should be considerations prior to undertaking an Inspection for Assessment.

Increasing or Decreasing the Inspection Interval

2.2.20 Formal guidance on increasing or decreasing the aforementioned inspection intervals is provided in the Code of Practice [1] and BD 63 [2]. Particular regard should be given to the need for a formal risk assessment when increasing the interval between Principal Inspections.

2.3   OBJECTIVES OF THE INSPECTION

2.3.1 Inspections are an integral part of the maintenance and management of highway structures and are carried out for a variety of reasons. It is important for the Supervising Engineer and the inspector to have a clear understanding of the objectives of the inspection and how the work fits into the management plan for the structure.

2.3.2 All structural maintenance and management activities have the ultimate aim of ensuring the continued safety and functionality of the structure. However, the work may be subdivided into three phases, each with different objectives:

- **Phase 1: Condition Monitoring** – the process of inspecting, testing and recording the condition of the structure;

- **Phase 2: Diagnosis** – the process of deciding the causes of any defects that are observed;

- **Phase 3: Solution Development** – the process of determining what remedial measures may be required to address the detected faults, defining the scope of work and carrying out maintenance.

2.3.3 Inspections may be required during each phase. Most inspections fall clearly into Phase 1, e.g. Safety, General and Principal Inspections, others may have objectives that cover two phases, e.g. Special Inspections fall into Phases 1 and 2.

2.4   CRITERIA THAT INFLUENCE THE INSPECTION SCHEDULING

2.4.1 There are a wide range of criteria that can potentially influence the date of an inspection. Some of the more common issues that should be taken into account during the scheduling of inspections include:

- Inspection requirements

- Efficient use of resources

- Availability of resources

- Traffic management

- Structures near or on railways or watercourses

- Weather conditions

- Tidal locations

- Environmental issues

- Coordination of works

- Scheduling tolerances

2.4.2   The following sections provide guidance on each of the above. It is important that this is used in conjunction with any structure specific or local criteria that may influence the inspection date.

Inspection Requirements

2.4.3   The inspection schedule should align with nationally recognised inspection frequencies (see Table C.3 and Table C.4). Where authorities deviate from these recognised frequencies, it is important that they fully document their reasons for the departure and the supporting rationale for the alternative timings (this is particularly the case for longer intervals between General and Principal Inspections). Careful consideration should be given to increased inspection intervals as this could lead to defects becoming more severe before they are detected, possible danger to the public and may leave the authority exposed to liability claims.

Efficient Use of Resources

2.4.4   The inspection schedule should seek to make efficient use of resources. For example, a significant factor in inspections is the time taken to travel to and from the structure. As such it may be prudent and efficient to aim to undertake inspections along particular routes or in the same area at the same time, thereby minimising the time wasted in travelling.

2.4.5   Authorities should also consider the most efficient arrangement for inspecting structures, e.g. in-house inspectors or consultants/contractors. This will depend on arrangements in place (e.g. term maintenance contact) and the characteristics of the structure stock. However, it is fundamental that the quality of the inspection staff is not comprised by drives for efficiency savings.

Availability of Resources

2.4.6   The inspection schedule must be commensurate with the availability of resources, primarily inspection staff, i.e. there is little point in preparing an inspection schedule if the internal/external staff are not available at the specified times. Issues to consider are:

- The size of the structure stock to the number of inspectors.

- The average time required for different inspection types, e.g. General and Principal Inspections.

- The average number of inspectors that need to be present for different inspection types, e.g. General and Principal Inspections. This should include consideration of any shadowing/training needs for new/existing inspectors.

- Other commitments of in-house inspection staff, for example, are they also responsible for building, depot, carriageway, etc. inspections and/or other duties, and what constraints do these have on their availability.

- Other commitments of external/term maintenance contractors/consultants, for example, are they committed to undertake inspections for several clients.

2.4.7   Consideration should also be given to the availability of inspection equipment, be this internal (e.g. ladders, data logging devices, pool vehicles) and external equipment (e.g. mobile platforms).

Traffic Management

2.4.8   The inspection programme may be constrained by traffic management requirements, for example, on busy roads there may be only be limited periods when lane closures are permitted. In some cases, inspections may need to be carried out at night; however, it must be remembered that night work is inherently more dangerous and has its own environmental impacts which should be taken into consideration (see paragraphs 3.5.55-3.5.61 and 3.10.1-3.10.6).

Structures Near or on Railways or Watercourses

2.4.9   Inspections near or on railways or watercourses should be arranged with, and carried out in full accordance with the requirements of, the relevant body. Severe restrictions usually apply to the timing of work affecting railways. Track possessions and electrical isolations are often only available overnight or on Sundays and generally need to be arranged well in advance. Similarly, access to bridges over canals can be seriously restricted by navigation requirements during the busier period of the year, from March to October.

Weather Conditions

2.4.10   When programming inspections, due regard should be given to prevailing weather conditions, for example it may be preferable to inspect the majority of structures on a specific route during the summer when there is longer daylight and more clement weather. However, the practice of inspecting structures during the summer and then writing up reports in the winter should not be adopted. The inspection reports should be prepared and submitted soon after the inspection, without exceeding the maximum durations specified by the authority.

2.4.11   Some defects, such as leaking joints, blocked drainage or cracks in concrete elements, are more prominent during or just after rainfall. While it is difficult to programme work to coincide with wet weather, the opportunity should not be overlooked. For example, bridges inspected during dry weather could be revisited (as part of a Safety Inspection) during the next spell of rain to check whether expansion joints or drainage are leaking.

Tidal Locations

2.4.12   For structures at tidal locations, planning inspections to coincide with tidal conditions can be advantageous. For instance, spring tides may afford access to more parts of the structure whereas neap tides may afford slower tidal flow and less change in water level.

Environmental Issues

2.4.13   When scheduling inspections, due consideration should be given to environmental issues that may influence the timing of inspections, such as nesting birds, hibernating bats, etc. If access is required over agricultural land, the timing of the inspection could be affected by growing crops or other farming activities.

2.4.14   Particular consideration should be given to protected species of flora (e.g. fen orchid, shore dock and meadow clary) and fauna (e.g. bats, otters, water voles and great crested newts). The Supervising Engineer may be familiar with some species but expert advice may be required to identify particular environmental issues at a specific site. As well as the advice included in Section 3.11, details of suitable sources of environmental guidance may be obtained from the organisations listed in Volume 1: Part F: Appendix A.

Coordination of Works

2.4.15   Wherever advantageous and practicable, inspections should be co-ordinated with maintenance or other activities, forming an integral part of the maintenance strategy for the route. On busy roads it will usually be necessary to book the roadspace with the engineer responsible for co-ordinating maintenance activities on that length of road. The Supervising Engineer should liaise with other staff, owners, organisations and the public, as necessary.

2.4.16   If use can be made of equipment or traffic management required for other maintenance works, then cost efficiencies can be made and the disruption to road users reduced. At the very least, the Supervising Engineer should consult other colleagues within the authority, and/or view planned schemes on the Asset Management System, to find out what else may be planned. When inspections are carried out in conjunction with other maintenance work, care must be taken to ensure that any interference between the two activities is taken into account when preparing method statements and risk assessments (see Sections 3.2 and 3.4). Two safe methods of working may become dangerous if carried out close to each other or at the same time.

2.4.17   Similarly, when a transport link (e.g. road, rail or water) managed by another authority is affected the Supervising Engineer should consult with the appropriate body. They should obtain the necessary permissions with regards to any restrictions or obstructions and co-ordinate the inspection with any work that may be undertaken on the other authority's network.

Scheduling Tolerances

2.4.18   General and Principal Inspections should ideally be scheduled in accordance with the intervals defined in Section 2.2. Given the aforementioned influencing criteria, in some circumstances it may be appropriate to alter the scheduled date of an inspection in order to produce a more rational and deliverable schedule e.g. bringing forward or delaying inspections. In such circumstances it may be appropriate to vary the scheduled date by up to ± 6 months provided the Supervising Engineer is satisfied that any marginally increased inspection interval is acceptable. The timing of the following General or Principal inspection should remain as per the original schedule, and should not be changed to accommodate the altered timing of this inspection.

3 Planning and Preparing for Inspections

3.1 INTRODUCTION

3.1.1 Adequate planning is fundamental to undertaking inspections safely and efficiently and ensuring that the requisite information is obtained. Appropriate planning should consider the factors described in the following sections and summarised in Table C.5.

Table C.5 – Planning and Preparing for Inspections			
No	Items to be considered	Comments	Section
1	Method statement	A method statement should be prepared in conjunction with the risk assessment.	3.2
2	Health and Safety	The appropriate health and safety considerations should be taken into account e.g. review the structure H&S file, check if CDM regulations apply, consider personal and public safety.	3.3
3	Risk assessment	A risk assessment including mitigation measures for dealing with hazards should be undertaken before work starts.	3.4
4	Methods of access	How will access be gained to the required part(s) of the structure?	3.5
5	Equipment needed	The equipment required for the inspection needs to be determined and its availability and serviceability checked.	3.6
6	Structure records	What records are available? e.g. drawings, previous inspection reports, etc.	3.7
7	Type and extent of testing	Types of tests, their location, extent and intensity.	3.8, Part E
8	Competence of inspection staff	Inspectors of the appropriate calibre for the work are essential.	3.9, Part A: 4.2
9	Notification of the work	There may be a need to notify other parties and, in some cases, to obtain their approval of the proposed inspection.	3.10
10	Environmental impacts	Is the scope and nature of work likely to affect the public or nearby sensitive wildlife habitats?	3.11

3.2 METHOD STATEMENT

3.2.1 As part of the planning process a method statement that summarises all relevant information should be prepared and agreed before undertaking an inspection. The method statement should take into account access requirements, environmental considerations and Health and Safety checks.

The method statement should be developed in parallel with the risk assessment (see paragraph 3.4), for identifying and assessing potential risks and formulating and refining the safest method of working. If the work falls within the scope of the *Construction (Design and Management) Regulations* [16], the appropriate procedures must be followed and a Health and Safety file prepared or updated. The content and level of detail should be commensurate with the circumstances and the type of inspection. However, the following information should normally be included in the method statement:

- Details and programme of the work to be undertaken during the inspection.

- Equipment required.

- Methods of access to be used.

- Traffic management details.

- A risk assessment including safe procedures for dealing with hazards.

- Resources and competence of staff that will carry out the inspection.

- Planned working times.

- Temporary works to be provided.

- Protection from highway, railway or other traffic.

- Requirements for action by others.

- Any co-ordination or notification required.

- Any environmental impacts of the work.

3.2.2   Copies of the method statement and risk assessment should be retained for future reference. In many cases it will be appropriate for these to be added to the structure Health and Safety File, which is required for any work within the scope of the CDM Regulations. Generic method statements and risk assessments may be appropriate for groups of similar structures.

3.2.3   The inspection team should familiarise themselves with the method statement and risk assessment (see paragraph 3.4) to ensure that they are aware of what is required, how work is going to be conducted, what their respective roles are and how identified risks would be mitigated.

3.3   HEALTH AND SAFETY

3.3.1   Inspections of highway structures, including any testing should be managed to comply with the requirements of the *Health and Safety at Work Act* [17], the *Management of Health and Safety at Work Regulations* [18], the *Construction (Health, Safety and Welfare) Regulations* [19] and any associated regulations contained in the *Approved Codes of Practice* [20, 21, 22] and in the *Construction Health and Safety Manual* [23]. In addition to these requirements, inspection personnel should comply with the authority's internal health and safety procedures when planning and undertaking inspections of highway structures. These often include guidelines and check lists for staff working in a variety of situations. An example of a check list for personal safety of staff is provided in Volume 1: Part F: Appendix B.

3.3.2   The *Health and Safety at Work Act* [17] provides a comprehensive framework to minimise risks to people arising from workplace activities, including the public and others who may be affected by the work activities, as well as those actually carrying out the work. A Supervising Engineer would be defined as an employer, under the *Act* [17], of organisations they instruct to carry out work on their behalf as well as members of their own staff. They have a basic duty of care to act, as far as is reasonably practicable, to minimise health and safety risks to organisations they employ and the employees of those organisations. Fulfilment of the duty may involve monitoring or provision of:

- safe plant and systems of work;

- safe use, handling, storage and transport of articles and substances;

- necessary information, instruction, training and supervision;

- a safe place of work and the means of access to and egress from that place;

- a working environment which is safe and healthy and which includes adequate welfare facilities.

3.3.3   Health and Safety requirements that should be given particular regard in relation to the inspection of highway structures include, but are not restricted to:

- Presence of asbestos (see paragraphs 3.3.7-3.3.8);

- Personal Safety (see paragraphs 3.3.9-3.3.25);

- Public Safety (see paragraphs 3.3.26-3.3.29);

3.3.4   Health and Safety requirements that should be given particular consideration in regard to the methods of access that may be employed during inspections are described in detail in Section 3.5 and include, but are not restricted to:

- Working on or adjacent to live highways (see paragraphs 3.5.4-3.5.20);

- Working on or over railways (see paragraphs 3.5.21-3.5.22)

- Working in or over water (see paragraphs 3.5.23-3.5.30);

- Working underwater (see paragraphs 3.5.31-3.5.41);

- Working in confined spaces (see paragraphs 3.5.42-3.5.54)

- Night time working (see paragraphs 3.5.55-3.5.61);

- Encountering toxic mould (see paragraph 3.5.47);

- Working at height to access elements of structures to be inspected, using scaffold, mobile elevating work platforms, etc. (see paragraphs 3.6.9-3.6.27);

3.3.5   The *Construction (Design and Management) Regulations* [16, 20, 24, 25] place specific duties on those involved in construction work, which includes investigations (but not site survey), maintenance and repair as well as new construction. These duties relate to the ways that health and safety aspects are managed during the design, planning and construction phases. Some inspections fall within the scope of these Regulations, especially Special Inspections and Inspections for Assessment, which involve extensive investigation and testing. However, it is unlikely that Acceptance, Safety, General, or Principal Inspections will do so.

3.3.6   As part of the preparation for an inspection, it will be necessary for the Supervising Engineer to ascertain whether the work will be within the scope of the *CDM Regulations* and if so, the required procedures should be complied with [16, 20, 24, 25]. These will include the preparation of a Health and Safety File, which should be retained for future reference.

Presence of Asbestos

3.3.7   Asbestos in bridges is commonly found within the waterproofing system, as permanent formwork, in drainage pipework or as insulation to water pipes passing through the bridge. *The Control of Asbestos at Work Regulations (CAWR)* [26] require authorities or the person controlling premises to have an Asbestos Management System in place, which should be accessible to anyone who will work on or in the premises. Consultation between the Health and Safety Executive and the Highways Agency has indicated that bridges and other highway structures are covered by the concept of premises. However, this may not be the case for other authorities, and should be checked with each individual authority.

3.3.8   The Supervising Engineer has a duty to have in place an Asbestos Management Plan for each structure, unless it has been shown that there is none, and for controlling access to the structure. There must also be a named 'Duty-holder' who is responsible for operation of the Asbestos Management Plan and revising it as necessary. The Asbestos Management System and in particular the individual Asbestos Management Plans should enable the inspector to check for the presence of asbestos containing materials prior to the inspection, thereby enabling appropriate action to be taken if or when asbestos is present.

Personal Safety

Personal Care

3.3.9   Under the *Health and Safety at Work Act* [17], all inspection personnel have a duty to take reasonable care of their own health and safety and that of others who may be affected. Inspectors are also required to co-operate with the Supervising Engineer, the authority's Safety Officer or other competent individuals on health and safety matters. An example of a check list, reminding staff of personal safety matters when working on or off the public highway, is provided in Volume 1: Part F: Appendix B.

3.3.10   Inspection of highway structures can be a hazardous operation and, when concentrating on the task in hand, it is easy to be momentarily unaware of danger. Stepping backwards out of a working zone into a live traffic lane is but one example. Inspectors should therefore be alert at all times, developing the ability to concentrate on the task in hand while being aware of surrounding hazards.

3.3.11 Working while under the influence of alcohol or drugs or when tired or ill can impinge on the ability to concentrate which may adversely affect the work and cause danger. When working on or over railways there are strict regulations under the *Transport and Works Act* [27].

3.3.12 Although a high degree of physical fitness is not required of an inspector, a reasonable level of agility and mobility is required. Sufferers of respiratory problems should be aware that they may be in dusty or dirty environments. Those prone to black-outs should not be involved in inspections.

3.3.13 Inspection work may involve contact with dirty surfaces; personal hygiene is therefore important. A supply of soap, water and disposable paper towels is advised. A first aid kit appropriate for the number of staff and nature of likely injuries should be readily available.

3.3.14 The areas under and around highway structures are frequently used for dumping rubbish and considerable debris can collect within a culvert. This ranges from abandoned vehicles to industrial or household waste or discarded hypodermic needles, which may present a wide variety of hazards. Inspectors should therefore avoid handling or disturbing rubbish. Where possible, arrangements should be made for the waste disposal authority to remove it, however, if it is necessary to handle rubbish, the appropriate Personal Protective Equipment (see paragraphs 3.6.28-3.5.30) should be worn. Particular care should be exercised in areas where hypodermic needles or syringes may be encountered.

3.3.15 In such work environments, there is an element of incidental exposure to biological agents, i.e. harmful micro-organisms such as bacteria, fungi, viruses, internal parasites, and other infectious proteins. For example, handling waste contaminated with human/animal waste or working with equipment or in an environment that is contaminated e.g. sewers. Inspectors should be aware that they may be harmed by biological agents by being infected with a micro-organism, by being exposed to toxins produced by the micro-organism, or by having an allergic reaction to the micro-organism or substances it produces. The advice contained in the Health and Safety Executive publication *Infection at work: Controlling the risks* [28] should be followed for the prevention of incidental exposure.

3.3.16 The classification of biological agents is documented in the *Approved List of Biological Agents* [29] which was produced under Section 15 of the *Health and Safety at Work Act* [17]. The Approved List also gives a separate indication of which biological agents are capable of causing allergic or toxic reactions or where there is an effective vaccine available. The *Approved List* should be read in conjunction with *The Control of Substances Hazardous to Health (COSHH) Regulations 2002* [30] and, in particular, Schedule 3 – additional provisions relating to work with biological agents.

3.3.17 Pigeons and other birds sometimes roost or nest on bearing shelves or other remote ledges on bridges, causing an accumulation of bird droppings. Such droppings can harbour parasites or fungal spores and should therefore be regarded as toxic. Suitable respirators should be worn in areas where bird droppings have accumulated.

3.3.18   All staff that may be required to work in or near water or enter confined spaces (especially long culverts) should be warned of the dangers of and be issued with a Leptospirosis (Weils Disease) Card or issued with the HSE Leaflet INDG84 08/06 C500 and be familiar with the relevant control measures.

3.3.19   Inspectors should not undertake work which they consider to be a danger to themselves or others. Thus, they should not attempt work which they believe to be inherently dangerous, nor should they attempt work for which they are not competent through limitations of training, experience or ability, or through ill health. Likewise, they should not attempt work which has not been properly planned, for which an adequate risk assessment has not been prepared, or where there is a lack of the required equipment.

3.3.20   Inspectors should be aware of the procedures to be followed and people to contact in the event of an accident to themselves, their colleagues, or members of the public. It is advisable to consider the need for a member of the inspection team to be trained in first aid. They should also be alert at all times to any situations which may create a hazard and take whatever steps are necessary to ensure that hazards are removed or minimised.

Electricity

3.3.21   Inspections and investigations may involve working with electrical equipment – power tools, floodlights, etc. All such work should be undertaken in accordance with the *Electricity at Work Regulations* [31]. Where the electrical equipment operates at 50 or more volts Alternating Current (AC) or 75 or more volts Direct Current (DC), the equipment should comply with the *Electrical Equipment (Safety) Regulations* [32].

3.3.22   In addition, many bridges and other highway structures have electrical equipment attached to them in the form of electricity supply cables, lighting, matrix signals, etc. Inspection work on such structures should comply with the *Electricity at Work Regulations* [31]. Wherever practicable, electrical equipment attached to structures should be isolated or disconnected before carrying out inspections in the vicinity. Inspectors should assume that cables and equipment are live unless they have clear authoritative information that they are not. If an inspector encounters an unexpected cable or item of electrical equipment, they should cease any work which might affect the item and arrange for a suitably qualified and experienced person to identify it. Work should not resume until the suitably qualified and experienced person either has been able to ensure that the item is not live or can confirm that it is safe to proceed with the proposed work. Similarly, if any damaged electrical apparatus is discovered, the inspector should arrange for it to be made safe before proceeding.

3.3.23   The locations of any overhead lines in the vicinity of the bridge should be noted, so that the appropriate precautions can be made to keep all working areas and access equipment safely away from them. The regulations of the relevant authority should be complied with when working near any overhead lines or apparatus.

3.3.24   A further electrical hazard which can occur during inspections is damage to underground cables. Work requiring excavation should not be undertaken without checking for the presence of cables or other buried apparatus, utilising the appropriate cable detection equipment. The structure records (see Section 3.7) should be consulted but the relevant organisations should also be contacted for advice on the location of their cables as the structure records may not be accurate. Should it be necessary to excavate near cables or ducts or any other buried apparatus, hand digging should be employed until the services have been exposed. Guidance on working near overhead and underground services is given in the *Construction Health and Safety Manual* [23].

3.3.25   The risk assessment and method statement (see paragraphs 3.2 and 3.4) prepared for an inspection should take account of any potential electrical hazards at the highway structure site.

Public Safety

3.3.26   When planning and carrying out inspections at any location, the safety of the public and other third parties must be considered [33]. Traffic management will normally be used to provide safe conditions for motorists, but the needs of pedestrians, cyclists, equestrians and occupiers of adjacent land or property must also be catered for.

3.3.27   It will not always be possible to avoid working alongside or over a route used by pedestrians or cyclists. In these circumstances additional care must be taken and, if appropriate, a lookout should be provided. The lookout's responsibility would be to warn both the public and inspectors of the possible danger. Alternatively, and if appropriate, the working area may be identified by means of stakes and warning tape to fence off the area. Suitable warning signs may also be required.

3.3.28   Pedestrians should not be diverted from their normal routes unless a diversion is clearly designated. It should be checked to ensure it is at least as safe as the normal route. When working above or adjacent to pedestrian routes, great care should be taken to avoid dropping tools or any other item.

3.3.29   In any situation where an inspector considers there will be a danger to third parties if work proceeds, the work should be stopped and further advice sought as to how to proceed safely.

3.4   RISK ASSESSMENTS

3.4.1   The *Management of Health and Safety at Work Regulations* [17, 34] require that an assessment is carried out identifying the risks to workers and others who may be affected by undertaking work and to record the significant findings of that assessment.

3.4.2   A Risk Assessment is the systematic process of:

- defining the activity, procedure or situation to be assessed;

- identifying all significant hazards;

- determining the likelihood of harm occurring due to the presence of the identified hazard;

- assessing the consequences of the harm;

- forming a judgement based on the possible consequences that the level of risk is either acceptable or unacceptable; and

- recording the conclusions arrived at and the information on which they were based and, if the risk is unacceptable, revising the method of working until the risk is acceptable or can be adequately mitigated or controlled.

3.4.3 Further advice and guidance may be found in the HSE publications *Five Steps to Risk Assessment* [35].

3.4.4 A risk assessment requires concentration on those risks that are liable to arise because of the work activity. Normally, trivial risks and risks arising from life in general do not need to be recorded unless they are increased by the work activity. A risk assessment should be an aid to the development of safe working practices and not an end in itself. It should not be a task which is out of proportion to the actual work activity.

3.4.5 The assessment should be appropriate to the nature of the work. Where the work activity or work place changes, the risk assessment may need to concentrate more on the broad range of risks that may arise. Risk assessments should be continually reviewed, particularly in the light of comments from persons involved in carrying out the work process.

3.4.6 A risk assessment should always be carried out before commencing work. For example, a risk assessment for a Safety Inspection may be brief and simple, but it must not be omitted. A risk assessment does not easily lend itself to objective numerical analysis. For the purpose of assessing risks associated with inspections, a simple qualitative approach may often be appropriate.

Preparing an Inspection Risk Assessment

3.4.7 The main objective of a risk assessment is to establish and record the measures that are to be taken to eliminate or reduce risks to inspection personnel and to members of the public during the proposed activity, and thereby to ensure that those managing or working on the project are provided with the information that they need to work safely. Its secondary objective is to provide evidence of compliance with legislation. There are no official requirements as to the form that a risk assessment should take. However, the *Approved Code of Practice* [21] provides guidance.

3.4.8 Some inspections will require staff with very specific safety training. Examples are access to railways, working in confined spaces and working at heights. Any such requirements should be identified and specified in the risk assessment.

3.4.9 There is bound to be some similarity between inspection projects and it is often appropriate to use generic or earlier risk assessments as guidance. However, it is not sufficient to just change the heading, date, address and the names. If the procedure is to have any real value, original thought should go into each project. For this reason, check lists should be used with care: although they provide a convenient reminder of most risks, they will not necessarily identify unusual hazards. The assessment should cover risks to the workforce and the public, with particular attention given to pedestrians and to vehicle, waterway and railway traffic as appropriate. Where the assessment identifies particular problems for health and safety, a detailed statement should be prepared covering the procedures to be adopted to minimise risks.

3.4.10   The best results are often achieved when at least two people work on a risk assessment, especially when identifying hazard and developing mitigation or control measures, as discussion often triggers new thoughts. It is also helpful to involve the people working on the project by allowing them to produce a draft version. Where appropriate, the highway authority's Safety Officer, the Health and Safety Executive, the police and other interested, expert or competent parties should be consulted.

3.4.11   It may often be appropriate to combine the risk assessment with some other document or report such as a health and safety plan or method statement. These three documents are closely related and in some cases there may be a considerable degree of overlap. The most appropriate format will depend on the nature and size of the inspection.

3.5   METHODS OF ACCESS

3.5.1   When planning any inspection, careful thought should be given to the methods of access required. This would generally require consideration of the types of access equipment to be used, the restrictions or obstructions caused by the equipment, traffic management requirements, and the routes to be used to and from the highway structure site. Consideration should be given to the items listed in Table C.6 with regards to where and when will access be needed and how will it be achieved.

Table C.6 – Access Considerations
Where
•   from highways (see paragraphs 3.5.4-3.5.15);
•   away from highways (see paragraphs 3.5.16-3.5.20);
•   at or over railways (see paragraphs 3.5.21-3.5.22);
•   in or over water (see paragraphs 3.5.23-3.5.30);
•   underwater (see paragraphs 3.5.31-3.5.41);
•   in confined spaces (see paragraphs 3.5.42-3.5.54);
•   in environmentally sensitive areas (see paragraph 3.11);
When
•   during short term possessions (see paragraphs 3.5.5-3.5.15 and 3.5.21);
•   between tides (see paragraphs 3.5.31-3.5.41);
•   to avoid disturbance to agriculture (see paragraphs 3.5.16-3.5.20);
•   to avoid disturbance to wildlife (see Section 3.11);
•   at night (see paragraphs 3.5.55-3.5.61)
How
•   what access equipment will be needed (see Section 3.6).

3.5.2   Safety of the inspectors and the public must always be considered as an integral part of these deliberations. A pre-inspection visit to the site is recommended prior to undertaking Principal Inspections and, where appropriate, Special Inspections.

3.5.3   Some of the following sections provide guidance on the selection of the appropriate access equipment and on safe methods of working. Further advice is given in *Temporary Access to the Workface* [36].

Working on Highways

3.5.4   Many inspections require some of the work to be carried out from the carriageway, hard shoulder or verge. Particular care must be taken to ensure the safety of both the inspectors and the public. All personnel working within the highway boundary should normally wear high visibility outer clothing except when in a vehicle. Particular access issues that should be considered when undertaking inspections on or near the highway include traffic management, mobile land closures, single vehicle works, and inspection from the verge; guidance on these is provided below.

Traffic Management

3.5.5   Restrictions to any part of the carriageway or motorway hard shoulder should be set out and signed in accordance with the relevant provisions of Chapter 8 of the *Traffic Signs Manual* [37, 38], *Approved Code of Practice for Safety at Street Works and Road Works* [39] and Volume 8 of the *DMRB* [40]. Recommendations on factors that should be taken on board when planning, executing and removing of traffic management are given in the Chapter 8 of the *Traffic Signs Manual* [37, 38] and should be considered where appropriate.

3.5.6   Where reasonably practicable, inspections should be designed to avoid or minimise the need for traffic management. However, the safety of the inspection team should be balanced against the convenience of the motorist. On busy roads, inspections may need to be carried out at times of minimal traffic flow, such as at night or early on Sundays. When designing traffic management for all-purpose roads, the needs of motorists, pedestrians and other users should be catered for.

3.5.7   Where access to undertake an inspection requires traffic management on a road owned by another authority (be it major road, minor road, or a private road or track), the traffic management arrangements should be agreed with the owner. Even on minor tracks, warning signs and protection may sometimes be needed (also see paragraphs 3.3.26-3.3.29).

3.5.8   While working within a lane closure or other form of traffic management, the inspector should ensure that an adequate safety zone is maintained between working and trafficked areas. Vehicles and equipment should be sited as far from the live carriageway as practicable – preferably off the highway altogether. Where it is necessary to park vehicles or plant on a closed area of carriageway, it is advisable to site them 'upstream' from the work site. Front wheels should be turned towards the verge. Access equipment should never be allowed to overhang any live traffic lane or safety zone; particular care should be taken when using equipment with articulated booms. Safety zones should be arranged around the area of working, not only to protect any aerial equipment but also to avoid injury to passers-by if a hammer or other item should fall.

Mobile Lane Closures

3.5.9   Mobile lane closure is a technique for use on dual carriageway roads whereby an operative works, stopping if necessary, ahead of suitably equipped lane closure and blocking vehicles. The blocking vehicles are fitted with direction arrows, beacons and crash cushions. Mobile lane closures should be carried out in accordance with Chapter 8 of the *Traffic Signs Manual* [37, 38].

3.5.10 Mobile lane closure is suited to short duration activities at carriageway level. Without mobile lane closure it would often be necessary to cone off a length of carriageway. Training of all personnel involved is essential, as is the need to adhere exactly to the required procedures.

Single Vehicle Works

3.5.11 Inspection work requiring the inspector to work on or close to the carriageway may sometimes be carried out using a single vehicle standing or operating on the carriageway. However, this method of working should only be adopted after a thorough risk assessment including consideration of alternatives. Vehicles should be positioned 'upstream' of the working area and all personnel should wear high visibility outer clothing.

3.5.12 Single vehicle works should never be used on unrestricted dual carriageways without hard shoulders; instead, works should be carried out by static traffic management or a mobile lane closure.

Inspection from the Verge or Adjacent Earthworks

3.5.13 Inspections are frequently carried out using personnel or equipment on the verge or adjacent earthworks – within the highway boundary but off the carriageway. The work ranges from visual inspections on foot to detailed testing requiring access and other equipment. The duration of the work may vary from a few minutes to several days.

3.5.14 The risks of working close to live traffic lanes should always be considered. Even when working clear of the carriageway, there may be a need for some encroachment onto the hard shoulder or traffic lanes. Examples include parking the inspectors' vehicles, while gaining access or egress, manoeuvring plant into position, or while moving from one side of the structure to the other. Appropriate traffic management should be installed in all cases; this will normally require a lane closure (see paragraphs 3.5.5-3.5.8).

3.5.15 When inspecting on foot from the verge or adjacent earthworks on a motorway, the inspector's vehicle may stop on the hard shoulder as described for single vehicle works (see paragraphs 3.5.11-3.5.12). On roads without hard shoulders, stationary vehicles should be parked off the carriageway wherever possible. A convenient lay-by or field gate entrance should be used if available, or the vehicle may stop on the verge. If it is necessary to stop on the carriageway, the vehicle should be parked as close to the edge as possible. Providing visibility is adequate, it is generally preferable to stop on a minor road rather than the main road. Vehicles should generally be positioned 'upstream' of the working area. Where possible, the inspector should seek to avoid the need to work from or to cross the carriageway.

Off-Highway Working

3.5.16　Wherever practicable, inspection work should be carried out off the highway (i.e. outside the highway boundary), with access equipment and vehicles also sited off the highway. Permission for access should be obtained from landowners or occupiers, and ground conditions may need to be checked.

3.5.17　Off-highway, pedestrian or vehicular traffic may use the area, so suitable barriers or warnings should be considered. Similarly there may be a need to fence the working area against livestock. The requirements of vehicles, pedestrians and livestock should be taken into account; accesses, rights of way or entrances should not be obstructed without first consulting and agreeing with the appropriate individuals or organisations (also see paragraphs 3.3.26-3.3.29).

3.5.18　Access routes to and from the highway structure site should be chosen so as to minimise damage or disruption; agreement with the landowner or occupier will generally be required. Where it is necessary to cross, or work on or adjacent to, agricultural land, special requirements may be needed to avoid the spread of animal or plant diseases. These should be ascertained and abided by. Care should also be taken to avoid permitting livestock to stray. Where appropriate the *Countryside Code* [41, 42, 43, 44] should be followed.

3.5.19　At some sites there may be features of ecological or landscape significance. Good working practices should be adopted at all times, in particular to avoid pollution of watercourses, and special procedures followed whenever required (see Section 3.11).

3.5.20　Despite being close to service roads and possibly populated areas, some locations around highway structures should be regarded as 'remote areas'. For example, a person lying injured by a culvert headwall at the foot of a highway embankment could remain undiscovered for a long time. Inspectors should be aware of areas which may effectively be remote and should follow the authority's procedures when entering such areas. These procedures will usually prohibit anyone from entering a remote area alone without first notifying others. Mobile telephones can be particularly useful for obtaining help in an emergency. In some rural areas, however, it may not be possible to use mobile phones due to lack of network coverage. Therefore as part of planning an inspection, the coverage of the particular phone that will be used by the inspector should be checked.

Working on or Over Railways

3.5.21　Railways are hazardous places. Consequently, the railway authorities have strict rules governing access onto and over their property; unauthorised entry is illegal. Early consultation with the railway authority is essential, as obtaining permission for entry or a track possession can be a lengthy process. Track possessions can normally be obtained only for short periods, usually at night or during a Saturday night and Sunday morning. Track possessions for bridge inspections are often scheduled to coincide with other maintenance work on the line.

3.5.22　Railway authorities produce their own safety procedures to be adopted when working on or near their lines. In general, work on or near the line is carried out in an area safeguarded by either stopping all trains, fencing off from lines in use, or being separated by a minimum distance from lines in use. Outside of these areas work is only undertaken when absolutely necessary and when lookout protection is provided. The specific procedures of the relevant railway authority must be adhered to in all cases.

Working in or Over Water

3.5.23　Working in or over water presents the risk of drowning; people can drown even in small watercourses. Therefore, precautions should be taken to prevent persons from tripping, falling or being swept into water. Platforms and gangways must be secure and protected with guardrails and toe boards. Allowance should be made for tidal range or any other variation in water level.

3.5.24　Safety harnesses and life jackets or buoyancy aids should be worn where appropriate. Lifebuoys or rescue lines should be readily available close to a working area over water and a rescue boat should be on active stand-by in deep water.

3.5.25　Inspection of some low headroom structures over watercourses may be carried out using a dinghy or other suitable craft. It is often most appropriate to use an inflatable, flat bottomed craft as these provide a reasonably stable platform. If inspections are carried out from a boat the inspector should be accompanied by a second person who is in sole charge of the craft. The craft should be positioned under the structure by means of fixed lines attached to the structure or bank or anchors. The craft should not normally be maintained in position during the inspection through the use of its engine. The standard of control of the craft is crucial to its safety and in most cases should only be undertaken by experienced boatmen.

3.5.26　Structures over navigable waterways may often best be inspected from an access platform mounted on a barge. This is a specialised means of access, which should be undertaken with great caution and only under the instruction of a person or organisation experienced in this form of access. At all times during the inspection, a rescue craft should be in attendance with fully equipped and trained personnel on board. The need for a rescue boat may be forgone if a risk assessment shows that this is unnecessary, e.g. on quiet canals.

3.5.27　It may be appropriate for the inspector to enter shallow water wearing wellington boots or waders. However, great care should be exercised due to the possibility of unseen hazards (glass or other debris, tree roots, scour holes, etc.), soft mud or slippery surfaces. The difficulty of keeping firm footing in fast flowing water should not be underestimated. A wading pole should always be used to test the depth of water. Waders should be used with special caution; in the event of falling over in the water, the waders may float, inverting the wearer and making it difficult to regain an upright posture. For this reason, thigh waders are safer than waist or chest waders. In deep water – over 1m deep – or if it would be necessary for the inspector to put his face underwater, the inspection of areas underwater should only be carried out as an underwater inspection (see paragraphs 3.5.31-3.5.41).

3.5.28　Wet or muddy surfaces can be extremely slippery and venturing onto such surfaces should be avoided if possible.

3.5.29　Waterways are administered by a number of bodies including the Environment Agency, British Waterways, Waterways Ireland, local authorities and harbour authorities. The appropriate body should be consulted in case there are any special requirements to be considered.

3.5.30　Inspection of culverts should always be undertaken with great care with the possibility of flash floods and accidental discharges into the watercourse being taken into account. Where significant water flow is unavoidable, the water is deep or there is low headroom, the inspection should be treated as an underwater inspection (see paragraphs 3.5.31-3.5.41). In rare cases, the inspection of the culvert above water level may be undertaken from a boat.

Working Underwater

3.5.31　Underwater parts of structures will require inspection from time to time. Typical reasons are to investigate the presence or extent of scour, either as a routine precaution or following a flood, or to ascertain the condition of underwater parts of the structure. In many cases the inspection will include structures such as retaining walls and adjacent areas of the bed of the watercourse.

3.5.32　Underwater inspections are potentially hazardous and can be very expensive, so it is important to plan adequately. Much of the planning and associated risk assessment will require similar consideration to any other inspection; however, the following aspects (inspection information, scheduling, lowering the water level, diverting the watercourse and diving) should be subject to particular consideration. Further guidance on underwater inspections is provided in the *Guide to Inspection of Underwater Structures* [45].

Inspection Information

3.5.33　The Supervising Engineer should provide those undertaking the inspection with as much information as is reasonably possible concerning the underwater parts of the structure and the watercourse. Sources of information may include, but are not restricted to:

- Previous inspection records and where available risk assessments indicating known or suspected hazards such as restricted access, fast currents or high turbulence, contaminated water, vessel traffic, low visibility and low temperature.

- Drawings of abutments, piers, etc., especially those showing depth and shape of foundations, type of construction, etc.

- A sketch of the structure identifying abutment and pier numbers, directional identification (from-to), flow direction and limits of examination e.g. all parts of the structure which are at bed level or which extend over the bed. These include inverts, aprons, revetments, scour protection works, etc.

- Details of all parts of the structure between bed level and High Water level. This will include parts of the structure which are accessible without divers at certain states of tide or time of year, but which are considered to be included as 'underwater parts' of the structure.

- Location and value of reference datum for a scour survey.

- Data of previously measured bed levels, either original or from previous examinations.

Scheduling

3.5.34   Unless the inspection is required urgently, it may be possible to time the inspection to coincide with favourable conditions such as low spring tides or low river flow. For example, inspection of culverts should normally be undertaken at the end of a long dry spell such that the flow of water is at its minimum. Refer to Section 2.4 for guidance on scheduling inspections.

Lowering the Water Level

3.5.35   The possibility of lowering the water level should be considered, so as to expose part, or all of the areas, which require inspecting. In canals or some rivers it may be possible to drain the water or reduce levels, however, this should only be undertaken following consultation with the relevant organisations since draining can have a variety of harmful effects along the stretch of watercourse. Local dewatering, using a cofferdam, flexible dam or limpet dam, may be a practicable alternative. If, after any reduction in water level, the inspector would still need to enter water deeper than 1 m or immerse their face underwater, qualified divers should carry out the inspection.

Diverting the Watercourse

3.5.36   It may be practicable to dam the watercourse and divert the flow. This approach may be beneficial for enabling the inspection of some culverts. The environmental implications associated with diverting the watercourse should be carefully considered, and the potential for causing flooding on the diversion course. Diversions are best undertaken in times of low water flow.

Diving

3.5.37   Carrying out effective underwater inspections requires considerable experience on the part of the diver. Underwater it is rarely possible to view the structure as a whole; orientation and location can be difficult to determine. The divers should preferably have relevant civil or structural engineering experience in addition to diving qualifications.

3.5.38   All diving operations in the UK are covered by the *Health and Safety at Work Act* [17, 46], the *Diving at Work Regulations* [47], and the *Approved Code of Practice for Commercial Diving Projects* [48]. All divers involved in commercial operations are required to hold valid diving medical certificates, a completed log book, a Health and Safety approved diving qualification and a valid HSE Diving First Aid Qualification; The Statutory Instrument lays down four main categories of diver qualifications. For most cases of underwater inspection for highway structures the minimum qualification would be HSE Part IV which covers diving to 30m using air alone.

3.5.39   Diving contractors undertaking underwater inspections should hold a HSE approved Diving Procedures Manual and be registered with the Health and Safety Executive. The *Diving at Work Regulations* [47] require diving contractors undertaking underwater inspections to prepare a job specific diving plan. A separate risk assessment for non-diving activities may need to be prepared in accordance with the *Construction (Design and Management) Regulations* [16].

3.5.40   The diving team should normally comprise a minimum of three HSE qualified divers: a diving supervisor, a working diver and a stand-by diver. However, depending on the nature of the diving operation, the use made of mechanized equipment and hazards, additional divers or other personnel may be required.

3.5.41　A wide range of inspection, measurement and testing techniques may be used underwater, including both still and video photography. However, the detail which can be determined and the accuracy of location are both hampered by the conditions. Even in good visibility, a close visual inspection cannot be relied upon to detect cracks less than 3mm in width especially where marine or aquatic growth is present. Where there is little visibility, the diver may have to rely on touch instead of sight: in these conditions only gross faults such as voids or cracks greater than 25mm width would be detected.

Working in Confined Spaces

Types of Confined Spaces

3.5.42　A 'confined space' is not readily defined as it does not always have to be enclosed on all sides; for example an open topped access shaft is a confined space. Confined spaces are typified by restricted access and space, poor ventilation and the lack of oxygen or presence of harmful gases. Workers in confined spaces may not be aware of the danger, for example when oxygen is gradually replaced by other gases. Examples of confined spaces are:

- interior of box girders and bridge towers;

- cofferdams and caissons;

- pumping station wells;

- sewers and manholes;

- culverts, pipelines, headings, tunnels, and shafts;

- bored piles, wells, tanks, ducts, silos, valve pits, excavations and trenches.

3.5.43　Excavations and trenches are included in the examples of confined spaces in the above list as particularly in the case of chalk excavations there is a possibility of oxygen deficiency due to acidic attack. In addition excavations and trenches may have restricted means of egress.

3.5.44　Confined spaces may be attractive to children, vandals and others. It is therefore essential that measures are taken as far as reasonably practicable to prevent unauthorised entry. Further advice is given in the *Approved Code of Practice to the Confined Spaces Regulations* [49].

Hazards

3.5.45　Hazards which may be present in confined spaces include:

- oxygen deficiency or excess;

- toxic fumes, including gases and vapours;

- flammable gases or liquids;

- fire;

- dust;

- bacteria or fungal spores;

- vermin;

- water;

- restricted and delayed egress in an emergency.

3.5.46 Hazards associated with a confined space often can be exacerbated by the work being undertaken within them. For instance the adverse effect on health due to fumes or gases produced by welding, flame cutting, painting and use of solvents are all greater in a confined space, an example being the removal of lead paint from surfaces in a box girder. In the case of a visual inspection it should not be assumed that there will be no additional hazards, as for example disturbance of stagnant deposits when moving about in a culvert or sewer may lead to the release of gases. Inspectors should also be aware that the 'normal' hazards encountered in inspection work can be more difficult to deal with in a confined space.

3.5.47 Whenever mould growth is encountered in a box girder or similar location, it should be treated as toxic and all inspection work should cease until the level of toxicity has been established as being within safe limits. Advice on safety should be sought from the Health and Safety Executive.

Confined Space Working

3.5.48 All work in a confined space should be considered as potentially dangerous and a carefully organised approach is therefore required, using a permit to enter/work system. A permit to work sets down in detail the work to be done and the precautions needed before entering and during any time spent in the confined space. As part of the permit to work it is usual for a permit to enter to be specified for the control of entry. However, the fact that such a system is adopted does not guarantee safety.

3.5.49 A permit to work system should include such topics as work activities, sequence of work, responsibilities, atmospheric testing and monitoring, personal protective equipment and other equipment, ventilation, communication, methods of entry and exit, lighting and, most importantly, emergency rescue procedures. The permit should be prepared by a competent authorised person familiar with the hazards and risks associated with the work. A sample Permit to Work is provided in Volume 1: Part F: Appendix C.

3.5.50 Whenever possible additional hazards should not be introduced into confined spaces. As an example, if a petrol driven generator is required this should be kept outside in a well ventilated area with the exhaust directed away from the entrance. Likewise where substances which give off vapours are proposed, alternative materials should be considered. Anyone who does not have a genuine need to enter a confined space should not do so as their presence may increase the risks involved.

3.5.51 The Supervising Engineer should compile and maintain a list of structures, or parts of structures, which are considered to be confined spaces. The list should be agreed with appropriate personnel and all relevant parties made aware of the list. Adequate risk assessments and method statements (including emergency procedures) should be prepared before any work requiring entry is undertaken (see Sections 3.2 and 3.4). Following the assessment of the risks involved in entering a structure, it may be decided to limit entry only to those staff issued with a permit-to-work (see Sample Permit to Work, Volume 1: Part F: Appendix C).

3.5.52   A safe system of work must be prepared and an assessment of the risks involved made for the entry requirements for each confined space and for each mode of entry. Requirements may be different, in the same space, at different times or for different purposes. All parties should be made fully aware of the requirements. Staff should be given appropriate training before they enter confined spaces.

3.5.53   Staff should be provided with personal protective equipment (e.g. helmets, goggles, gloves, overalls, masks) and safety equipment (e.g. safety harness, rope and attachment, gas tester, torch, breathing apparatus) appropriate to the requirements of each structure. Communication equipment should also be provided where necessary. Periodic checks should be made to ensure the equipment is functioning correctly.

Classification of Confined Space Working

3.5.54   In order to assist in the identification of confined spaces and the development of the required system of work, confined space working can be divided into three main classes as described below.

- **Class A Working** – Class A working is for use for inspection work where it is possible to see the inspector working and there is ready communications. In some situations personnel winching may be required. A typical arrangement for inspection of a manhole chamber is shown in Figure C.2. A minimum of two personnel are required for Class A working. Depending on the confined space depth and associated hazards, it is recommended that equipment should include but not be limited to: gas detector; personal protective equipment; communications; harness; life lines; lighting; riding winch (dependent on depth); escape breathing apparatus; resuscitator.

- **Class B Working** – Class B working is for use for inspection work where an intermediate person is required to maintain audible or visual communications. In some situations personnel winching may be required. A typical arrangement for the inspection of a culvert accessed from a manhole is shown in Figure C.3. A minimum of three personnel are required for Class B working. Equipment required would be the same as for Class A working. Particular consideration should be given to the provision of escape breathing apparatus which would be required if the total distance to safe atmosphere may be of some distance.

- **Escape breathing apparatus for Class A & B Working** – Consideration should be given to the type of escape breathing apparatus to be used. For example, would a 10 minute escape set be sufficient, or should alternative temporary equipment (sufficient for an escape) be provided? It should be noted however, that Class A and B Working excludes the provision for the use of 30 minute breathing apparatus. This is subject to sub-section 4 and 5 of Section 30 of the *Factories Act* [13].

- **Class C Working** – Class C working is for use for inspection work where Class A & B are not appropriate and for other types of more hazardous activities such as testing, investigations and minor reinstatements. Typical arrangements are shown in Figure C.4 for situations where it is possible to see the inspection team and there is ready communications and in Figure C.5 where this is not possible. A minimum of four personnel are required for Class C working and typical equipment would include gas detection (minimum 2); personal protective equipment; communications; harness; life line; lighting; riding winch (dependent on depth); resuscitator; first aid kit; air mover; escape breathing apparatus; 30 minute working breathing apparatus, etc. However, the number of personnel and type of equipment to be provided will be dependent on the risk activity. In order to determine resources, this type of work should be carefully planned, with the risks being assessed and written safe system of work produced. For example, on a large project, such as extensive investigation and testing in a long box girder, consideration should be given to the placing of kits of additional rescue, breathing and resuscitation equipment at strategic points and arranging for these to be suitably marked and signposted.

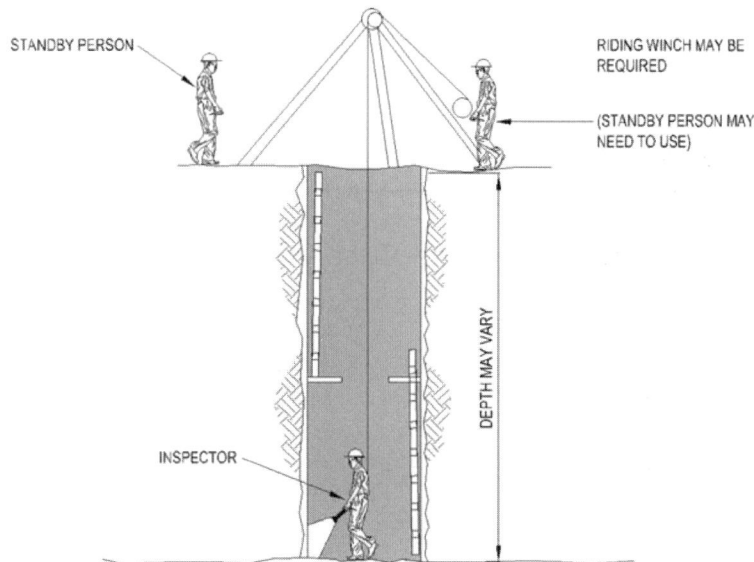

Figure C.2 – Confined space working: Class A

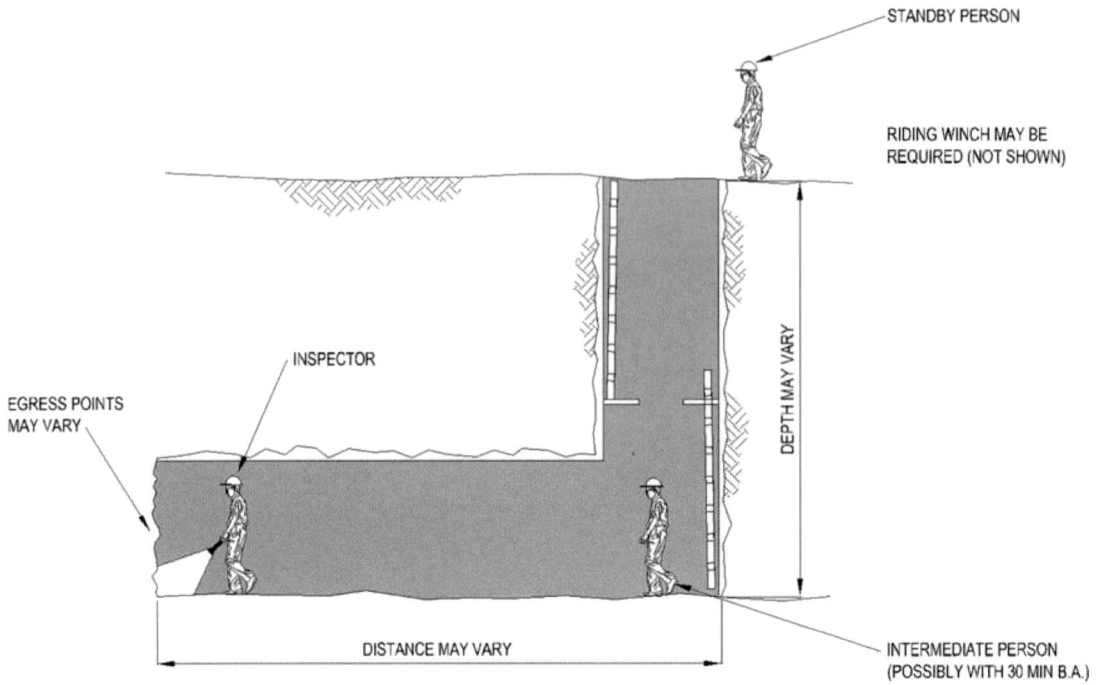

Figure C.3 – Confined space working: Class B

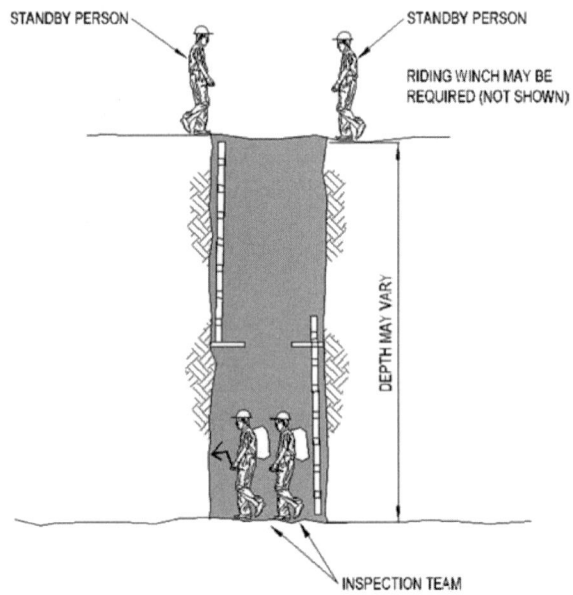

Figure C.4 – Confined space working: Class C1

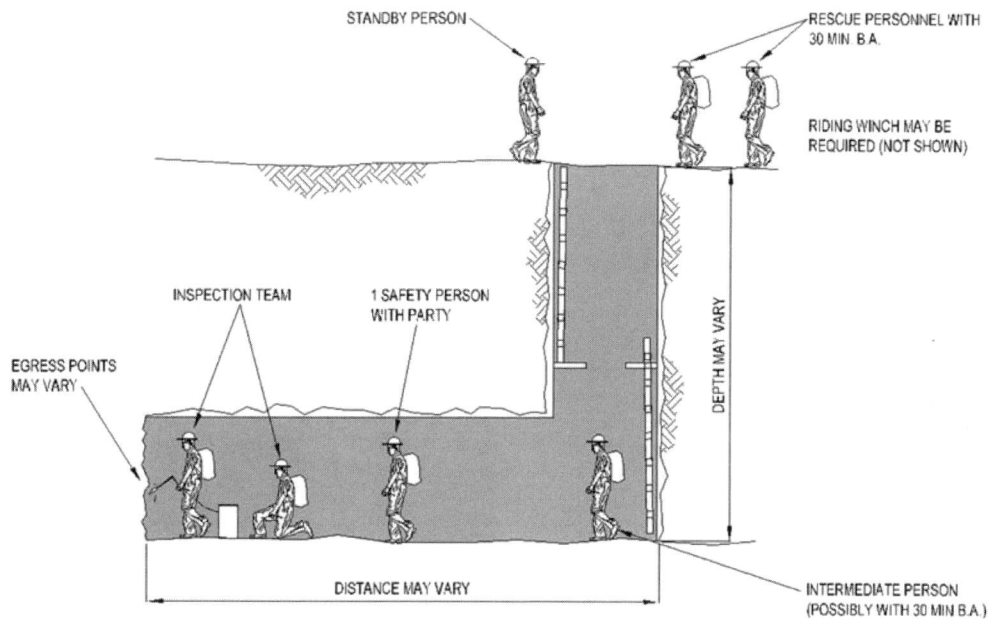

Figure C.5 – Confined space working: Class C2

Night-Time Working

3.5.55   Inspecting at night, on carriageways which are heavily trafficked during the day, can greatly reduce the delays and disruption caused by traffic management. With night-time only working, lane closures are introduced after the evening peak traffic flow and removed before the build-up of the morning peak flow. Chapter 8 of the *Traffic Signs Manual* [37, 38] contains useful advice on night-time working.

3.5.56   Night time working is not generally appropriate for General Inspections as normally at such inspections no traffic delays are incurred and there may be practical difficulties in providing the required luminance to surfaces not closely accessed.

3.5.57   Working at night is inherently more dangerous than working during the day. The additional risks may include working in darkness or restricted lighting, fatigue, un-natural work cycles and working at low temperatures. Also, although traffic flow is reduced, lighting and signing of the traffic management needs special care to reduce the risk of accidents due to darkness. A risk assessment (see paragraph 3.4) that takes into account these additional factors should be prepared and appropriate mitigation measures identified.

3.5.58   It is important to ensure that staff working at night has properly rested before starting work. It is dangerous for staff to do a day's work then, after only a short break, carry on working at night.

3.5.59   When planning night-time working it is essential to provide sufficient resources to ensure that the work can be completed and the road reopened by the due time. The direct costs of the inspections can be higher than daytime working due to reduced productivity and the costs of additional lighting or other equipment. However, the reductions in traffic delay provide substantial benefits on busy roads.

3.5.60   Night-time working should be considered wherever it is a safe, practicable option, for example on dual two-lane carriageways with a daily flow of over

40,000 vehicles and dual three-lane carriageways with a daily flow of over 60,000 vehicles.

3.5.61 Night-time working can be highly disruptive for local residents as the public perception of noise and light is often considerably more acute at night (see Section 3.11). The Environmental Health Department of the authority should be consulted in appropriate cases. Advance publicity can significantly reduce potential distress and eliminate complaints.

3.6 EQUIPMENT

3.6.1 Part of planning an inspection involves ensuring that the inspector is appropriately equipped for the type of inspection to be undertaken. The inspection team should ensure that all required access equipment, personal protective equipment, data recording and measuring equipment is available and is functioning correctly. Table C.7 lists typical equipment to consider.

Table C.7 – Suggested Checklist of Inspection Equipment	
Access equipment	
• ladders • scaffolds	• mobile elevating work platforms
Personal protective equipment	
• personal protective clothing and other equipment • simple hand washing equipment – water, soap, paper towels • first aid materials	• haversack or shoulder bag for equipment (to leave both hands free) • mobile telephone or other means of communication
Data recording equipment	
• clipboard with waterproof covering • inspection sheets/pro-formas or report forms • pocket tape recorder • mobile phone camera	• writing and marking materials (pen, pencil, eraser, chalk, permanent marker) • data logger or notebook computer • digital camera and spare memory • video camera

Continued

Table C.7 – Suggested Checklist of Inspection Equipment (continued)	
Measuring or inspection equipment	
• cover meter – for depth to reinforcement	• access keys
• surface temperature thermometer	• manhole key
• steel rule or straight edge	• steel thickness gauge
• tell-tales or gauge points – with adhesive	• pen-knife
• Demec gauge	• hand held GPS device which can display coordinates
• underwater probing rods or ranging rods	• laser measuring equipment
• feeler gauges	• binoculars
• endoscope	• small inspection mirror
• steel tape – 3m or 5m	• light tapping or chipping hammer
• plumb bob and string line	• screwdriver – for prodding
• visual crack-width gauge	• paint thickness gauge
• spirit level	• torch, handlamp or helmet mounted lamp
• level and staff	• measuring wheel
• hand drill for concrete dust sampling	• telescopic staff – for measuring headroom
• phenolphthalein indicator for carbonation tests	

Access Equipment

3.6.2   A preliminary visit to the highway structure site may be necessary in order to determine what form of access equipment is required; what restrictions or obstructions there are; and how the equipment will be transported to and from the site. The area around, over or under the highway structure should be observed, to identify any obstructions such as overhead lines, trees, lighting columns, etc., which could restrict the use of equipment. Ground conditions should also be considered, since poor ground can affect the mobility of access equipment or vehicles and may not be capable of supporting vehicle or outrigger loads.

3.6.3   In some circumstances it may be necessary to check for the presence and location of underground services (see paragraph 3.10.3). Underground services may limit the positioning of equipment and restrict the methods used for digging trial excavations.

3.6.4   Overhead electrical equipment, such as overhead line equipment on electrified railways and high and low voltage electricity distribution lines, present special hazards. Any work in the vicinity of such equipment should comply with the requirements of the relevant authority, e.g. railway authority, electric company.

3.6.5   Equipment to gain access for inspection includes ladders, scaffolds, mobile elevating work platforms (MEWPs), abseiling and walkways or mobile gantries permanently fixed to the structure. All involve working at height above ground level and, as such, present hazards but, with sensible care, the risk of accident can be reduced to an acceptable level. Inspectors should never use any access equipment alone. The guide *Temporary Access to the Workface* [36] provides practical information on the various means of access.

3.6.6   The means of access has an important influence on the safety of an inspection; it is the need for personnel to gain access to a particular point which creates most of the risks associated with inspection. It can be difficult to gain access safely to some parts of a highway structure. In such situations the risks involved need to be balanced against the value of the information. The Supervising Engineer should consider:

- Is the information really necessary?

- Can it be obtained in some other way that reduces risks?

- Will it influence decisions on the maintenance required and/or management strategy?

3.6.7   Deciding who needs to have access may influence the method chosen. For example, if the Supervising Engineer or a specialist needs to see a particular defect personally, scaffolding or a MEWP may be required, whereas, if a photograph will suffice, it may be better to use abseiling.

3.6.8   The following sections outline the principal forms of access equipment, commenting on matters relating to inspections. They do not contain full guidance for the safe use of the various kinds of equipment as such guidance can be obtained from a variety of sources, including those listed in Section 7. When working at height all inspection personnel should comply with the Work at *Height Regulations* [50].

Ladders

3.6.9   Ladders are the simplest form of access equipment. They should normally be used as a means of gaining access to an inspection area rather than inspecting from the ladder. However, ladders can be appropriate for 'point' inspections, where there is a need to look at a specific small location and/or defect.

3.6.10   If using a ladder to gain access, it should project a minimum height of 1.05m above the landing unless there is another secure handhold immediately adjacent. When using a ladder, inspectors should never lean out from it and should never carry objects while ascending or descending. It is recommended that any equipment required for the inspection is carried in a haversack or shoulder bag leaving both hands free for the ladder.

3.6.11   Stepladders and trestles are not generally suitable for access purposes and should be used with caution, although they can be appropriate for inspecting structures such as pedestrian subways. Further guidance on the safe use of ladders and step ladders in contained in the *Health and Safety Executive Guide INDG 402* [51].

Scaffolds

3.6.12   Although a properly erected scaffold can give good access to a highway structure it can also obstruct sight of, or access to, defects in the structure. Ideally a scaffold should be tied to the structure. The erection of scaffolding should be planned properly to ensure that it meets working requirements, is designed to carry the necessary loadings and complies with the requirements of the *Construction (Health, Safety and Welfare) Regulations* [17]. Members of the public should not be put at risk. All scaffolding should be erected, altered or dismantled by trained and experienced persons, who must be under competent supervision. Barriers or warning devices must be placed around the base of a scaffold to prevent collision either by vehicles or by persons.

3.6.13   Scaffolds should be checked before every use for vandalism or unauthorised removal of parts and tagged (to specify the safe working load). Responsibility for the inspection and maintenance of scaffolding should be assigned to a suitably experienced and qualified person, whose duty should include the completion of inspection reports as required by the *Construction (Health, Safety and Welfare) Regulations* [17].

3.6.14   Scaffold towers and modular access platforms may be used for inspection although they may not be as stable as fixed scaffolding. The maximum height of the topmost platform is 3.5 times the minimum dimension of the base. The base may include outriggers if they are firmly in contact with the ground on a firm base or sole plates. In high winds kentledge or guys and ground anchors should be attached.

Mobile Elevating Work Platforms

3.6.15   Mobile elevating work platforms (MEWPs) are lorry mounted or self-propelled articulated or telescopic boom type machines. Most types are designed to stand below a structure and reach up to give access to the underside and sides. However, some machines are designed to stand on a structure with booms which telescope out, down and under the same structure.

3.6.16   All MEWPs used for the inspection of highway structures should comply with *BS EN 280* [52]. Operators should be competent and hold a relevant certificate of training achievement issued by either the Construction Industry Training Board or the International Powered Access Federation. The recommendations contained in the *HSE booklet HS(G)19* [53] should also be complied with.

3.6.17   Each MEWP should be inspected by a suitably qualified and experienced person in accordance with clause 7.1.1.5 of *BS EN 280* [52], within the preceding 6 months and following any modification, maintenance or repair that could affect its stability, strength or performance. Each MEWP should also be load tested within the preceding 12 months in accordance with clause 6.1.4.3 of *BS EN 280 [52]*.

Facelift (GB) Ltd

3.6.18   Each MEWP should be fitted with dual controls such that it can be operated through its full working envelope from either the work platform or the ground.

At each location where the MEWP is to be operated and prior to its use, it should be tested using ground level controls to demonstrate that it operates smoothly for all practicable motions and that all safety devices work correctly.

3.6.19　The work platform of each MEWP should be fitted with clearly marked safety harness attachment points that are sufficient for the number of persons that could be carried by the platform.

3.6.20　Inspectors should wear a climber's helmet and safety harness when working on a MEWP (see paragraphs 3.6.28-3.5.30). The safety harness should be connected by lanyard to one of the platform's attachment points and to no other position. If the inspector needs to exit from the platform at height a second lanyard should first be secured to a suitable part of the structure before releasing the original lanyard. The work platform should be positioned such that one side is reasonably parallel and close to the structure and at such a height as to facilitate movement from the platform to the structure. The MEWP should never be allowed to overhang live traffic lanes or the safety zones.

Facelift (GB) Ltd

3.6.21　Self propelled scissor-type MEWPs are not appropriate for use in many outdoor situations. Most are designed for use only on hard level ground and are suited for use inside buildings, e.g. operating on floor slabs. Self-propelled scissor lifts are not generally as stable as lorry-mounted MEWPs: there can be a serious danger of overturning if they are operated out of level or in windy conditions. Ground conditions must be good enough to support the wheels without settlement. If outriggers are fitted, it may be necessary to use suitable strong packing to spread the load. Any rough terrain performance of scissor lifts is intended only for travelling to the work position.

Abseiling

3.6.22　The use of abseiling, or roped access, allows inspections to be carried out where the provision of more conventional means of access would be difficult or prohibitively expensive. Access is gained from above by means of suspended ropes, which provide means of support and positioning. In addition to inspecting vertical faces of a bridge it is also possible to inspect the soffit using ropes or light weight metal structures slung under the bridge.

3.6.23　Abseiling is a specialised task and must only be undertaken by trained and experienced firms. An inspector using this method must have a support team at all times and sufficient, well-maintained equipment.

3.6.24 Due to the nature of the equipment it may not always be possible to inspect all parts of the structure at touching distance. It may be necessary to attach eye bolts and other fixing points on the structure to anchor or support ropes. Further information is provided in *Guidelines on Use of Roped Access Methods for Industrial Purposes* [54].

Permanently Installed Access

3.6.25 Some large or unusual bridges have permanently installed access equipment in the form of ladders, walkways, cradles or gantries or provision for such equipment such as gantry rails. Before using any permanently installed access equipment the inspector should ensure that the equipment has been maintained and is safe to use: it should be clear who is responsible for regular inspection and maintenance. Where appropriate, the equipment should be examined and tested before use.

3.6.26 The inspection work should be planned to ensure the design/safe working loads are not exceeded and any other limitations on use are adhered to. Where the safe working load of the equipment is not known this should be determined before use. Only competent and trained operatives should use the equipment.

3.6.27 A mobile gantry including any fixings or runway beams requires its own programme of inspection and maintenance to ensure its safety. Consideration should be given to the requirements of the Institution of Structural Engineers' report on *The Operation and Maintenance of Bridge Access Gantries and Runways* [12]. Hoists, winches and associated cables should be inspected in accordance with the requirements of the relevant orders under the *Factories Act* [13].

Personal Protective Equipment

3.6.28 Inspectors should wear personal protective equipment (PPE) to suit the operations being undertaken. Commonly required PPE includes:

- *Safety helmet* – The Construction (Head Protection) Regulations [55], as amended by *The Personal Protective Equipment at Work Regulations* [56], require suitable head protection to be provided and worn when there is a risk of head injury. Only Sikhs wearing turbans are exempt. For working at ground level an industrial safety helmet is suitable. However, for working at height, where there is a significant risk of falling a climber's helmet is advisable. A climber's helmet has no peak and is securely attached by adjustable straps.

- *Eye protection* – Safety spectacles with toughened lenses and side screens or safety goggles should be worn when drilling or undertaking any process for taking samples of hardened material (e.g. concrete, masonry, metal) for testing or analysis. Eye protection is also needed when using a chipping hammer to remove any material which has become partly detached.

- *Hearing protection* – The Control of Noise at Work Regulations 2005 [57] require that suitable hearing protection such as earmuffs, earplugs, semi-inserts/caps, should be provided and worn when carrying out noisy work, e.g. drilling or when entering hearing protection areas. Hearing protection should be worn properly (inspectors should be trained on how to do this), and at all times when undertaking noisy work, and/or when working in hearing protection areas. Further information is contained in the Health and Safety Executive's leaflet *Noise at work* INDG362(rev1) [58] and pocket card *Protect your hearing or lose it!* INDG363(rev1) [59].

- **Safety footwear** – Safety boots or shoes with steel toe caps and steel mid-soles should be worn. For working in shallow water or corrosive materials, Wellington boots with steel toe caps and steel mid-soles should be worn. Waders should be used with caution; in the event of falling over in water, the waders may float, inverting the wearer and making it difficult to regain an up-right posture (see paragraphs 3.5.23-3.5.30).

- **High visibility clothing** – High visibility outer clothing of approved type should be worn at all times, except while in a vehicle, when working within the highway boundary or on railway property.

- **Clothing** – Inspection staff should wear clothing appropriate to the weather conditions, where appropriate, this should include standard industrial overalls.

- **Gloves** – It is important to select suitable gloves for the job. In general, gloves of cotton or leather are intended for protection against handling concrete and steel, while rubber or PVC laminated gloves are for protection against chemicals, cement, solvents and glue.

- **Full body safety harness with lanyard and shock absorber** – A full body safety harness fitted with an energy absorber, lanyard and connectors should be worn at all times when working where there is a significant risk of falling. All mobile elevating work platforms should be equipped with safety harness attachment points.

- **Dust mask** – In dusty conditions or when using equipment which creates dust, appropriate respiratory protective equipment should be worn. In most situations a filtering face mask will suffice. However, in confined spaces or when using certain solvents, other equipment, for example breathing apparatus, may be needed

- **Buoyancy aid** – Inspection staff should wear a buoyancy aid when working in or over water or in danger of falling into deep water.

3.6.29 Other specialised equipment will be required in certain situations; for example, breathing apparatus for use in confined spaces. The actual requirements should be specified in the risk assessment.

3.6.30 All PPE should be handled, stored and used carefully. It should be protected and maintained as recommended by the manufacturers. All PPE should be fitted correctly and securely, badly fitted equipment may not perform as anticipated and could even be a hazard in itself.

Data Recording Equipment

3.6.31   The method used to capture defect data on site is at the discretion of the Supervising Engineer and will depend on the equipment in use both on site and in the office. Typical methods include:

- Traditional clipboard with waterproof cover and pencil with standard pro-forma or pro-forma printed specifically for each structure, to record text, measurements and sketches.

- Pocket tape recorder to record the findings.

- A robust data logger or notebook computer, entering the information directly onto screens, which mimic paper pro-forma.

- Digital cameras, mobile phone cameras and/or video cameras.

3.6.32   Digital cameras provide an effective means of recording defects and other features of a structure. However, good quality pictures require considerable memory capacity, so an adequate supply of spare memory is essential on site. Most digital cameras enable the picture to be checked immediately to see if it is of suitable quality. Mobile phone cameras may be used to send photographs to the Supervising Engineer for comment while on site.

3.6.33   Video cameras are rarely used in inspection as they generally require a reasonable level of proficiency from the inspector, combined with subsequent editing and referencing. However, video cameras can be useful in particular circumstances such as recording movement or at locations of difficult or expensive access. In this latter scenario a video camera could be mounted on a boom or robot to be sent into areas where human access is expensive, difficult or impossible, although specialist equipment is likely to be required for controlling the camera remotely and for viewing the image.

3.6.34   In order to identify which structure and what part of the structure is being shown, it is essential to provide a means of referencing for all forms of pictorial records, i.e. sketches, photographs, digital pictures and video recordings.

Measuring Equipment

3.6.35   Depending on the type of inspection and the data required to be collected, the inspector may also need relevant measuring equipment such as that listed in Table C.7.

3.7   STRUCTURE RECORDS

3.7.1   When preparing for inspections, a review of the structure records should be undertaken, to obtain a thorough understanding of the characteristics of the structure and of any features or defects which may require special attention, such as the condition of the structure at the time of the last inspection and any significant maintenance or modifications since the last inspection.

3.7.2   The number and type of records to be reviewed would depend on availability and the nature of the inspection, but should generally cover the appropriate records held in the CDM Health and Safety File, such as those as defined in *BD 62* [60]. This should contain relevant information associated with the whole life management of the highway structure and include records under the following generic headings:

- Inventory Data

- Drawings

- Design Data

- Construction and Demolition Methods

- Materials, Components and Treatments

- Certification and Test Results

- Operation Requirements

- Inspection Schedules and Records

- Maintenance Records

- Structural Assessment and Load Management Data

- Legal Documentation

- Environmental Information

- Supplementary records

3.7.3   It is desirable that the records listed above, and described in more detail in BD 62 [60], are held for all existing structures as these are extremely useful and may contribute to reducing the work of preparing for inspections. For example, records of the methods of access and traffic management used for previous inspections, together with notes on any problems encountered, will eliminate the need to plan the work from scratch.

3.7.4   For some existing structures there may be gaps between the records listed above and those currently held. The Supervising Engineer should seek to identify these gaps and close them in a cost effective and efficient manner by combining record reviews, data collection and record creation with on-going management activities. For example, these activities may be combined with General Inspections, Principal Inspections and/or routine maintenance activities, when records could be verified by inspectors whilst on site.

3.7.5   Where the structure is old it may be helpful to obtain information concerning the construction materials and methods in use at the time of construction. Several publications [61, 62 and 63] provide useful background information on historical construction procedures for concrete, steel and masonry structures.

3.8   TYPE AND EXTENT OF TESTING

3.8.1   Site testing is normally identified during or following a General, Principal or Special inspection or it may be required to obtain additional data to support a structural assessment. It is therefore essential to plan testing operations, i.e. the range of tests and their location, and their extent and intensity, to suit the specific objectives for which testing may be required. Volume 1: Part E of the Manual provides a summary of testing methods relevant to highway structures.

3.9 COMPETENCE OF INSPECTION STAFF

3.9.1 As outlined in Volume 1: Part A: Section 4.2, all inspections should be undertaken by personnel that are judged by the Supervising Engineer to satisfy the minimum health, experience and, where appropriate, qualification requirements for the particular inspection type.

3.10 NOTIFICATION OF OTHER OWNERS AND THIRD PARTIES

3.10.1 Highway authorities do not necessarily own all the land under or near a highway structure and may only retain a right to access. Where inspections require access to land under different ownership, either at the highway structure or adjacent to it, the records should be checked and any landowners and/or tenants consulted to agree arrangements.

3.10.2 For bridges over railways, canals and navigable waterways, it is essential to consult with the relevant authority well in advance and agree details of possessions and safe working practices (see paragraphs 3.5.21-3.5.22 and 3.5.29 for further advice). For underbridges or culverts over non-navigable watercourses, the appropriate drainage or environmental authority may need notification (local Drainage Board, Environmental Agency, etc).

3.10.3 The presence of services in or near the highway structure should be considered and the service authorities consulted for details. Underground services or drains may restrict the placing of access equipment or the location of trial excavations. Overhead lines may also restrict the types of access equipment to be used.

3.10.4 Consultation with environmental or conservation bodies may be necessary where the inspection work is liable to have a significant environmental impact (see Section 3.11).

3.10.5 In some circumstances, it may be necessary or advisable to notify the general public of the proposed inspection work. Appropriate situations that should be considered include:

- where severe delays to road users are unavoidable;

- where the inspection work would result in the temporary closure of vehicular or pedestrian routes; or

- when noisy or night-time working is required in residential areas. In such cases the Environmental Health Department of the local authority should be consulted and prior consent obtained.

3.10.6 Prior notification and advance publicity could help reduce distress that may be caused by the work, by informing people of the need and duration of the work and demonstrating that they have been considered.

3.11 ENVIRONMENTAL CONSIDERATIONS

3.11.1 Due consideration should be given to the environment when inspections are planned and undertaken at highway structures, including access and working operations at adjacent areas. The Supervising Engineer should ascertain whether any significant environmental impacts are likely to occur and, if so, seek expert advice to identify and implement the appropriate working practices and/or mitigation measures. Details of suitable sources of environmental guidance may be obtained from the organisations listed in Volume 1: Part F: Appendix A. Particular attention should be placed on maintaining the population of the country's characteristic fauna and flora and the communities they comprise [64].

3.11.2   Some plant and animal species and their habitats are given special protection by UK and European legislation [65] and a list of specially protected species is provided in Volume 1: Part F: Appendix D. Highway structures, although man made, are part of the landscape and quickly become habitats used by wildlife. Also, some structures are situated in areas of special environmental interest or designated areas protected by statue, which could be damaged by inspection activities. The following sections outline the principal environmental aspects, also summarised in Table C.8, which should be considered and include recommendations for planning and carrying out inspections.

Table C.8 – Environmental Considerations			
Impacts on	**Aspects to be considered**	**Action**	**Paragraph**
1. People			
in vehicles	Safety and disruption of traffic	Provide traffic management	3.5.4-3.5.15
pedestrians	Safety and convenience	Possible diversionary routes	3.5.4-3.5.15
		Fence off works	3.5.17
nearby residents and businesses	Noisy operations	Adopt quiet methods Avoid night working	3.11.3-3.11.4
	Night time working	Obtain consent from EHD	3.11.3-3.11.4
		Advance publicity	3.10.5-3.10.6
landowners and tenants	Convenience and access	Check records for right of access and agree access arrangements	3.10.1
2. Farmland			
Livestock	Access, timing, etc	Consult farmer	3.5.16-3.5.20
	Straying	Fence off works	3.5.17
		Close gates etc.	
	Disease	Adopt necessary procedures	3.5.16-3.5.20
Crops	Crop damage	Consider timing of inspection	3.5.16-3.5.20
		Keep to arranged locations	
	Disease	Adopt necessary procedures	3.5.16-3.5.20

Continued

Table C.8 – Environmental Considerations (continued)			
Impacts on	**Aspects to be considered**	**Action**	**Paragraph**
3. Habitats	Ecological or landscape significance of adjacent land	Obtain advice/consent from relevant body. Keep to arranged locations	3.11.22-3.11.24
4. Watercourses	Pollution	Adopt good practice Avoid spillage of fuel or other pollutants	3.11.25-3.11.28
5. Plants	Destruction of rare or protected species	Adopt necessary procedures Keep to arranged locations	3.11.29-3.11.31
6. Wildlife Bats	Disturbance of bats or roost sites	Consult the SNCO and arrange bat survey where required Time inspection to avoid disturbance	3.11.29-3.11.31 3.11.8-3.11.13
Nesting birds	Disturbance of birds when nesting	Time inspection to avoid disturbance	3.11.14-3.11.17 3.11.29-3.11.31
Other wildlife	Disturbance of otters or other protected wildlife	Obtain general licence to remove pest species Consult relevant organisations	3.11.29-3.11.31 3.11.18-3.11.21
Bird droppings	Accumulations can be a health hazard	Use respirators and other PPE	3.3.9-3.3.17

People

3.11.3 Some highway structures are sited in sensitive areas, such as residential areas or adjacent to hospitals, where inspection work or noise from plant may cause inconvenience or nuisance to the public. The adoption of good working practices and sensible timing may help to reduce the effects of noise. Nevertheless the Environmental Health Department of the local authority should be consulted if there are likely to be problems. The Environmental Health Officer may set limits to the noise which will be permitted.

3.11.4 Night-time working can be particularly disruptive. If work at night is planned near residential properties or other sensitive areas, the Environmental Health Department must be consulted and prior consent obtained. This consent will stipulate noise limits.

Plants

3.11.5 There is currently approximately 39,000 plant or plant-like species, including algae, fungi, mosses, liverworts, ferns, horsetails, and flowering plants (trees and shrubs), native or naturalised in the UK. Plants, and the diversity of plants in a particular area, reflect the influence of many interrelated factors such as soil, hydrology, climate and type of management (either natural or by humans). Some species have very specific requirements, while others have adapted to almost any situation.

3.11.6 Plants can be found in almost any habitat, wherever sufficient resources can support their requirements. Some plant species have evolved very complex life cycles or behaviour based on an interaction with other species. Plants form the basic material on which most other species survive.

3.11.7 Some plant or plant-like species are protected under the *Wildlife and Countryside Act* [66] and the *Conservation (Natural Habitats, etc.) Regulations* [67, 68]. Under Section 13 of the *Wildlife and Countryside Act* [66], it is an offence to intentionally pick, uproot, or destroy any wild plant. The general procedure to be followed for the consideration of protected species during inspections is outlined in paragraphs 3.11.29-3.11.31. Further guidance for avoiding the destruction of protected or rare plants is contained in *HA 84* [65] and the *Ecology and Nature Conservation Advice Note* [64].

Bats

3.11.8 Bats in Britain and Ireland have declined in numbers due to habitat degradation, widespread use of pesticides, persecution, and destruction of their roosting sites. The roosts are of great importance to their survival. Due to the decline and numerous threats they face, bats and their roosts are legally protected in the United Kingdom under the *Wildlife and Countryside Act* [66], the *Conservation (Natural Habitats, etc.) Regulations* [67, 68] and the *Wildlife (Northern Ireland) Order* [69].

3.11.9 Under the above legislation, it is an offence to deliberately capture, injure or kill a bat, and to possess or control any live or dead bat. Most important so far as bats and highway structures are concerned, the legislation affords protection to any structure or place the bat uses for shelter, protection or breeding, by making it an offence to damage, destroy or obstruct access to such a place, as well as to deliberately disturb the bats whilst they are using it.

3.11.10 A wide range of bridge types have suitable crevices for bats, offering safety, cool and stable temperature conditions, high humidity, and nearby drinking water and feeding areas for bats roosting in spring, summer and autumn. Bridges in river valleys are especially liable to be used as bat roosts. Some studies have indicated that up to 16% of bridges in England and Wales are used as bat roosting sites. Stone bridges are more likely to provide suitable crevices than concrete or steel structures. However, a survey in Cumbria indicated that 25% of concrete bridges had suitable roost crevices and 5% had bat roosts [70].

3.11.11 Bats may be found in bridges at any time, and may use different parts of the same bridge depending on the time of year, as they require different roost conditions with the changing seasons, i.e. bats often move between summer roosts, hibernacula and transition roosts. Nursery roosts where the females give birth and rear their young (approximately June to August) would be found in warm sites. Bridges with deep crevices may offer good hibernation sites (approximately October to March) being isolated from external temperature fluctuations and have stable, cool temperatures. Partially blocked arches are particularly suitable as hibernacula. Male roosts and transition roosts tend to be more variable in nature. Many bridges contain suitable crevices which could be used as night roosts for resting, eating large prey, or socialising.

3.11.12 Most roosts are located in bridge superstructures, but roosts have also been found in abutments, spandrel walls and parapets. Bats can roost in almost any type of crevice that is greater than 100mm deep, 12mm wide and protected from the elements. The wide variety of crevice types that have been used include:

- Crevices between stones where mortar has fallen out, or in damaged stonework;

- Drainage holes and pipes, including ceramic and steel pipes;

- Expansion joints;

- Construction joints;

- Gaps between beams and slabs in bridge decks;

- Gaps created where the span overlies the piers or abutments;

- Box voids in concrete structures; and

- Behind thick growths of ivy on bridges.

3.11.13 A guide for signs of bats is given in Volume 1: Part F: Appendix E. The general procedure to be followed for the consideration of protected species during inspections is outlined in paragraphs 3.11.29-3.11.31. Further guidance on avoiding the disturbance of bats or their roosts in bridges is provided in *HA 80* [71].

Birds

3.11.14 Birds often nest in bridge structures and several species use bridges as night-time roosts. Some ground nesting birds may nest on the ground near or adjacent to bridge structures. All birds, their nests and eggs are protected under the *Wildlife and Countryside Act* [66] and the *Wildlife (Northern Ireland) Order* [69]. Under this legislation, it is an offence, with certain exceptions, to intentionally:

- kill, injure or take any wild bird;

- take, damage or destroy the nest of any wild bird while it is in use or being built;

- take or destroy the egg of a wild bird; or

- disturb any wild bird listed in Schedule 1 of the *Wildlife and Countryside Act* [66] (see Volume 1: Part F: Appendix D) while it is nest building, or at a nest containing eggs or young, or disturb the dependent young of such a bird.

3.11.15 The most notable exceptions to the above provisions are:

- An authorised person (e.g. a landowner or occupier) may kill or take 'pest species' (see Table C.9) and destroy or take the nest eggs of such a bird. This is permissible under the terms of general licences issued by Government departments.

- A person charged with killing or attempting to kill a wild bird, other than one included in Schedule 1 of the *Wildlife and Countryside Act* [66] (see Volume 1: Part F: Appendix D), is not guilty of an offence if they can demonstrate that this action was necessary to preserve public health or air safety, prevent spread of disease, or prevent serious damage to agriculture.

Table C.9 – 'Pest Species' of Birds
Carrion crow
Collared dove (not in Northern Ireland)
Feral pigeon
Great black-backed gull
Herring gull
House sparrow
Jackdaw
Jay (not in Northern Ireland)
Lesser black-backed gull
Magpie
Rook
Starling
Wood pigeon

3.11.16 Game birds are covered by separate Game Acts, which generally provide protection during the close season. Close seasons vary, but normally run from the beginning of February to the end of August.

3.11.17 The general procedures to be followed to avoid illegal disturbance of birds or their nests are outlined in paragraphs 3.11.29-3.11.31. It should be noted that accumulations of bird droppings at roost or nest sites can be a health hazard; appropriate precautions should be taken (see paragraphs 3.3.14-3.3.17 and Section 3.4 and the *RSSM* [25]).

Other Wildlife

3.11.18 Bridges or their immediate surroundings may be inhabited or frequented by other protected wildlife. Where the presence of protected wildlife is known or suspected, the Supervising Engineer should contact the Statutory Nature Conservation Organisation (SNCO) or county or regional wildlife trust to seek advice on mitigating any disturbance. Details of suitable sources of environmental guidance may be obtained from the organisations listed in Volume 1: Part F: Appendix A.

3.11.19 As an example, bridges or culverts over watercourses may be frequented by otters. Their presence can sometimes be detected from droppings (spraints) left to mark their territory on ledges, rocks or sediment under bridges. Hollows in bridge abutments or culverts and drains may be used as lying-up sites or holts. The adjacent riverbank may also be used. Otters and their holts are also protected under the *Wildlife and Countryside Act* [66], the *Conservation (Natural Habitats, etc.) Regulations* [67, 68] and the *Wildlife (Northern Ireland) Order* [69].

The Supervising Engineer should contact the SNCO or county or regional wildlife trust to ascertain whether otters are likely to be present at specific bridges. This information should be updated every few years since the range of otters is increasing in some parts of the country. Further information concerning otters is provided in *HA 81* [72].

3.11.20 Water voles may also be found along certain watercourses. Although it is unlikely that they may be found within a structure their burrows may be located nearby. At the time of publication of this Manual water voles themselves were not protected by law (it is anticipated that this may change in future), however, limited legal protection through Schedule 5 of the *Wildlife and Countryside Act* [66] (as amended) in respect of Section 9(4) is given to their habitats. It is an offence to intentionally damage, destroy, or obstruct access to any structure or place which water voles use for shelter or protection or disturb water voles while they are using such a place.

3.11.21 Badgers are known, if the ground conditions are right, to burrow next to structures especially within embankments found near or adjacent to bridges. They as well as their habitat are protected under the *Badgers Act* [73] and one requires a licence to remove them or close their setts. Further information concerning badgers is provided in *HA 59* [74].

Surrounding Habitats

3.11.22 The land at and adjacent to the structure, or along the route used to gain access to the structure, may be a designated site [64], i.e. areas of high nature conservation value protected to varying degrees by statute, international conventions, or local authority planning controls. They form a network of habitats which may be of global, international, European, national, regional or local importance. Generally, the priority for protection of designated sites is as follows:

- Sites of international importance:

- World heritage sites;

- Biosphere reserves;

- Ramsar Sites – wetlands of international importance;

- Special Protection Areas (SPA);

- Special Areas of Conservation (SAC)

- Sites of national importance:

- National Nature Reserves (NNR);

- Marine Nature Reserves (MNR);

- Sites of Special Scientific Interest (SSSI);

- Areas of Special Scientific Interest (ASSI);

- Areas of Special Protection for Birds (AOSP);

- Ancient Woodlands;

- Natural Heritage Areas

- National Parks;

- Areas of Outstanding Natural Beauty (AONB);

- Environmentally Sensitive Areas (ESA);

- Sites of Regional or Local importance:

- Local Nature Reserves (LNR);

- Regional Parks;

- Non-Statutory Sites of Importance for Nature Conservation;

- Non-Statutory Nature Reserves

- Forest Nature Reserves

- County Sites of Biological Interest;

- Other wildlife sites

3.11.23 In any of these areas there may be advisory or statutory restrictions on intended activities of the bridge inspection team. It will therefore be necessary to contact the SNCO or the county or regional wildlife trust or biological records office (see Volume 1: Part F: Appendix A) to establish whether there are any areas of significance at or near the structure. Should there be, the methods of working should be drawn up taking this into account.

3.11.24 Access to the structure should be arranged so as to minimise damage to the environment. Vehicles and equipment can cause rutting or ground compaction as well as direct damage to the vegetation. The impact can be minimised by careful routeing of vehicles, by the use of temporary access ways or mats, or by the timing of the work. Wherever practicable, vehicles should be confined to existing tracks or hardstandings.

Pollution

3.11.25 Care should be taken to avoid pollution. This is most likely to occur from fuel spillage or from effluent such as the lubricating water used during core drilling. The pollution of controlled waters is covered by the *Water Resources Act* [75], the *Control of Pollution Act* [76] and the *Water (Northern Ireland) Act* [77] and it is an offence to cause or knowingly permit any poisonous, noxious or polluting matter or any solid waste to enter controlled waters. Controlled waters are all watercourses (i.e. both surface and ground water).

3.11.26 Drip trays should always be used under petrol or diesel powered plant where there is any chance of polluting controlled waters. Oil powered pumps, generators and the like should be positioned on impervious drip trays surrounded by earth or sand bunds and located at least 10 metres from any controlled waters. Any accidental spillages of hydrocarbons or other pollutant into a watercourse should be reported immediately to the Environment Agency, or other appropriate organisation, using the emergency hotline (see Volume 1: Part F: Appendix A).

3.11.27 Some testing operations can give rise to other forms of pollution, such as dust. This can normally be avoided or minimised by the use of best practice procedures. In some sensitive areas noise can cause a nuisance, but good practice and sensible timing should alleviate the problem.

3.11.28 In the *national Air Quality Strategy* [78] eight substances were identified as requiring closer control due to their effects on people and the environment. As a result, some authorities have designated Air Quality Management Areas (AQMA) located around major roads, and have produced, or are in the process of producing, action plans identifying how the air quality issues in these areas could be controlled or tackled. Inspection personnel should be aware of AQMAs and have a duty to ensure the air quality within these areas is not adversely influenced by any operations undertaken as part of the inspection.

Consideration of Protected Species During Inspections

3.11.29 Due consideration should be given to protected species during inspection planning. Inspections should be divided into two classifications to identify those inspections that are likely and unlikely to impact on protected species. Two classifications that may be used are:

- Non-invasive inspections – non-invasive inspections are essentially those requiring no, or limited, physical contact with the structure, e.g. a bridge is inspected visually, either from the ground level, the deck or from permanently installed access equipment. Safety and General Inspections are of this type. Non-invasive bridge inspections may be undertaken at any time, subject to the constraints of access.

- Invasive inspections – invasive bridge inspections are those requiring in-situ testing and the use of hand, probe or torch to investigate deep (i.e. greater than 100mm) crevices. Potential types of crevice are listed in paragraphs 3.11.8-3.11.13. Some Principal and most Special Inspections are 'invasive'.

3.11.30 When undertaking a *non-invasive* inspection it may be beneficial for the inspector to be aware of protected species before a undertaking a (non-invasive) General Inspection, therefore if evidence of different protected species are found during the inspection this should be reported to the Statutory Nature Conservation Organisation (SNCO), but the inspection may proceed, see Figure C.6. Figure C.6 also presents a procedure that may be followed for *invasive* inspections, key points of which include:

- The structure records should be checked to ascertain whether or not protected species are associated with the structure and if there is a need for any special measures related to the protected species.

- If the structure is a known protected wildlife or other species site, the SNCO and any other relevant organisations (see Volume 1: Part F: Appendix A) should be able to provide information and advice on when, and possibly how, work should proceed. In general, work may be undertaken at specific times of the year depending on, for example, how or when wildlife uses the structure, see Table C.10.

- If the structure is not a known protected species site, the inspection may proceed. However, if the structure has potential for the presence of protected species then a pre-inspection survey by a competent environmental specialist should be arranged. This would prevent any delays to the work if protected species are found.

- During the inspection, if any evidence of protected species is found it should be reported to the SNCO, but the inspection may proceed.

- During the inspection if protected species are found, then 'invasive' work should stop and the SNCO contacted. The SNCO should make provision for a visit by a competent environmental specialist and then advise on how the work should proceed.

3.11.31 If a structure is used all year by wildlife, it is unlikely that the same part of the structure will be used all year round. Where this is the case, inspection could proceed on the parts of the structure not currently occupied. However, appropriate advice relevant to the specific situation should be sought.

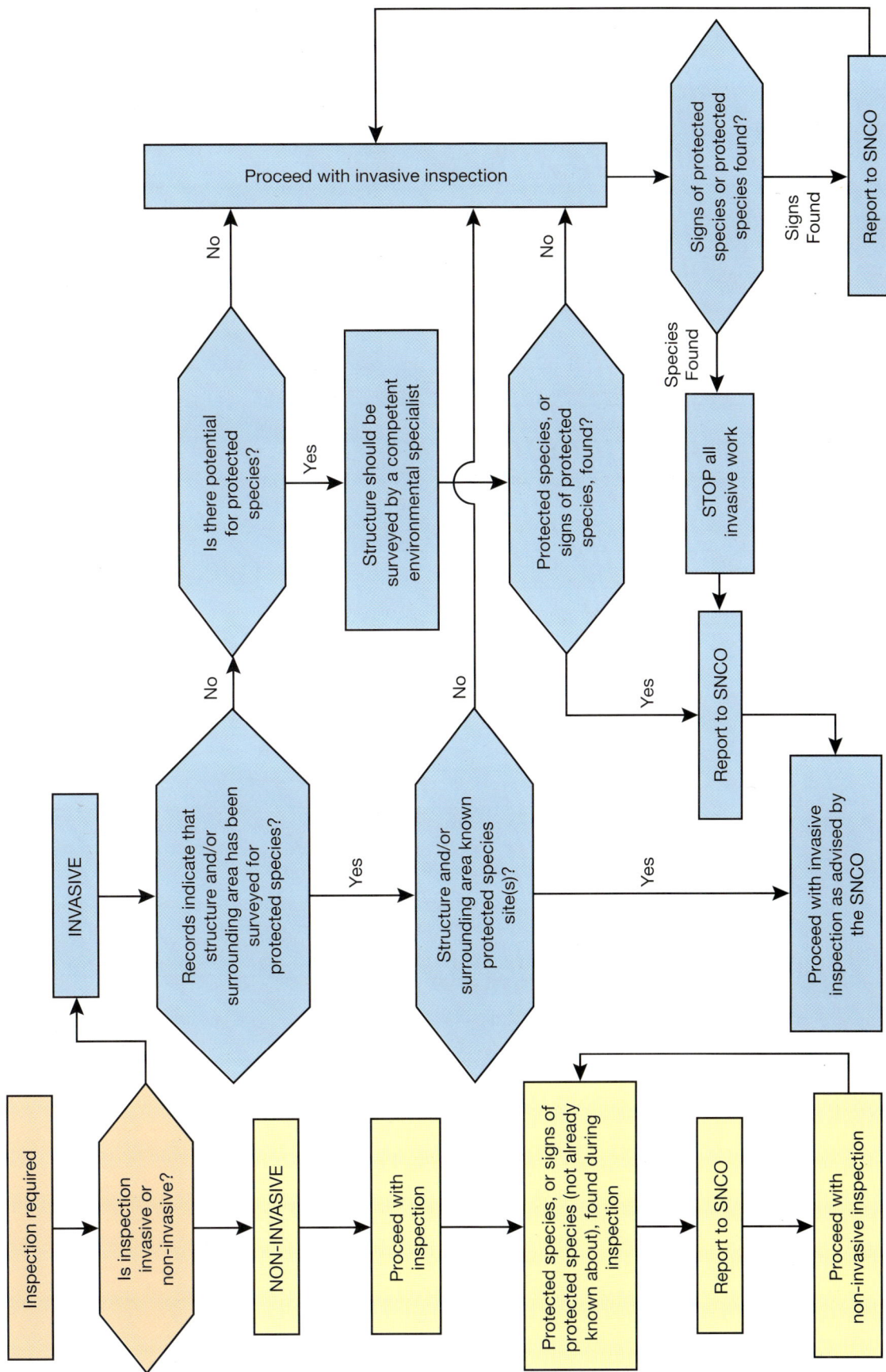

Figure C.6 – Procedure for consideration of protected species at highway structures

Table C.10 – Seasonal Constraints: Animals

Type	Month	Jan	Feb	Mar	Apr	May	Jun	Jul	Aug	Sep	Oct	Nov	Dec
Birds	Breeding			◼ Start of nesting dependent on local variations →									
	Over wintering	◼	◼	◼						◼	◼	◼	◼
Amphibians	Breeding		◼	◼	◼	◼							
	Hibernation	◼										◼	◼
Reptiles	Breeding				◼	◼	◼	◼					
	Hibernation	◼	◼									◼	◼
Invertibrates	Breeding			◼	◼	◼	◼	◼	◼	◼			
	Over wintering	◼ Egg stages									◼ Egg stages →		
Bats	Breeding				◼ Breeding			◼ Nursery roosts					
	Over wintering	◼ Hibernacula										◼ Hibernacula	
Badgers	Breeding	◼ Females in setts with young					◼						
Otters	Breeding	◼	◼	◼	◼	◼	◼	◼	◼	◼	◼	◼	◼
Dormice (*Muscardinus avellanarius*)	Breeding						◼	◼	◼	◼			
	Hibernation	◼	◼	◼							◼	◼	◼
Fish	Adult sea trout						◼ Ascend rivers to spawn			◼			
	Adult salmon	◼	◼							◼ Spawning →			
	Salmon smolts				◼ Migration to sea	◼							
	Salmon eggs or fry										◼ In river bed →		
	Coarse fish			◼ Spawning	◼	◼	◼						

Notes	
	1. The bars indicate the main periods to avoid when undertaking works which affect these species. Consideration should be given to climatic and geographical variations, which may bring forward or delay the period.
	2. Some animals and their habitats have statutory protection. Where protection is given by legislation, advice should be obtained from the relevant Statutory Consultees.
	3. Badgers should not be disturbed between 1 December and 30 June as females are in setts with young.
	4. In respect of fish, the constraints indicated provide a guide only and variations due to species and season should be considered on a site-by-site basis.

4 Performing Inspections

4.1 INTRODUCTION

4.1.1 Upon arrival on site, a careful check should be carried out to confirm the identity of the structure, i.e. to ensure it is the correct one. Mistakes can easily occur and in some instances records may be misleading. Unless the inspection team is already familiar with the structure, a quick look around is advisable to make sure that elements such as movement joints, bearings, etc. may be correctly identified, that the orientation is understood by the team and that any drawings confirm the observed layout. Where possible the co-ordinates of the structure should be checked.

4.1.2 Inspectors should be aware that the appearance of some structural materials may sometimes be misleading. For example, a retaining wall with a masonry face may be solid masonry or it may be masonry cladding to some other material such as mass concrete, reinforced concrete, steel sheet piling or even reinforced earth. Similarly some structures may have been subjected to repairs or alterations which are superficially similar to the rest of the structure but which may conceal a different form of construction. In the majority of cases the structure records should contain sufficient information to clarify such situations; however, inspectors should be aware that some details may have not been recorded or that the structure records are in the process of being up-dated.

4.1.3 Before commencing work, inspectors must ensure that any necessary traffic management measures (see paragraphs 3.5.5-3.5.20) are in place, that they have been correctly set up and that they are diverting the traffic in a safe manner. At intervals during the work, checks should be made to ensure that the traffic management is still operating correctly, for example signs may have fallen over or become obscured; cones may have been displaced; signals may have ceased to function, etc. Traffic management is put in place for the safety of the inspectors as well as that of the public and as such inspectors have a duty of care to ensure that it operates satisfactorily.

4.1.4 The inspection should proceed in a logical and systematic manner within the constraints imposed by any safety, traffic management and access considerations. The first stage of any inspection should be to review the overall condition of the structure paying particular attention to any evidence of structural movement, e.g. settlement. During this initial 'overview' of the structure, the inspector should focus on identifying the effects of structural defects and not the defects themselves.

4.1.5 Inspections should be thorough, i.e. conscientious and to the full requirements of the specific brief for the type of inspection being undertaken. The work should never be skimped and if problems are encountered, these should be discussed with the inspection team leader and, if necessary, referred to the Supervising Engineer. For example, a typical problem might be difficulty in obtaining access to some part or parts of the structure. Every effort should be made, without compromising safety, to obtain all the required information during the inspection. Where there are restrictions on the working hours, such as on busy motorways or during railway track possessions, it is particularly important to work efficiently so that the work may be completed within the allotted time.

4.1.6   In some instances, the lead inspector may be required to consider whether the value of the inspection would improve by undertaking work additional to the original brief while on site. For example, the results of early tests or the initial discovery of unforeseen movements should be used to review the scope of the work. Any proposed changes should be agreed with the Supervising Engineer prior to being implemented.

4.1.7   Inspectors should always be alert to anything unusual while on site and focus on any part of the structure that may cause particular concern. Possible examples include incomplete or missing secondary elements, clearances that are too small or too large, or elements which are over or under-sized. Any such observations or concerns should be brought to the attention of the Supervising Engineer.

4.1.8   Any damage or disturbance caused on or adjacent to a highway structure during an inspection should be made good. This frequently includes reinstating drill or core holes in concrete or masonry, painting exposed steel surfaces or refilling trial pits. The correct materials and good workmanship are essential as poor repairs may result in accelerated deterioration or affect the appearance of the structure.

4.1.9   On completing an inspection, the team should verify that all the information required has been captured. The site must be cleared thoroughly of all equipment, materials and rubbish. If working on a highway, before a stretch of the highway is reopened to traffic, the lead inspector, or another responsible person, should ensure that the area is safe for public use. On railways, a formal checking and reporting procedure should be followed to ensure safety at the end of a possession.

4.1.10   The following sections give general advice and guidance on carrying out inspections for structures constructed of different materials and certain special structures. Detailed advice on the defects that may occur on these structures is included in Part D of the Manual. The level of activity and information acquired should be commensurate with the type of inspection being undertaken.

4.2   CONCRETE STRUCTURES

4.2.1   The main cause of deterioration of reinforced concrete structures is corrosion of the reinforcement. Inspectors should pay particular attention to the presence of reinforcement corrosion or the risk that corrosion may occur in the future. Areas particularly at risk are those subjected to leakage of de-icing salts through joints, and concrete subjected to salt spray from passing traffic or from the sea for structures in a marine environment. Vulnerable areas on bridges may include bearing shelves, half joints, piers and abutments, crossheads, ballast walls, deck ends and areas around defective or blocked drainage.

4.2.2   Where cracking of concrete due to reinforcement corrosion or corrosion of prestressing tendons is suspected, in addition to visual examination it may be appropriate to carry out some simple testing during a Special Inspection such as measurement of chloride content, carbonation depth, reinforcement cover or electrode-potential (half-cell); details of theses techniques are provided in Part E. This would enable a better assessment of the condition of the reinforced concrete to be made. The results obtained should be recorded in the Structure Records for future reference. Further guidance may be obtained from *BA 35* [5], *BA 88* [79], *Diagnosis of Deterioration in Concrete* [80] and T*echnical Guide 2: Guide to Testing and Monitoring the Durability of Concrete* [81].

Highways Agency

4.2.3   Concrete structures suspected of suffering from alkali-silica reaction (ASR) or any other form of chemical degradation should have a Special Inspection to check the cause and extent of any deterioration. Further information on ASR can be found in *Structural Effects of Alkali-Silica Reaction: Technical Guidance on Appraisal of Existing Structures* [82].

4.2.4   Prestressed concrete structures (pretensioned or post-tensioned) can suffer from any of the defects described above for reinforced concrete. However, particular attention should be paid to cracks in the concrete or any other indication, e.g. rust staining, that the prestressed elements may be subject to corrosion and therefore at risk of loss of prestress.

4.2.5   Post-tensioned concrete bridges with grouted tendon ducts are particularly vulnerable to corrosion and severe deterioration in segmental construction and/or where internal grouting of the ducts is incomplete. Such bridges may have been subjected to a Special Inspection in accordance with *BA 50* [6]. The findings of the Special Inspection should be taken into account when planning and undertaking an inspection. Where such an inspection has not been undertaken previously, a Special Inspection should be carried out. The purpose is to establish whether there are voids in the grouted ducts and the extent of any tendon corrosion or other deterioration, so that the vulnerability of the bridge and its residual strength may be assessed. It is important to determine the form of the bridge and its load-carrying system as this can have a considerable influence on its vulnerability to tendon corrosion.

4.3   STEEL STRUCTURES

4.3.1   Steel is particularly vulnerable to corrosion when exposed to wet conditions or to aggressive ions, such as chlorides from de-icing salt, or when exposed to a marine environment. Most steelwork on highway structures is therefore protected with paint or some other protective coating. Corrosion is usually associated with the breakdown of protective systems, which is probably the most common defect associated with steel superstructures. It is important to assess the magnitude, location and form of corrosion and, if possible, identify its cause. Inspectors should assess and record any loss of structural section. Special Inspections of the protective system using specialist inspectors may be required to identify the cause of any deterioration of the paint system and to identify the need for maintenance painting. There are also circumstances when Special Inspections are required in order to identify if corrosion is taking place and to monitor it over a period of time.

Highways Agency

4.3.2   The steelwork in some structures, particularly bridges, has been enclosed to reduce the rate of corrosion and to provide access for inspection. Such enclosures should be inspected during all General and Principal Inspections of the structure. Although enclosures should have long service lives some components or seals may have shorter lives. Further guidance on enclosures for bridges is contained in *BA 67* [83].

4.3.3   Older bridges may be at risk of fatigue-induced failures, although fatigue susceptible details may also be present on more recent bridges. Deformation or distortion of members may reduce the load carrying capacity of the structure. Sighting along flanges may aid checking of members, taking measurement of the maximum deformation if necessary.

4.3.4   Weathering steel is particularly vulnerable in wet/dry situations and at web flange joints, where settled rust deposits may retain water like a sponge. Weathering steel should be visually inspected for irregularities in the appearance of the patina, at critical areas and in particular at fixed joints and expansion joints. Any irregularity of appearance should be reported. Where irregularity occurs, a Special Inspection may be needed to ascertain the cause. Steel thickness measurements on weathering steel are also required generally at six year intervals at predetermined locations to check for loss of steel, normally at the time of a Principal Inspection. Where structures contain such material, the authority should follow the procedures given in *BD 7* [9].

4.3.5   Steel/concrete composite bridges rely on interaction between the steel and the concrete provided by shear connectors. Failure may be indicated by separation between the top flange and the concrete slab. The inspector should examine this interface for evidence of separation.

4.3.6   Corrugated steel buried structures (CSBS) used as culverts, deteriorate mainly through hydraulic wear in the invert and along the wet/dry line. The hydraulic action removes protective coatings and exposes the steel substrate to corrosion. Deterioration of CSBS is also caused by exposure to water laden with de-icing salts or sulphur compounds present in the backfill and surrounding soil. Deterioration of CSBS used as cattle creeps, pedestrian underpasses, etc. will also occur due to this cause. Deterioration is often localised and in extreme cases results in perforation of the steel shell, which might require strengthening works or, if in an advanced state, replacement of the structure. Inspection of corrugated steel buried structures is generally limited to exposed surfaces. Inspectors should look for signs of bulging and deformation in the shape or line of the steel arch or ring and for signs of the settlement of fill in areas above or adjacent to the arch. An overview of the immediate surrounds should be made to identify changes such as erection of structures, subsequent to the construction of the corrugated structure that may, for example, alter the loading or the level of the water table in the vicinity. Further advice on the inspection of these structures is given in *BA 87* [84].

4.3.7   Connections are points of weakness in steel construction, whether welded, riveted or bolted, and may have material or loading defects. Inspectors should be aware that structural movement or failure may initially propagate as movement at connections. Therefore, all connections should be checked for defects. Welds, particularly those between deck plates, and stiffeners should be inspected for cracking, which may require the use of non-destructive testing techniques. Bolts and rivets should be checked to establish that none are loose or missing.

4.3.8   Older structures often have details which are susceptible to corrosion so inspectors should give particular attention to areas such as:

- Small gaps between components which are not adequately sealed.

- Where components are built into concrete or masonry.

- Water traps and areas where debris can build up.

- The insides of unsealed hollow members that are not readily accessible, e.g. look for external indications of corrosion and/or use specialist techniques during a Special Inspection.

- Areas subject to leakage of de-icing salts, e.g. members below deck joints, joints in trough or plate decking.

4.4   CABLE SUPPORTED STRUCTURES

4.4.1   Particular factors that may affect the performance of cable supported structures include excessive vibrations, corrosion, fatigue, and the general inability to reliably ascertain the condition of the cables, especially in the critical anchorage zones. Depending on the type of inspection performed, currently available methods that may be used for cable supported structures include conventional visual inspections and some non-destructive testing techniques such as magnetic, ultrasonic, x-ray, laser, acoustic, and remote or contact-based vibration methods.

Warwickshire CC

4.4.2   During Principal Inspections of cables, the entire surface of the cable should be inspected at close range, followed by an inspection of neoprene boots and rings, visible surfaces of guide pipes, and accessible anchorage surfaces. Visual inspections of cables typically involve the following (the relevance of the following will depend on the specific arrangement of the cables and the size of the structure):

- Identification of longitudinal or transverse cracking or excessive bulging in the sheathing, as well as damage at connections to dampers or cross cables.

- Inspection for cable alignment irregularities including waviness or excessive sag. Cable sag may be estimated or measured using optical devices or through video or photo image processing. Cable angle may be measured with an inclinometer at specific points.

- Identification of changes to bridge deck elevations.

- Examination of protective tape wrapping, e.g. tears, cracks, and delaminations.

- Examination of the sheathing, particularly any evidence of cracking in the sheathing located at high stress areas.

- Identification of damage to connections between anchorage pipes and cable sheathing.

- Inspection for damage, loosening, lack of water tightness, and deterioration of neoprene boots and band clamps.

- Inspection for damage or dislocation of neoprene rings and keeper rings, if applicable.

- Identification of gaps between the neoprene rings and the sheathing.

- Examination of sheathing surface inside the guide pipe through a boroscope or other means, looking for damage or deformation to the sheathing near the anchorage.

- Identification of cracking or damage to guide pipes or evidence of the impact of cable components on guide pipes.

- Examination of surface conditions on the visible anchorage components including ring nuts, end caps, and bearing plates.

- Examination of visible parts of saddles for damage, corrosion, and cracking.

- Review of evidence of moisture or fillers (such as grease) exiting the anchorage components. If there is an access port at the end cap (ideally at the lowest point), it can be opened and examined for moisture or moisture contaminated grease.

- Removal, in some cases, of the end caps on the sockets to allow for visual inspection of the anchorage plate and anchorage devices and to see if there is moisture or corrosion inside.

- Inspection of the cross tie cables for sagging, i.e. losing their tension force and require to be retensioned.

- Inspection of damage or cracking on components of cross tie cables. Evidence of fretting and fatigue, especially at connections, are of particular interest.

- Examination of dampers, if any, as per recommendations of manufacturer.

4.4.3   Due to the nature of these bridges and because no two cable supported structures are identical, it is recommended that inspection procedures for major cable supported bridges are tailored to each specific structure. Special Inspections entailing the use of non-destructive methods should be led by a specialist familiar with both the testing techniques and these particular types of structures. Further guidance for the use and effectiveness of appropriate testing methods for cable supported structures are contained in *Synthesis 353: Inspection and Maintenance of Bridge Stay Cable Systems* [85].

4.5   MASONRY STRUCTURES

4.5.1   Inspection of masonry structures relies on visual inspection rather than testing. The main defects found on masonry structures are: cracking, arch ring separation, bulging and deformation, loss of mortar, loss of bricks or stones, seepage of water through the structure and deterioration of the bricks or stones. Cracking arises from a variety of causes including overloading, vibration or impact from traffic, settlement, foundation failure, temperature and humidity changes i.e. cycles of freeze/thaw activity or wetting and drying. It may be necessary to initiate a Special Inspection in order to determine the cause of the cracking.

Cambridgeshire CC

4.5.2   Cracks in masonry may affect the appearance only or be indicative of a more serious defect. Recent or progressive cracks are more serious than those which may have occurred soon after the structure was constructed. Evidence that cracks are recent may include clean faces to the crack and loose fragments of masonry or mortar. Cracks formed in the mortar only may be indicative of joint deficiencies. Inspectors should map the extent of cracking in order that comparisons can be made with previous inspections.

4.5.3   Inspections should generally seek to take into account the age of the structure, the type of masonry, local knowledge (many masonry structures are very old) and the exposure environment. Some types of masonry (e.g. sandstone) deteriorate more readily than others (e.g. granite) and this can be exacerbated by the severity of the environment they are in. Further information on the inspection of masonry arches is given in *Masonry Arch Bridges: Condition, Appraisal and Remedial Treatment* [86].

4.6   CAST IRON AND WROUGHT IRON STRUCTURES

4.6.1   Cast iron may be found in older bridges, being first used in the United Kingdom in 1779 at Ironbridge. It has only rarely been used since 1914. There are several types of cast iron, the type usually found in structures is known as grey, or flake graphite cast iron, from the dull grey appearance of a freshly fractured surface.

Surrey CC

4.6.2   Wrought iron may be found in older bridges, being first used in the United Kingdom in 1840 and rarely after 1914. The manufacturing processes placed practical limitations on the size of elements so larger elements had to be built up from relatively small components, using wrought iron rivets and bolts.

Wrought iron was also commonly used for cables and forged links, especially in 19th century suspension bridges. Other applications include trusses and lattices, handrails and balustrades.

4.6.3   The homogeneity and purity of cast iron and wrought iron in the aforementioned structures is below the standards of present day materials. This variability should be taken into account in the inspection process. The only certain method for distinguishing between wrought iron, cast iron and steel is chemical analysis and/or metallographic examination of a sample sawn (not flame cut) from the member. However, there are a number of other characteristics of wrought iron elements which give indications of the type of material, and some of these are listed in Volume 1: Part F: Appendix F and described in the *Appraisal of Existing Iron and Steel Structures* [87].

4.6.4   Corrosion of both cast iron and wrought iron is relatively slow but it may reach significant proportions particularly for wrought iron because of the age of the structure and the composition of wrought iron. In general, the corrosion products cause expansion and can be readily detected. Corrosion occurs along lines of slag inclusions, which run parallel to the longitudinal axis of the element and causes the material to delaminate. Since this occurs within the element, deterioration of the element may be greater than is apparent at the surface. Tapping with a hammer by an experienced inspector can provide useful qualitative information.

4.6.5   Areas of severe corrosion, graphitisation of cast iron or delamination of wrought iron, identified during a Principal Inspection may need a more detailed Special Inspection to establish the severity of the defect and identify its cause. Where there is a build-up of rust, a visual inspection is not sufficient to evaluate section loss. A Special Inspection is normally needed which includes the removal of rust to base metal and the measurement of section thickness using callipers, ultrasonic thickness meters (for cast iron) or other appropriate methods. Ultrasonic thickness meters are not recommended for wrought iron as they are unreliable due to the laminar nature of the material.

4.7   ADVANCED COMPOSITES STRUCTURES

4.7.1   The surface of advanced composites should be inspected for signs of crazing, cracking or delamination and for signs of local damage such as impact or abrasion. Where there is a protective layer, it should be checked to ensure that it is intact. Bonded plates should be checked to ensure that they are not becoming detached. It is recommended that this should generally be carried out by inspectors with experience of the delamination of such materials. Further guidance is given in *Strengthening Concrete Structures with Fibre Composite Materials: Acceptance, Inspection and Monitoring* [8] and *Repair and Maintenance of FRP Structures* [88].

4.8   TIMBER STRUCTURES

4.8.1   The main problems for timber structures/elements are decay, insect attack, splitting and separation of laminated layers. The principal forms of decay are dry rot and wet rot with the latter more likely on highway structures. Timber attacked by dry rot looks dry and brittle, developing deep cracks across the grain and breaking into brick-shaped pieces. Wet rot can only attack wood with high moisture content; it does not spread into dry wood. Affected wood becomes soft, pulpy and wet, with the structure of the wood progressively breaking down. Prolonged dampness and vegetation growing from crevices are also signs that the timber may be decaying. Areas which are particularly susceptible to decay are those which are in contact with both water and air.

4.8.2   Chemical treatment to prevent decay will not penetrate to the middle of the timber so even if the outside is sound, decay may still be occurring below the surface. Signs of hidden decay include water stains on the timber or soft areas on the surface.

4.8.3   Insect attack may occur anywhere and can seriously weaken a timber structure. Insect holes usually have dust in them or near them. A few small holes (less than 5mm in diameter) are not usually serious. If there are more small holes or much larger holes, the problem is serious.

4.8.4   Evidence of possible decay or insect attack can be detected using a sharp instrument to check the condition below the surface. Where deterioration has occurred, samples may be taken for examination and testing. Sampling in this way is usually only done in exceptional circumstances.

4.8.5   Splitting commonly occurs in timber as it dries out, and does not necessarily seriously affect the structure. Splitting defects that should be treated more seriously include:

- Splits across the grain of the wood.

- Splits orientated so that water can accumulate in them.

- Splits around connections such as bolt holes.

- Splits that are observed to be increasing in size.

4.8.6   Loose or damaged joints can seriously affect the strength of the structure, and in some cases can also cause serious accidents. Steel connection members, such as plates, bolts, pins and cables, may also be subject to corrosion, particularly in saline environments. Additionally, oak when wet gives off acids that can corrode ferrous connectors.

4.8.7   In glued-laminated timber elements, separation of the laminations may occur due to degradation of the adhesive. Delamination may be seen at the edges of the timber, where the edges of laminations are exposed, or on top or bottom surfaces as blistering.

4.9   CULVERTS

4.9.1   The general condition of trash screens and other ancillary items such as restraint systems or handrails should be noted. It may be appropriate for inspectors to comment on the apparent effectiveness of the trash screens or even suggest improvements as these items require frequent maintenance and, when blocked, can cause flooding. In some cases minor alterations to the layout and operation of the screen may improve effectiveness and reduce maintenance requirements.

4.9.2   The difficulties of providing access and ensuring safe working conditions may make the cost of Principal Inspections of some small diameter culverts disproportionate to the information gained. In appropriate cases it may therefore be acceptable to reduce the frequency of inspections requiring access to the barrel of the culvert. Formal guidance on decreasing inspections intervals is provided in the *Code of Practice* [1] and in *BD 63* [2] (also see paragraphs 2.2.20 and 2.4.18).

4.9.3 For a culvert to be considered for a reduced frequency of Principal Inspections, the following criteria should normally be satisfied:

- previous inspections have not detected significant defects or rapid deterioration;

- the culvert is functioning as intended;

- there is no evidence of settlement or other instability in the adjacent embankment or overlying carriageway;

- there is a clear line of sight through the culvert, so any blockages or gross deformations could be seen;

- inspections of either end of the culvert are carried out at the frequencies recommended in this Manual; and

- the culvert is not subject to frequent flooding (i.e. at or near full bore).

4.10 RETAINING WALLS

4.10.1 The principal defects which may occur in a retaining wall, are excessive movement of the whole wall (tilting, sliding, etc.) or of part of it (bulging, differential settlement, etc.) and problems arising from water seepage. These defects can also occur in other earth retaining structures such as bridge abutments and wing walls. Structural defects leading to excessive movement or misalignment which may be overlooked during close inspection may be apparent from a distance. Sighting along restraint systems, handrails, string courses or other features is a good method for detecting misalignments.

Highways Agency

4.10.2 The form of construction of the retaining wall may influence the location and types of defects. Cracks, for example, on the face of a wall may correlate with the location of steps constructed in the rear of a wall or bulging of a face may occur between adjacent counterforts.

4.10.3 Defects in flexible masonry retaining structures may stem from changed conditions in the vicinity of the structure, for instance adjacent construction works or clogged drainage paths.

4.10.4   Reinforced concrete retaining walls can suffer from corrosion due to the ingress of chloride ions from de-icing salts. Where walls face onto carriageways, they can become contaminated with chloride from the spray of passing traffic. Walls which support highways can also become contaminated if run-off from the carriageway is allowed to reach the top of the wall and either trickle down the face or seep down the back of the wall.

4.10.5   Inspectors should be particularly alert to changes in the loads imposed on retaining walls. These can frequently be caused by raising the ground level or storing materials behind the wall. Where there is vehicular access along the top of the wall, any changes in use should be noted.

4.11   MASTS, GANTRIES AND CANTILEVER SIGN STRUCTURES

4.11.1   Masts, gantries or cantilever sign structures are constructed from a variety of materials and are susceptible to the same forms of deterioration as other structures made of the same materials. Masts and gantries are relatively light structures, generally designed to be as slender as practicable. Consequently masts and gantries are more susceptible than other highway structures to structural failure arising from damage or deterioration. Inspectors should therefore be particularly alert for evidence of significant damage or deterioration.

4.11.2   The lower sections of supports are more vulnerable to corrosion because they are within the traffic splash and spray zones. Fixing brackets and straps for signs and electrical conduit on steel structures need careful inspection to confirm that they have not damaged any protective coating or impeded drainage. Since fixings may be of a relatively small cross-section, the amount of steel loss which can be tolerated may be small.

4.11.3   Particular care should be given to identifying signs of foundation failure. The vertical alignment of the structure should be checked in both planes. Lack of verticality in any direction may indicate a foundation or fixing problem and its cause should be investigated.

4.11.4   Most masts and gantries support electrical equipment. Particular care is therefore required to ensure that the inspection can be undertaken safely (see paragraphs 3.3.21-3.3.25). Where it impinges on the inspection work, and whenever practicable, electrical apparatus should be isolated before inspection work proceeds in the vicinity. Any damaged electrical equipment should be reported immediately to the appropriate authority and work nearby should not proceed until the damaged item has been certified as safe.

4.11.5   The signs, lighting and other equipment supported by masts and gantries also require inspection and maintenance at regular intervals. This includes cleaning, bulk lamp changing and electrical inspection and testing. Whenever practicable, inspections of the structure should be co-ordinated with other inspections or routine maintenance. If an inspector observes what appears to be a significant defect on an item of equipment not within the scope of the structural inspection, he should report it without delay to the Supervising Engineer.

5 Recording Inspection Findings

5.1 INTRODUCTION

5.1.1 The majority of the inspection work involves collecting relevant data and describing defects in terms of their type, location, extent, severity and, if possible, cause. Accurate reporting is essential in enabling the Supervising Engineer and others to make appropriate decisions concerning the safety and maintenance of the structure. In addition, reporting should be consistent so that inspectors, visiting the structure at a later time in the inspection cycle, can determine whether specific defects changed with regards to their extent or severity. Appropriate written notes should be captured as soon as an observation is made. After noting details of the defect, the notes should be checked to ensure that the supervising engineer will gain correct information on the overall condition of a structure.

5.1.2 Significant changes in the overall condition of a structure between distinct inspections may sometimes be attributed to different weather conditions at the time each inspection was undertaken. For example, rain, wet surfaces or poor light may affect observations and the width of a crack or joint may be dependent on temperature fluctuations. Therefore, it is important that the weather conditions at the time of the inspection and where appropriate during the previous few days should always be recorded.

5.1.3 Whenever appropriate photographs should be taken to record both individual defects and the general condition of the structure. To illustrate the scale, inclusion of an item such as a coin or ruler in the photograph, would be very useful when photographing details. Photographs taken from a distance to show a substantial part of the structure can also be very useful, especially in conjunction with close-ups of individual defects. Where access is difficult, it can be useful for future reference to photograph (or record on video) the access or traffic management arrangements adopted. These should supplement, not replace, the relevant parts of the method statement.

5.2 DATA CAPTURE AND INSPECTION PRO-FORMA

5.2.1 The purpose of recording defects is to both help decide on and prioritise any maintenance actions needed, and to monitor any development in the defects. Although there may be differing defect recording systems utilised by particular owners, the system should nevertheless support these two broad principles. A defect could be rated in terms of its influence on the condition of the element in which it is located, such as loss of functionality of the element, or it could be rated in terms of its action upon the element, for example the probability of causing damage to the element or (for an appearance related defect) the visual tolerability of the defect. Inspectors should check with the authority the form of the defect reporting system to be used.

5.2.2 Generally, inspection data should be recorded in a format that gives a clear and accurate description of the condition of a structure. The observed defects along with their associated severity and extent levels (see Section 5.3) should be recorded on an appropriate inspection pro-forma during General and Principal Inspections and, where relevant, Special Inspections.

5.2.3   An inspection pro-forma should be drawn up before an inspection is undertaken to specify the relevant information to be collected. The pro-forma should accommodate information on the form and materials of the structure, the referencing system, the span/panel and elements being inspected, the extent, severity and location of any defects, the recommended action and its priority, and inspector's comments. Some of the aforementioned information may be entered onto the pro-forma before the inspection.

5.2.4   The pro-forma should provide a simple and effective way of creating a consistent inspection reporting system that can be adjusted to suit the needs of individual owners. Typical pro-forma layouts that may be used during inspections are included in Volume 1: Part F: Appendix G. For example, the CSS inspection procedure [89 and 90], recommends that a generic pro-forma is used to perform the first inspection on a structure. After the inspection, the data collected should be used to create a structure specific pro-forma comprising all the general structure data and only element types relevant to that structure, with the remaining elements deleted or their fields blanked-out.

5.2.5   Inspection reports should be signed by the inspector and dated in paper format as evidence in case of future potential claims by the public. The Supervising Engineer may review and sign General and Principal Inspection reports periodically and also identify maintenance needs.

5.3   ELEMENT CONDITION RATING

5.3.1   Some inspection recording methods rely on direct evaluation of element condition, whilst others rely on rating and locating individual defects. One recommended element condition rating process is contained in the *CSS Guidance Documents* [89, 90] and this is summarised Volume 1: Part F: Appendix G. Some authorities use alternative methods of deriving element condition ratings and as necessary, reference should be made to the particular owner guidance.

Significance of Defects

5.3.2   Inspectors should immediately inform the Supervising Engineer, if in their opinion, a defect may compromise the integrity of the structure and the safety of the public.

5.3.3   Defects of a structural nature may in time cause dangerous or costly loss of fabric of the structure, and their significance may be influenced by the reserve structural capacity of the affected element. This type of defect therefore requires special consideration by both the inspector and the Supervising Engineer. It is recognised that the inspector may not know the reserve capacity of the various elements; therefore, the severity rating should be based on the inspector's observations. However, the inspector, should also indicate, in their opinion, whether consideration should be given to undertaking a structural assessment of the element with the defect in order to confirm its structural significance. The inspector should base this comment on their knowledge of structural behaviour and professional judgement.

5.3.4   The Supervising Engineer will ultimately decide on the significance of a defect, and assessing the cause of the defect is an important step in determining this. It is important therefore that the inspector has an appreciation of likely causes and wherever possible relays this to the Supervising Engineer.

5.3.5   The Supervising Engineer may also need to assess, often by analytical means, whether a defect is affecting the required load carrying capacity of a highway structure. For example a defect may appear minor but could be the first indication of a serious problem. Diagonal cracks in a beam may be very fine and appear insignificant, whereas they could be the only sign that the beam may have inadequate shear capacity.

5.3.6   Where testing is undertaken in addition to or following on from a visual inspection, the severity rating of the defect should be reviewed once the results of these tests are known.

5.4   INSPECTION RECORDS

5.4.1   All inspections should result in a record, in a format appropriate to the inspection type. Standardised formats should be used for inspection records. The format should be clear, follow a logical sequence and incorporate all the necessary information. The inspection records support maintenance planning and management and should assist this process by adopting a relatively consistent format from one inspection cycle to the next.

5.4.2   It is recommended that, as a minimum, the information described below is provided in the relevant inspection report. In addition to this information all inspection records should also contain the date of the inspection, those responsible for undertaking the inspection, general information about the structure (e.g. name, reference and location) and details of the prevailing weather conditions at the time of the inspection (and where appropriate the weather during the previous few days).

General Inspection

5.4.3   A completed inspection pro-forma (see Section 5.2) may be sufficient as the General Inspection record; this should include as a minimum an indication of the location, severity, extent and type of any defects.

Principal Inspection

5.4.4   The record of a Principal Inspection should comment on the significance of any defects, include a completed inspection pro-forma (normally as an appendix) and give a broad statement on the overall condition of the structure. The report should state if a Special Inspection is required, and where attention should be given to particular elements during the following General or Principal Inspection.

5.4.5   It is recommended that a Principal Inspection should also include a review of the completeness and accuracy of the inventory records. Any deficiencies in the records should be rectified as part of the Principal Inspection.

5.4.6   The following provides a checklist of records that should be considered for creation/updating by a Principal Inspection:

- The location, severity, extent and type of all defects on the structure, including, where appropriate, detailed descriptions and/or photographs (or sketches) of the defects that clearly identify their location and illustrate the severity/extent of damage.

- For bridges over roads the relevant headroom information based on measurements taken during the inspection.

- Any significant change (e.g. works carried out or deterioration) since the last Principal Inspection.

- Any information relevant to the integrity and stability of the structure.

- The scope and timing of any remedial or other actions required before the next inspection.

- The need for a Special Inspection, additional investigations and/or monitoring.

- A description of any testing that was undertaken, details of the information collected and an interpretation of the information.

5.4.7 Additional requirements specific to the authority or the structure characteristics may also be required from the Principal Inspection.

Special Inspections

5.4.8 The following may be used as a checklist for the minimum set of records created by a Special Inspection:

- Background and reasons for the Special Inspection.

- A detailed description of the condition of those parts of the structure that have been inspected including, where appropriate, photographs and/or sketches.

- Any significant change (e.g. works carried out or deterioration) since the last maintenance inspection to those parts of the structure that have been inspected.

- A description of any testing that was undertaken, details of the information collected and an interpretation of the information.

- Any information relevant to the integrity and stability of the structure.

- The scope and timing of any remedial or other actions required before the next inspection.

- The need for any additional investigations and/or monitoring.

- All aspects identified and/or required by the Monitoring Specification for structures managed in accordance with *BD 79* [4].

5.4.9 Additional requirements specific to the authority or the structure characteristics may also be required from the Special Inspection.

Diving Inspections

5.4.10 Where diving operations are required for Underwater Inspections a Diving Report Form should be filled in. An example of a completed form and accompanying sketches are shown in Volume 1: Part F: Appendix H.

5.5 DATA STORAGE

5.5.1 Inspection procedures are more effective when used in conjunction with a suitable computerised database or asset management system. The principal information obtained from all inspections should be entered onto the database, thus providing an up-to-date record of the condition of each structure. Information from previous inspections would be retained, thereby building up a profile of the change of condition over time.

5.5.2 In addition to providing for the entry of principal information, some databases have the facility to attach additional documents. Entire reports including comments, general text and photographs can be linked to the database. The use of electronic data storage, browsing and retrieval methods can improve working efficiency.

5.6 EVALUATION OF INSPECTION RESULTS

5.6.1 The results of an inspection should be sufficient to determine whether a structure is safe for use and fit for purpose. The inspection results should trigger urgent action if necessary and enable the identification of current and future maintenance, prioritisation of work and an approximate estimation of the cost. The *Code of Practice* [1] provides guidance on how inspection results should be used in the maintenance planning process.

5.6.2 The *Condition Performance Indicator* [91], formerly known as the *CSS Bridge Condition Indicator* [92], can be used to provide a condition score for an individual structure, a group of structures and a stock of structures. The condition scores should be monitored over time to assess whether the condition is declining, improving or remaining constant as maintenance is carried out.

6 Input to Maintenance Planning Process

6.1.1   Maintenance planning and management is an on-going activity and as such requires up-to-date and relevant information on structural condition and performance to ensure the correct work is being planned and to assess the effectiveness of previous work.

6.1.2   Inspections, primarily General, Principal and Special Inspections, generally provide the most up-to-date and comprehensive data on the condition of highway structures and as such are a key input for maintenance planning. The *Code of Practice* [1] provides guidance on how inspection results and other structural performance information should be used to inform the maintenance planning process.

7 References for Part C

1. *Management of Highway Structures: Code of Practice*, TSO, 2005.

2. *BD 63 Inspection of Highway Structures*, DMRB 3.1.4, TSO.

3. *BA 74 Assessment of Scour at Highway Bridges*, DMRB 3.4.21, TSO

4. *BD 79 The Management of Substandard Highway Structures*, DMRB 3.4.18, TSO.

5. *BA 35 Inspection and Repair of Concrete Highway Structures*, DMRB 3.3.2, TSO.

6. *BA 50 Post-tensioned Concrete Bridges: Planning, Organization and Methods for Carrying Out Special Inspections*, DMRB 3.1.3, TSO.

7. *BA 30 Strengthening of Concrete Highway Structures Using Externally Bonded Plates*, DMRB 3.3.1, TSO.

8. *Technical Report 57: Strengthening Concrete Structures using Fibre Composite Materials: Acceptance, Inspection and Monitoring*, Concrete Society, Crowthorne, 2003.

9. *BD 7 Weathering Steel for Highway Structures*, DMRB 2.3.8, TSO.

10. *BS EN 1337 Structural Bearings – Part 10: Inspection and Maintenance*, British Standards Institution, 2003.

11. *BD 65 Design Criteria for Collision Protector Beams*, DMRB 2.2.5, TSO.

12. *The Operation and Maintenance of Bridge Access Gantries and Runways*, Institution of Structural Engineers, London, 2007.

13. *Factories Act 1961*, HMSO.

14. *Well-lit Highway: Code of Practice for Highway Lighting Management*, TSO, 2005.

15. *BD 21 The Assessment of Highway Bridges and Structures*, DMRB 3.4.3, TSO.

16. The *Construction (Design and Management) Regulations 2007* (SI 2007, No. 320), HMSO.

17. *Health and Safety at Work Act etc 1974*, HMSO.

18. *Management of Health and Safety at Work Regulations 1999* (SI 1999, No. 3242), HMSO.

19. *Construction (Health, Safety and Welfare) Regulations 1996* (SI 1996, No. 1592), HMSO.

20. *Managing Construction for Health and Safety – Construction (Design and Management) Regulations 1994: Approved Code of Practice*, Health & Safety Executive, HSE Books, Sudbury, 2001.

21. *Management of Health and Safety at Work – Management of Health and Safety at Work Regulations 1999: Approved Code of Practice and Guidance L21*, Health & Safety Executive, HSE Books, Sudbury, 2000.

22. *The Personal Protective Equipment at Work Regulations 1992 – Approved Code of Practice L25*, Health & Safety Executive, HSE Books, Sudbury, 1992.

23. *Construction Health & Safety Manual (2 Volumes)*, Construction Confederation, Construction Industry Publications, Birmingham, 1998.

24. *Construction (Design and Management) Regulations (Northern Ireland) 1995* (SI 1995, No. 209), HMSO (Amended SR 2001, No. 142).

25. *Roads Service Safety Manual (RSSM)*, Department of the Environment for Northern Ireland, Belfast, 1999.

26. *The Control of Asbestos at Work Regulations 2002* (SI 2002, No. 2675), HMSO.

27. *Transport and Works Act 1992*, HMSO.

28. *Infection at Work – Controlling the Risks: A Guide for Employers and the Self Employed on Identifying Assessing and Controlling the Risks of Infection in the Workplace*, Health & Safety Executive, Advisory Committee on Dangerous Pathogens, 2003.

29. *The Approved List of Biological Agents*, Health & Safety Executive, Advisory Committee on Dangerous Pathogens, 2004

30. *The Control of Substances Hazardous to Health Regulations 2002: Approved Code of Practice and Guidance L5*, Health & Safety Executive, HSE Books, Sudbury, 2002.

31. *Electricity at Work Regulations 1989* (SI 1989, No. 635), HMSO.

32. *Electrical Equipment (Safety) Regulations 1994* (SI 1994, No. 3260), HMSO.

33. *HS(G)151 Protecting the Public – Your Next Move*, Health & Safety Executive, HSE Books, Sudbury, 1997.

34. *Management of Health and Safety at Work Regulations (Northern Ireland) 2000* (SR 2000, No. 338), HMSO (Amended SR 2006, No. 255).

35. INDG163 (rev2) *Five Steps to Risk Assessment*, Health & Safety Executive, HSE Books, Sudbury, 2006.

36. SP121 *Temporary Access to the Workface: A Handbook for Young Professionals*, CIRIA, London, 1995.

37. *Traffic Signs Manual, Section 8: Traffic Safety Measures and Signs for Road Works and Temporary Situations, Part 1: Design*, Department of Transport, TSO, London, 2006.

38. *Traffic Signs Manual, Section 8: Traffic Safety Measures and Signs for Road Works and Temporary Situations, Part 2: Operations*, Department of Transport, TSO, London, 2006.

39. *Safety at Street Works and Road Works – A Code of Practice*, 2nd Edition, HMSO, 2002.

40. *Traffic Signs and Lighting,* DMRB Volume 8, TSO.

41. *Countryside Code*, Countryside Access: http://www.countrysideaccess.gov.uk.

42. *Countryside Code*, Countryside Council for Wales: http://www.ccw.gov.uk.

43. *Countryside Code*, Scottish Natural Heritage: http://www.snh.org.uk.

44. *Countryside Code*, Countryside Access and Activities Network for Northern Ireland: http://www.countrysiderecreation.com.

45. *Guide to Inspection of Underwater Structures*, Institution of Structural Engineers, London, 2001.

46. *Health and Safety at Work (Northern Ireland) Order 1978* (SI 1978, No. 1039), HMSO (Amended SI 1998 No. 2795).

47. *Diving at Work Regulations 1997* (SI 1997, No. 2776), HMSO.

48. Health & Safety Executive *Commercial Diving Projects Inland / Inshore: Approved Code of Practice L104*, HSE Books, Sudbury, 1998.

49. *Safe Work in Confined Spaces: Approved Code of Practice, Regulations and Guidance L101*, Health & Safety Executive, HSE Books, Sudbury, 1997.

50. *Work at Height Regulations 2005* (SI 2005, No 735), HMSO.

51. *INDG402 Safe Use of Ladders and Stepladders – An Employers' Guide*, Health & Safety Executive, HSE Books, 2005.

52. *BS EN 280 Mobile elevating work platforms: Design calculations, Stability Criteria, Construction, Safety, Examinations and Tests*, British Standards Institution, Milton Keynes, 2001.

53. *HS(G)19 Safety in Working With Power-Operated Mobile Working Platforms*, Health & Safety Executive, HSE Books, Sudbury, 1982.

54. *Guidelines on Use of Roped Access Methods for Industrial Purposes*, 2nd Edition, Rev 1., Industrial Roped Access Trade Association, 2002.

55. *Construction (Head Protection) Regulations 1989* (SI 1989, No. 2209), HMSO.

56. *Personal Protective Equipment at Work Regulations 1992* (SI 1992, No. 2966), HMSO.

57. *Control of Noise at Work Regulations 2005* (SI 2005, No. 1643), HMSO.

58. *INDG362(rev1) Noise at work – Guidance for employers on the Control of Noise at Work Regulations 2005*, Health & Safety Executive, HSE Books, 2005.

59. *INDG363(rev1) Protect your hearing or lose it!*, Health & Safety Executive, HSE Books, 2005.

60. BD 62 *As Built, Operational and Maintenance Records for Structures*, DMRB 3.2.1, TSO.

61. *Special Issue: Historic Concrete*, Institution of Civil Engineers (Ed.), Proc. Inst. Civ. Engrs Structs & Blgs, 116, Aug/Nov, 1996.

62. *Historical Structural Steelwork Handbook*, Bates W, British Constructional Steelwork Association, London, 1984.

63. *The Maintenance of Brick and Stone Masonry Structures*, Sowden AM (Ed.), E & F N Spon, London, 1990.

64. *Ecology and Nature Conservation*, DMRB 11.3.4, TSO.

65. HA 84 *Nature Conservation and Biodiversity*, DMRB 10.4.1, TSO.

66. *Wildlife and Countryside Act 1981*, HMSO (Amended SI 2004, No. 1487).

67. *Conservation (Natural Habitats, etc.) Regulations 1994* (SI 1994, No. 2716), HMSO.

68. *Conservation (Natural Habitats, etc.) Regulations (Northern Ireland) 1995* (SI 1995, No. 380), HMSO (Amended SI 2004, No. 435).

69. *Wildlife (Northern Ireland) Order 1985* (SI 1985, No. 171 (NI 2)), HMSO (Amended SI 1995, No. 761 (NI6)).

70. *The Conservation of Bats in Bridges Project: A Report on the Survey and Conservation of Bat Roosts in Bridges in Cumbria*, Billington GE & Norman GM, English Nature, Peterborough, 1997.

71. HA 80 *Nature Conservation Advice in Relation to Bats*, DMRB 10.4.3, TSO.

72. HA 81 *Nature Conservation Advice in Relation to Otters*, DMRB 10.4.4, TSO.

73. *The Protection of Badgers Act 1992*, HMSO.

74. HA 59 *Mitigating Against Effects on Badgers*, DMRB 10.4.2, TSO.

75. *Water Resources Act 1991*, HMSO.

76. *Control of Pollution Act 1974* (Amended by SI 1988, No. 818 and SI 1989, No. 1150), HMSO.

77. *Water (Northern Ireland) Act 1972*, HMSO.

78. *The Air Quality Strategy for England, Scotland Wales and Northern Ireland, Working Together for Cleaner Air*, Department of the Environment, Transport and the Regions, HMSO, 2000.

79. *BA 88 Management of Buried Concrete Box Structures*, DMRB 3.3.5, TSO.

80. *Technical Report 54 Diagnosis of Deterioration in Concrete Structures*, Concrete Society, 2000.

81. *Technical Guide 2: Guide to Testing and Monitoring the Durability of Concrete*, Concrete Bridge Development Group, Concrete Society, Slough, 2002.

82. *Structural Effects of Alkali-Silica Reaction: Technical Guidance on Appraisal of Existing Structures*, Institution of Structural Engineers, 1992.

83. *BA 67 Enclosure of Bridges*, DMRB 2.2.8, TSO.

84. *BA 87 Management of Corrugated Steel Buried Structures*, DMRB 3.3.4, TSO.

85. *Synthesis 353: Inspection and Maintenance of Bridge Stay Cable Systems – A synthesis of Highway Practice, Transportation Research Board*, National Cooperative Highway Research Programme, Washington D.C., 2005.

86. *Masonry Arch Bridges: Condition Appraisal and Remedial Treatment*, Publication C656, CIRIA, 2006.

87. *Appraisal of Existing Iron and Steel Structures*, Bussell M, Steel Construction Institute, 1997.

88. *BRE Good Repair Guide 34: Repair and Maintenance of FRP Structures*, S Halliwell & E Suttie, 2003.

89. *Bridge Condition Indicators Volume 2: Guidance Note on Bridge Inspection Reporting*, County Surveyors Society, 2002.

90. *Addendum to CSS Bridge Condition Indicators Volume 2*, County Surveyors Society, 2004.

91. *Guidance Document for Performance Measurement of Highway Structures, Part B1: Condition Performance Indicator*, UK Bridges Board, 2007.

92. *Bridge Condition Indicators Volume 3: Guidance Note on Evaluation of Bridge Condition Indicator*, County Surveyors Society, 2002.

Part D
Defects Descriptions and Causes

This Part provides background information and guidance on the essentials of describing and categorising defects. The principal defects that are likely to be encountered in concrete structures, steel and steel/concrete composite structures, masonry structures and structures built of other materials are described with the emphasis placed on identification and likely causes.

1 Introduction

1.1.1 This Part of the Manual provides general guidance on principal defects, and their causes, that may occur in all types of structures and structural elements. The inspector should be aware of these but also should be alert for anything unusual. Table D.1 provides an overview of the layout and content of this Part of the Manual.

Table D.1 – Layout of Part D	
Section	**Summary of purpose and content**
2. Principal Causes of Defects	This section provides a description of the principal causes that are likely to lead to distress of highway structures irrespective of their construction material. It also highlights the factors that may affect the functionality of specific elements of some structures such as bridges, culverts, retaining walls, signs/signal gantries and masts.
3. Concrete Defects	This section describes typical defects that occur in concrete structures and in the concrete elements of structures built mainly from other materials with particular emphasis placed on identification and likely causes.
4. Steel Defects	This section describes typical steel defects that occur in steel and steel/concrete composite structures and in steel elements of structures built mainly from other materials with particular emphasis on identification and likely causes.
5. Masonry Defects	This section describes typical defects that occur in masonry structures and in the masonry elements of structures built mainly from other materials with particular emphasis on identification and likely causes.
6. Defects in Miscellaneous Materials	This section describes typical defects that occur in structures and structural elements fabricated of materials other than concrete, steel and masonry. It examines the most commonly used alternative materials, namely stay cables, cast iron, wrought iron, aluminium, timber and advanced composites. The defects, which are likely to be encountered in these types of structures, are described with particular emphasis being placed on identification and likely causes.

1.1.2 All inspections require the inspector to observe and record the defects present in a structure. While diagnosis of the causes of the defects is not necessarily a requirement of the inspection, it is of great value for the inspector to have an appreciation of structural behaviour and of the defects that might occur. Such an appreciation will guide and alert the inspector to particular signs enabling attention to be focussed where it is most needed. This would ensure that, when a defect is observed, the necessary data are collected on site so that a correct diagnosis can be made; especially when defects occur due to a combination of causes.

1.1.3 Identification of structural defects and their causes requires considerable care as structural distress within an element may often have consequential effects on other elements and it may not be immediately apparent which element has caused the failure. For example:

- Failure in the bridge foundations, due to settlement, sliding or rotation, is often manifested as cracking, differential movement or other defect in the substructure. Such movements may be displayed as abnormal clearances between the abutment ballast wall and the end of the deck, or as out-of-range movements at the expansion joints or bearings.

- Settlement of embankments may affect the substructure, appearing as depressions in the road surface adjacent to the bridge or as discontinuities in the kerb line.

- Abutments, wing walls or other earth retaining structures can become unstable if supporting ground is removed from one face or if additional loads are imposed on the other.

- Weep holes through earth retaining walls are frequently used to avoid the build-up of hydrostatic pressure behind the wall due to trapped water. Seepage of water at cracks or joints away from the weep holes can indicate a build-up of water due to blocked drainage.

1.1.4 Review of the structures records prior to the inspection and careful recording and collation of the appropriate records after the inspection will usually allow identification of the failure mechanism and enable further inspections and/or maintenance actions to be planned effectively.

2 Principal Causes of Defects

2.1 OVERVIEW

2.1.1 A structure may exhibit signs of distress due to one or a combination of two or more of the following:

- Inadequate structural capacity or clearances, i.e. inadequate design, construction or maintenance, excessive loading or overstress, and sub-standard layout; Section 2.2.

- Naturally occurring damage, i.e. unforeseen movement, water seepage, scour, erosion, vegetation, debris and marine fouling; Section 2.3.

- Accidental or deliberate damage, i.e. fire or impact damage, graffiti, vandalism; Section 2.4.

- Structural materials deterioration, e.g. structural steel or concrete reinforcement corrosion; Section 2.5.

- Structural elements functionality, e.g. drainage or expansion joints not operating satisfactorily; Section 2.6 for bridge elements and Section 2.7 for elements and special features on other highway structures.

2.1.2 Diagnosis of the cause of distress can be difficult, especially if several effects are acting together, in possible combination with a material defect. Structural analysis will generally be required if the initial suspicion is overload or inadequate design. The inspector should endeavour to ensure that the Supervising Engineer is presented with as complete a set of information as possible for this purpose.

2.1.3 Although some structural defects may be identified during a close inspection (e.g. shear cracks in a beam), others may only be detected by standing back and observing the structure as a whole. For example, signs of distress, movement or misalignment, which may be overlooked during close inspection, may be more apparent when viewed from a distance (e.g. settlement showing up as a discontinuity in the line of the restraint system).

2.2 STRUCTURAL CAPACITY AND CLEARANCES

Inadequate Design, Construction or Maintenance

2.2.1 Inadequate design, construction or maintenance of a modern highway structure or one of its elements is fortunately rare in the United Kingdom. Where this does occur, the evidence of failure is likely to become apparent soon after construction. It is therefore very important to be especially vigilant during the early inspections of a new structure, i.e. the Pre Opening Inspection, Defects Liability Inspection and the first Principal Inspection defined in *BD 63* [1]. Inspectors should be aware that the visible evidence of inadequate design may be similar to that caused by excessive loading.

2.2.2   When the Supervising Engineer considers that a new structure may be inadequately designed, it is vital that a structural assessment be undertaken at the earliest opportunity (see *BD 21* [2] and *BA 16* [3]).

2.2.3   It is possible that a correctly designed structure is of inadequate strength due to constructional errors, i.e. utilisation of poor materials or bad workmanship. The effects of these defects may be of similar visual appearance to those caused by inadequate design.

Excessive Loading

2.2.4   Excessive loading or overstress, i.e. the actual applied loads exceed or have exceeded the design loads, can occur due to several reasons, such as passage of abnormal load vehicles, environmental loads (e.g. floods and earthquakes) or alterations to the surrounding ground. It is therefore likely that the cause may only be determined by long term monitoring unless it is obvious that the defects only occurred after a particular event.

2.2.5   As described by the guidance provided in the *Management of Highway Structures: A Code of Practice* [4], abnormal load movements should be carefully controlled, and all movements notified to the authority. Records of these load movements should, where possible, be recorded in the bridge management system and therefore be available to the Supervising Engineer to support establishing the number and types of vehicles, which have used the structure.

2.2.6   Overstress may also be caused by the passage of normal vehicles over sub-standard bridges. The Supervising Engineer should have identified any sub-standard structures and applied appropriate controls to ensure their safety in accordance with *BD 21* [2] and *BA 79* [5]. Long-term monitoring of sub-standard structures may be required.

2.2.7   Environmental effects and ground alterations will normally be obvious to the inspector and these should be accurately described in the inspection report. Changes in ground level due to excavation or the deposition of soil may cause problems through the removal of support or increase in loading. For example, a trench excavated in front of an earth retaining structure can lead to movement or even collapse since it will remove lateral ground restraint.

2.2.8   Changes in the level of the water table may result in a reduction in the foundation bearing capacity. This is a potential problem in areas where there have been changes in industrial production, reducing the level of borehole water abstraction and allowing the water table to rise. An example of this is the closure of a brewery, which had previously extracted large quantities of water from its own well system.

2.2.9   Signs of structural distress in a bridge, due to overloading, construction defects or inadequate strength, are more likely to be noticeable on the sides or soffits of beams. It can be difficult to detect structural distress in slabs, except that caused by excessive sagging moments, when transverse cracking may be evident at midspan.

Sub-standard Layout

2.2.10 The requirements for highway alignment, cross-section and headroom are contained in Volume 6 of the *DMRB* [6]; requirements for cross-sections and headrooms are given in *TD 27* [7]. However, existing layouts may not necessarily conform to these standards due to the age of a specific structure, changes in standards, alterations made since construction, resurfacing or site constraints.

2.2.11   Inspectors are not normally expected to be conversant with the layout standards. During Principal and Special Inspections they will often be required to measure the headroom or other critical dimensions, but otherwise a check of the alignment and clearances is not normally part of the inspector's duties. Nevertheless, inspectors should use their judgement to identify any features which appear to be significantly below normal standards or which give cause for concern. Such features might be creating a danger to the public or cause the risk of damage to the structure. Examples include:

- Poor visibility for road users, especially where there are road junctions or slip roads close to a bridge;

- Lack of, or inadequate, footways for pedestrians;

- Inadequate clearance or headroom between the carriageway and parts of the structure; or

- Missing, obscured, inaccurate or inappropriate traffic signs warning of hazards at the bridge.

2.2.12   In the majority of cases the highway authority will already be aware of the deficiency, but recent alterations (either permanent or temporary) may have been made to the roads over or under the bridge, or traffic signs may have become obscured. For instance, overgrown vegetation can cause serious obstruction to sightlines and can obscure signs. Impact damage may give an indication that clearances or headroom are too small or that the hazard warning signs are not effective.

2.2.13   If the inspector is concerned about a deficiency in the layout, this should be recorded on the relevant report form, supplemented with photographs, sketches or measurements as appropriate. The deficiency should be drawn to the attention of the Supervising Engineer, who will decide what action is required.

2.3   NATURALLY OCCURRING DAMAGE

Unforeseen Movement

2.3.1   Unpredicted movement is not always due to a fault in the design and may appear in any highway structure, even those that are relatively old and sparingly used. The inspector should be aware of the general layout of the structure and be alert to evidence of recent unusual movement at the ends of decks, bearings, construction joints, expansion joints, etc.

2.3.2   The most common problems are associated with ground movement in foundations or adjacent embankments and fill. For example, settlement of fill under the carriageway behind an abutment can result in forces on the abutment due to wheel impact in the depression of the surfacing. Such forces may damage the abutments.

2.3.3　Movement of the ground adjacent to the structure can cause foundation failure due to the application of increased or unexpected loadings. For example, settlement of embankments can lead to lateral loading on piles, especially those beneath bank seat abutments. Settlement, if severe or localised, may result in cracking of abutments, piers and other elements of the structure. Severe frost can cause certain soils to heave and apply unplanned forces on structures.

2.3.4　The settlement may be the result of geotechnical factors such as consolidation, shrinkage of clays or the presence of expansive soils. Consolidation will typically occur early in the life of a structure, becoming stable thereafter. Shrinkage and heave may occur at any time as a result of changes in groundwater levels. The inspector should take note of any evidence that may suggest that a change in groundwater level has taken place. Foundations constructed using timber piles or rafts, will rot if exposed to air and a drop in the water table may cause these to fail. Evidence of burst water mains, leaking canal basins, flooding, or the growth of large trees close to the structure should, for example, be noted.

2.3.5　Mining subsidence may cause large foundation movement or even failure, because it can instigate a ground wave, forcing relative movement between the different parts of a structure and compromising its stability. The effects will depend on the capability of the structure to accommodate movement. Normally this will have been considered at design stage or before mining is permitted. If mining subsistence is suspected the inspector should try to establish whether this is due to active or old mining workings.

2.3.6　Studying the line and level of the restraint system may reveal evidence of movement of other parts of the structure, for example the profile of a bridge deck may be checked for signs of sagging or unusual deflection by sighting along the top of the restraint system, taking measurements for determining the extent and location of hogging, sagging and other deformations. Evidence of movement may also be obtained by sighting along lines of mortar joints or other vertical and horizontal features built into the structure. Excessive vibration during the passage of a heavy vehicle may also indicate that the bridge is defective.

Water Seepage

2.3.7　Seepage of water is a common cause of damage to bridges, usually due to failure of the expansion joints, waterproofing system or to blockage of the drainage. Water penetration through a structure may be classified as:

- Damp – moist to the touch but not wet;

- Wet – drops falling regularly; or

- Running – a trickle or stream of water.

2.3.8　Most modern bridge decks are protected with a waterproof membrane laid under the carriageway surfacing. However, the membrane may not provide complete protection, e.g. it may have been damaged during resurfacing or other works or the membrane may not be sealed completely around drains, expansion joints or other items. It can be difficult to determine the source of leakage through a membrane since the water can travel a considerable distance from the source before being detected. Older bridges may not have waterproof membranes. The underside of bridges should be inspected for signs of leakage or staining, especially at the ends of deck slabs, at expansion joints, around drain outlets and at service troughs.

2.3.9   Although the water staining may appear unimportant, it can cause significant deterioration of reinforced concrete structures, since the water will often be chloride-contaminated from the use of de-icing salts. Leaking expansion joints result in water running down the deck end and the faces of the ballast walls. Where there is silt, debris or blocked drainage this water can lie on top of the pier or bearing shelf, creating ideal conditions for rapid corrosion. Water seepage also causes extensive damage on steel through corrosion of the steelwork. Water paths through masonry can cause leaching and degradation of the mortar and also deterioration of the stonework or brickwork. Damp masonry is liable to damage by freeze / thaw action. Inspectors should therefore try to establish the cause of any dampness, leaking, staining or even ponding so that effective remedial works can be undertaken.

2.3.10   Leakages or water staining can often be due to, or aggravated by, blocked gutters, drains or weep holes, or by the lack of adequate provision for drainage. Inspections should therefore include, where appropriate, a check that drainage appears to be clear and functioning properly.

2.3.11   The ideal conditions for checking whether leakage is occurring are when the bridge is drying out after wet weather. Such conditions will also highlight the presence of cracks. In dry weather water staining will often be visible, but it can be difficult to distinguish between fresh stains due to current leakage and old ones due to former repaired leaks. A return visit should be made after wet weather if necessary.

Scour

2.3.12   Scour is a natural phenomenon, resulting in the removal of material from the bed and banks of a channel or watercourse, caused by the flow of water over an erodible boundary. In a river, scour is normally most pronounced when the bed and river banks consist of granular alluvial materials but also occurs in cohesive materials, such as clay [8, 9].

2.3.13   Scour may occur for a number of reasons, e.g. natural changes of flow in a channel, long-term morphological changes to a river, or as a result of human activities, such as the building of structures in a channel or the dredging of material from the bed. It is therefore useful to classify scour into the following types:

- **General scour** – In any watercourse, erosion of the bed and banks occurs as the flow and velocities increase during a flood. The material eroded is transported by the flow and deposited again as the flow recedes. The resulting reduction of the bed level, whilst that material is in transport, is known as 'general scour' and is independent of the presence of structures encroaching on a channel.

- **Constriction scour** – The presence of a bridge across a watercourse will tend to restrict the waterway opening. Even if the bridge opening is wider than the main channel, it is likely to create a restriction to flood flows when the water level is above bank level. This restriction of flood flows causes an increase in the velocity through the bridge opening and a greater depth of scour results. This type of scour in the bridge waterway is referred to as 'constriction scour'.

- **Local scour** – This is the scour due to effects of the bridge structure on the local flow patterns. It is most likely to develop around piers or abutments in the main river channel but may also occur around piers and abutments set back on a floodplain when there is overbank flow.

- **General degradation** – This is the term used for the possible long term reduction of the bed level due to factors unconnected with the presence of structures. These might include morphological changes in the catchment, such as increased surface runoff due to urbanization, interruption of sediment supply or the effects of upstream flood alleviation works, or by works or changes downstream such as dredging activities, or the removal of a control structure. General long term changes to the bed level at a site may also occur as part of the general movement of sediment down a river. Sand and gravel bed transport, in particular, may occur in a series of 'waves' as banks of material are moved downstream by the larger occasional floods. This may result in cyclical changes of bed level at a site. A bridge constructed at a time when the bed level was high will be more susceptible to scour damage during the period when a trough passes the site.

2.3.14   When inspecting structures over watercourses, inspectors should check that the watercourse is not blocked with debris and that debris has not accumulated against the bridge. Even partial blocking of the watercourse may cause problems due to scour or extreme loadings during the next flood.

2.3.15   Structures built in or near rivers and other channels can be vulnerable to scour around their foundations. It is common for concrete aprons or scour protection works to be built around susceptible abutments and piers and/or for an invert slab to be placed in the bed of the watercourse. Whilst probing with rods sometimes provides useful information about the structure and the bed material, results are variable and a diving survey (see Volume 1: Part C: Section 3: paragraphs 3.5.31-3.5.41) may be required for more reliable information. Visible and accessible concrete above water level should be examined for typical defects, especially cracking and signs of movement.

2.3.16   If the depth of the scour becomes significant, the stability of the foundations may be endangered, with a consequent risk of damage or failure of the structure. In the past 15 years, there have been several bridge failures, resulting in significant transport disruption, economic loss and, on occasion, loss of life. The factors influencing scour are complex and vary according to the type of structure [8, 10]. For this reason regular inspection regimes should be set up for structures liable to scour. *BA 74* [9] provides guidance on the assessment of scour at highway structures.

Erosion

2.3.17   Erosion is the displacement (i.e. detachment, transportation and deposition to a new location) of solid material particles by the action of water, wind or ice.

2.3.18   Bridge substructures and associated elements such as aprons or training walls situated in watercourses with fast water flow may be subject to erosion. Erosion of concrete surfaces will eventually weaken the structures by reducing the section size and, in reinforced concrete, by exposing or reducing cover to the reinforcement. Erosion of corrugated steel structures or steel piling will damage or remove any protective coating. Inspectors should note any erosion damage and should estimate, where possible the depth and extent of any material loss.

Vegetation, Debris and Marine Fouling

2.3.19   Vegetation rooted in cracks, masonry joints or movement joints can cause spalling or other deterioration particularly if allowed to over-grow and also can impede the designed movement of a joint. Silt or debris can block movement joints or drainage and accumulations on or against a structure can create damp conditions leading to rapid deterioration. Tree root growth can severely locally disrupt wing walls and retaining walls, as well as affect the water table levels in the vicinity of structures.

2.3.20   Vegetation, debris and marine fouling may obscure the structure, hiding defects. The inspector may therefore need to clear selected areas to allow proper inspection.

2.4   ACCIDENTAL OR DELIBERATE DAMAGE

Fire Damage

2.4.1   Some fires are started deliberately (e.g. stolen vehicle fires) and others accidentally (e.g. ignition of debris or stored materials or as a result of traffic accidents under/or adjacent to the structure). Inspectors should therefore report evidence of fires and identify any potential fire hazards under or near the structure. Use of the area under a bridge by tenants for storage or working can be a significant fire hazard.

Atkins

2.4.2   Most construction materials are reasonably resistant to the effects of heat and it is unusual that a structure will be sufficiently affected so that major remedial works are required. The visible evidence of fire damage is often restricted to the obvious black smoke staining around the perimeter of the fire. However, severe fires can significantly reduce the strength of the structure through spalling and decomposition of concrete and masonry or through distortion and reduction in yield strength of steel. Further details of the effects of fire on concrete, steel, masonry and timber are given in subsequent sections.

Impact Damage

2.4.3   Impact damage can occur to vulnerable parts of a structure, with the definition of vulnerability depending on the type of structure. In some cases, the effects of impact can be prevented or reduced by, for example, providing collision protection beams for bridge decks, fenders for bridge piers in rivers and estuaries or safety fences for sign/signal gantries and masts. The condition and effectiveness of the protection equipment should be checked by a competent specialist at programmed intervals. The soffits of road bridges (particularly those with height restrictions) and sign/signal gantries are vulnerable to impact by over-height vehicles. Restraint systems and bridge kerbs are also regularly impacted. Damage to kerbs and expansion joints can be the result of snow ploughing.

2.4.4   Vehicle impact or other accidental damage may cause spalling, distortion, buckling or displacement of structural material. On concrete or masonry structures this usually results in scoring, spalling or displacement of masonry units, whilst on steel and advance composite structures the damage is usually visible as scoring or deformation. A structure subjected to vehicle impact may be further damaged if the vehicle catches fire.

2.4.5   Impact damage to a structure is usually obvious, although it is possible to incorrectly attribute it as a material defect e.g. reinforcement corrosion causing spalling. Severe damage may lead to reduced strength/functionality and in some instances complete failure of the impacted member or components such as bearings or movement joints, in which case partial reconstruction may be required. A Special Inspection should be carried out if there is any concern that there may be structural damage.

2.4.6   The inspector should bear in mind that the shock wave from the impact may travel a considerable distance into the structure and damage could occur some distance away from the impact site, e.g. impact at the vehicle restraint system of a bridge may cause damage at the bearings or impact on one leg of a sign/signal gantry may cause damage at the beam/leg connection on the opposite side of the carriageway.

Graffiti

2.4.7   Graffiti affects the appearance of the structure and is usually present in the most obvious and accessible areas of the structure. Graffiti does not usually cause damage other than to the appearance; however, some graffiti may cause offence to the public, e.g. racist and obscene graffiti. All graffiti should be described in the inspection report. Scrubbing, sacrificial coatings, specialist chemicals or even blast cleaning are utilised to remove the different types of paint and marker that might be used, although some removal methods can affect the substrate.

Highways Agency

Vandalism

2.4.8 Physical (manual) vandalism to the structural parts of a bridge is very rare and usually of a minor nature, as it almost always requires considerable time and effort to cause sizeable damage. Nevertheless, the structural elements of cable supported bridges are particularly vulnerable to this type of damage; for example one case of vandalism has been recently reported when vandals tried to hack thought some of the cables supporting one such footbridge.

2.4.9 It is far more common for the secondary elements of a structure to be subject to deliberate damage although complete restraint systems have been stolen for their scrap value. Pedestrian railings are commonly the subject of attack and may be either bent or completely removed, in which circumstances safety may be compromised. However, some vehicle restraint systems have been stolen, resulting in severe safety implications. In such cases early identification and repair is paramount.

2.4.10 Damage to paint systems or cathodic protection systems will not cause significant corrosion for several months. However, if not repaired, the damage may lead to more rapid deterioration, increasing the final amount of remedial work required.

2.4.11 Deliberately blocked drains can lead to seeping water finding an alternative path through the structure, causing leaching and increasing the risk of reinforcement corrosion.

2.5 STRUCTURAL MATERIALS DETERIORATION

2.5.1 Material defects are the most common problem in highway structures, usually in the form of long-term deterioration due to the severe exposure conditions to which most structures are subjected.

2.5.2 Material deterioration should not be dismissed as the potential cause of failure just because the degree of deterioration is not visibly great. Some forms of deterioration, such as pitting corrosion of the reinforcement or chemical degradation of the concrete, may not be visibly apparent. The loss of cross-section of an element may be critical but not appear to be particularly severe. It is important that all loss of section is accurately reported. The inspector should provide in the inspection report any sketches that will assist the Supervising Engineer to make an informed judgement about the residual capacity of the element.

2.5.3 Concrete structures may exhibit signs of cracking, spalling, reinforcement corrosion or other defects (see Section 3). Corrosion of reinforcement caused by chloride ingress from de-icing salts is a common problem in highway structures. Areas likely to be affected include the lower sections of sign/signal gantries and masts that are located within the splash and spray zones; and some parts on bridges, such as ballast walls, bearing shelves and the tops of abutments and piers, where damaged or faulty joints may have allowed the leakage of chloride-laden water. The lower sections of abutments and piers adjacent to carriageways are also likely to be affected by chloride attack, due to traffic spray. Corrosion of the reinforcement can sometimes occur in concrete foundations if run-off containing de-icing salts has percolated down to the foundations.

2.5.4   Concrete is also susceptible to attack by some naturally occurring minerals which may cause crumbling or disintegration of the material. The commonest incidence of this is caused by the presence of sulfates in the groundwater. Under certain conditions a thaumasite sulfate reaction can occur, leading to severe degradation of the concrete.

2.5.5   Structural steel elements of structures are vulnerable to corrosion, cracking and defective connections (see Section 4). For example, steel piles may corrode in aggressive ground conditions or if the piles have become exposed due to scour or erosion. A few old bridges are supported on timber piles, which may suffer from attack by insects or marine borers or from fungal decay. As with steel piles, such deterioration is most likely to occur where the piles are exposed to air or water.

2.5.6   Some steel structures, such as sign/signal gantries and masts, are generally protected from corrosion by galvanising or metal spray, with or without additional paint coatings. The life of a paint system is normally much less than that of the steel member it is protecting. The general condition of the protective system should always be recorded during inspections as it is beneficial to detect the early stages of breakdown, i.e. this may aid in substantially reducing the amount of subsequent maintenance required for correcting defects.

2.5.7   Masonry substructures should be checked for cracking, either in the joints or through the blocks, displaced masonry, loss of mortar from the joints and other defects (see Section 5).

2.5.8   The various forms of material defects that occur in concrete, steel and masonry and other structural materials are described in more detail in Sections 3-6.

2.6   BRIDGE ELEMENTS FUNCTIONALITY

2.6.1   Structures sometimes exhibit defects due to elements that do not operate satisfactorily. These may relate to any aspect of the structure, such as bearings, foundations, etc, but are often concerned with durability elements such as drainage or expansion joints.

2.6.2   While an inspector's primary duty is to inspect and report, they may also be in a good position to consider possible improvements or alterations to the structure. Inspectors should be encouraged to suggest minor alterations that could improve the durability of the structure or reduce the future maintenance requirements. For example, an ineffective drainage detail may allow chloride-contaminated water to spread across the surface of an abutment, while provision of a gutter may help in preventing this.

Structural Elements

Bearings

2.6.3 The condition of a bearing and its seating is an important indicator not only of the condition of the bearing itself but sometimes of other elements in the structure. If bearings do not function adequately, the structure may suffer excessive stress, causing cracking or deformation.

2.6.4 Debris may accumulate around the bearings and water from the road surface may leak through joints. Some bearings will deteriorate under these conditions; metal parts may corrode and sliding or rolling surfaces may become scored or jammed with debris. This may lead to excessive restraint against movement. Lubrication should be maintained where provision exists.

2.6.5 Faulty positioning or alignment may prevent the bearing from functioning correctly. For example, there may be incomplete contact at bearing surfaces on thrust plates and keys and gearing may bind or not engage properly. In skew and curved bridges, bearings and lateral shear keys may bind or suffer damage. Overstress or damage may occur if the bearing is set so that it reaches its limit of travel before the maximum or minimum temperature is attained.

2.6.6 Anchor bolts and other fittings may become loose allowing movement or vibration of the bearing. Rust or other deterioration can also limit the free movement of the bearing so that, in extreme cases, all the movement may take place between the bearing and its seating. Similarly, the bearings may seize completely forcing movement to be accommodated elsewhere in the structure.

2.6.7 The seating materials may crack or disintegrate and gaps may occur between the bearings and seating. On some bridges the mortar may not have been packed adequately under the bearings, leaving voids. It has been known for bearings to remain supported on temporary steel shims, which corrode with time. These are all potentially serious defects since they can reduce the support given to the superstructure.

Foundations

2.6.8 Evidence of foundation deterioration or failure is likely to be manifested as tilting, distortion or cracking of elements in other parts of a structure, for example excessive movement at joints between the deck and abutments. Occasionally, rotation of the foundation is so severe that it exceeds the capacity of the bearing. Similarly, expansion joints can open or close beyond their design limits. Differential movement of the substructure is usually easier to detect than the overall movement of a whole pier or abutment.

2.6.9 Movements due to foundation failure may be sudden (e.g. mining subsidence) or progressive over many years (e.g. water table changes). It is therefore important that the widths of expansion joints, the amount of differential movement (i.e. when a part of the structure moves more than adjacent parts) and positions of bearings are recorded at each Principal Inspection, for subsequent comparison.

Durability Elements

Drainage

2.6.10 Surface water should flow freely to outlets from all parts of the bridge deck without overflowing, ponding or leaking through the deck. Evidence of defective drainage on or adjacent to structures should always be recorded as this may result in serious damage. It is essential that routine maintenance is carried out on the drainage to ensure durability of the structure.

2.6.11 All drainage pipes, channels and gullies should be free from damage and all joints or connections should be properly sealed. Pipe hangers and fixings should be free from corrosion. Plugs and covers to traps and rodding points should be correctly in place. Pipes are often concealed and it may be necessary to trace the entire length of the drainage system. A substantial amount of debris can find its way into the drainage system and there may be interceptor traps at various strategic points. Traps should not be full and their outlets should be clear. Outlets should not discharge water where it may be detrimental to components of the structure, cause erosion of fill and embankment material or spill onto the road or railway below.

2.6.12 In addition to the blowing and spillage of water from drainage outlets that are inappropriately located, substantial amounts of water can be blown on to the soffit of beams and slabs by the wind aided by the funnelling effect of the abutments. Projections, grooves or drip features along the edge of the deck soffit should prevent the spread of water in this way. Water stains on beams, slabs, piers, columns and abutments may indicate leaky pipes, blocked gutters, inadequate drainage systems, leakage through decks or joints, or clogging of gullies and pipes. The gaps between beams on the deck soffit should be checked to ensure that water cannot escape from this location in order to eliminate the possibility of water to freeze and form icicles over the carriageway.

2.6.13 Drain holes are normally provided in box beams to remove water from the lowest point of the boxes. These should be checked to ensure that they are clear and protected against corrosion.

2.6.14 Inflammable and toxic materials may sometimes enter the drainage system, service ducts or troughs, creating a hazard. If contamination is suspected, it should be reported promptly to the Supervising Engineer.

Expansion Joints

2.6.15 Expansion joints are subjected to onerous performance requirements which include:

- accommodating movements of the bridge;

- withstanding traffic loadings;

- having good riding qualities and acceptable skid resistance;

- avoiding the generation of noise or vibration;

- being watertight, or providing for removal of water;

- being easy to inspect and maintain; and

- avoiding sudden deterioration which could cause a hazard to traffic.

Atkins

2.6.16   As a result of these onerous requirements, an expansion joint will generally
          have a shorter working life than the bridge itself. Indeed many of the materials
          used in joints are known to have much shorter lives: items such as split
          compression seals or detached sealants will need replacement, asphaltic plug
          joints may need levelling up and elastomeric joints may need partial
          replacement or resetting.

2.6.17   Loosening or movement of the joint is a common form of failure and may be
          accompanied by rattling and fracture of the bolts, joint components and
          seating. The onset of some form of loosening can often be detected by a
          crack developing between the joint and the adjacent surfacing. Eventually a
          series of cracks will develop in the surfacing itself or in the material forming the
          joint to the deck. Tapping with a hammer can sometimes detect failure in
          adhesion or anchorage of the joint. Joints may be damaged by snowplough
          blades.

2.6.18   There should be adequate space for the joint to function under the specified
          temperature range. For example, lateral displacement of a comb joint will lead
          to restraint of movement and damage to the joint. One section of a joint may

become vertically displaced relative to the other and, if this displacement is excessive, it will cause additional impact under traffic loading and may represent a hazard to the safety of small and two-wheeled vehicles.

2.6.19   Some joints are designed with open gaps through which water and debris can fall. These should be provided with a drainage system, but it may be blocked by accumulation of debris, allowing water to flow over the bearings and sub-structure. However, many joints are designed to exclude water and dirt using seals and sealants. The sealing does not always ensure watertightness over a long period of time. The quantity of water, which may leak through a joint, can be substantial and this can continue over a long period. It is essential that routine maintenance is carried out on the joints, including any cleaning of any transverse or through drainage. Where drainage is not provided at a joint, the joint may act as a dam causing a build-up of water pressure at the face resulting in deterioration of the surfacing. It is important that blocked drainage, silted-up-gaps or split compression seals are detected at an early stage since leakage of water containing chlorides can have very damaging effects on reinforced concrete parts of the bridge or on metal components. Silted-up gaps can also lead to the transmission of high forces into the joint fixing system.

2.6.20   On short span bridges the road surfacing is often laid over the joint, so as to provide better ride quality. Cracks may eventually appear over the buried expansion joint and these should be repaired or sealed.

2.6.21   Asphaltic plug joints are at risk from the formation of cracks, rutting or deformation. If the movement is excessive, the asphaltic plug is likely to have debonded from the surfacing. The life expectancy of such joints is not likely to be greater than that of the adjacent carriageway surfacing. The skid resistance of the joint surface should be checked as this can pose a problem at junctions and for motorcyclists on curves and bends.

2.6.22   Typical defects that commonly occur on the principal types of expansion joints are listed in Table D.2 [11].

Table D.2 – Expansion Joints Defects [11]	
Joint Type	**Defects**
Buried	• Surface cracks along joint • Breaking up of surfacing • Rutting and tracking • Leakage
Asphaltic Plug Joint	• Cracks and leakage • Debonding between joint and road surfacing • Loss of aggregate causing pot holes • Tracking and flow of binder onto road surfacing
Nosing	• Split and loose seals • Cracking along nosing • Break up of nosing material

Continued

Table D.2 – Expansion Joints Defects [11] (continued)	
Joint Type	**Defects**
Reinforced Elastomeric Joint	• Transition strips cracked, breaking up or debonded • Bolt plugs missing • Bolts loose or missing, causing movement of units • Joint surfacing worn, reducing skid resistance • Joint rubber worn or split, exposing steel plates • Debris in grooves • Leakage
Elastomeric in Metal Runners (Cast In)	• Split or dislodged seals • Cracking/break up of road surface/transition strip adjacent to metal runners • Wear and distortion to metal runners • Debris filling joint seal
Elastomeric in Metal Runners (Resin Encapsulated)	• Split or dislodged seals • Cracking along bonding system • Breaking up of bonding system • Wear and distortion to metal runners • Insecure anchorage • Debris filling joint seal
Comb/Tooth	• Transition strips cracked, breaking up or debonded • Wear to metal plates and cracked or missing teeth • Insecure anchorage • Gaps between teeth filled with debris • Damaged drainage membrane

Concrete Hinges

2.6.23   Where a structure incorporates horizontal transverse hinge joints, the hinge throat is often cracked and may require periodic monitoring of the long-term changes to the crack width, however, this may be influenced by seasonal temperature variation. Other defects such as leaching, or corrosion products leaking through the throat may also be observed below or adjacent to the joint with associated concrete delamination and spalling in the vicinity of the hinge.

2.6.24   Absence of visual evidence of defects should not be taken as assurance that no defects are present, as site investigations have shown the presence of corrosion damage without external indications. Leakage is not necessarily a true indicator of corrosion potential. Severe pitting corrosion can exist within outwardly 'dry' joints, where the carriageway joint filler soaks up chlorides in small but concentrated quantities. On the other hand severe leakage may dilute chlorides and wash out the joints.

Half Joints

2.6.25 Where a structure incorporates half joints, often failure of the expansion joint over the half-joint would lead to consequent leakage of water and possibly chlorides on to the bearing shelf of the half-joint. In some instances cracking will be present at the half-joint and may require periodic monitoring of the long-term changes to the crack width, which may be influenced by seasonal temperature variation. Other defects that may be present in the vicinity of the half joint include fretting of the concrete surface, leaching or corrosion products, concrete delamination and spalling. As with hinge joint, pitting corrosion may occur in these joints even in the absence of visual evidence.

Atkins

Cladding

2.6.26 Concrete cladding to structures is normally in the form of precast panels, which are attached to the underlying structure by discrete metal fixings, i.e. bolts, clips, etc. Alternatively, precast panels may have been used as permanent formwork when the structural concrete was cast, in which case these panels will be retained by means of ribs or other protrusions on the rear face. A third form of cladding is masonry facing using concrete blocks, clay bricks or stone.

2.6.27 Most concrete cladding panels are reinforced with steel and therefore susceptible to corrosion of the reinforcement due to carbonation or chloride ingress (see Section 3.4), unless stailess steel reinforcement has been used. The condition and integrity of the fixings securing the cladding to the rest of the structure should be given particular attention as failure can result in the panels falling off. In some instances corrosion of the fixings may cause cracking or rust staining on the face of the cladding, but this cannot be relied upon and it may be necessary to remove selected panels or inspect behind them using an endoscope.

2.6.28 The Supervising Engineer should determine the intervals and extent of the removal of cladding to permit the inspection of structural elements. It may be appropriate to remove cladding panels only in those locations where the risk of deterioration is highest. This is particularly important where the cladding was added after the original construction. In such cases, contamination of the element may have occurred before the cladding was fitted. Although no deterioration was evident at the time, defects may have developed since. The cladding itself should be inspected, on both faces where possible, for deterioration or defects.

2.6.29   Not all claddings or facings are designed to be removable. Brickwork or stone masonry facings to substructures are normally constructed as a facing wall tied at intervals to the structural element. Alternatively, a permanent concrete wall may have been cast against permanent formwork of precast concrete, fibre reinforced polymer, glass fibre reinforced panels or other material. In such cases inspection should normally be confined to observations of the condition of the cladding or facing supplemented where appropriate by endoscope inspection of the cavity, if any, between the facing and the structural element. It is important to ascertain the condition of ties connecting the facing masonry to the structure. Detailed investigations, requiring demolition of parts of the facing in order to gain access, may be required where there is cause for concern over the condition of the facing, ties or the structure behind.

2.6.30   Cladding panels often have special aggregates or other decorative facings. In time the adhesion of these facings to the body of the panel may fail, allowing individual pieces to become detached. The condition of any surface coatings, renders or decorative finishes and features should also be noted.

2.6.31   Nominal gaps are often left around cladding panels to allow for expansion, differential movement, minor adjustments, etc. These gaps may be sealed. Where the edges of panels come into contact, due to excessive movement, incorrect installation or slippage of fixings, for example, minor spalling may occur. Such damage is unlikely to be important in itself unless restraint of movement causes damage to the fixings, but the cause should be ascertained, as the relative movement may be a symptom of a more serious problem.

Safety Elements

Vehicle and Pedestrian Restraint Systems

2.6.32   Vehicle and pedestrian restraint systems, such as parapets and safety fences are constructed in reinforced concrete, steel, masonry, wrought iron, cast iron and aluminium. They are particularly liable to damage from traffic impacts and are also susceptible to the deterioration mechanisms described in Sections 3-6 often caused by the splash or spray from road vehicles.

2.6.33   Welds should be sound since much of the effectiveness of the restraint system in containing vehicles depends on continuity of rails and on the posts to rail connexions.

2.6.34   Holding down bolts and connecting bolts should be tight, taking account of any special expansion requirements in the parapet or between the parapet and the approach safety fence. Bolts or studs should be the correct length, with nuts properly engaged. Holding down bolts should also be watertight as they are often the cause of water penetrating the waterproofing system, leading to leakage through the bridge deck.

2.6.35   The tops of masonry parapets will normally be protected by coping stones or a similar device. Copings will prevent the penetration of water and additionally should direct water clear of the wall below. Missing coping stones will increase the rate of deterioration of parapets and wing walls by allowing water to enter the structure.

2.6.36   Masonry parapets may have been strengthened or rebuilt. Strengthening may have included the addition of reinforced concrete elements, and this can suffer from chloride or carbonation induced corrosion as discussed in Section 3.

2.6.37   Above deck inspection should include the examination of the horizontal and vertical alignments of each restraint system. Detailed surveys should record any significant distortion, as this may indicate a structural defect such as concrete creep of the deck or loss of prestress, which may be otherwise undetectable. However, a 'one-off' survey is of little use unless previous data is available for comparison. Therefore, where distortion in the deck profile is observed, Special Inspections should be carried out at frequencies determined by the Supervising Engineer. For guidance, it would be usual to carry out the second survey about 1 month after the initial measurements. The frequency of further surveys can then be established and will be dependent on the amount of movement recorded.

Ancillary Elements

Surfacing and Kerbs

2.6.38   Cracking of the bituminous carriageway surfacing can take many forms depending on the nature of the failure and the characteristics of the particular materials. In some cases, the cracking is an indication of failure of the surfacing material, while in others it indicates excessive movement or deterioration of the underlying deck. With time, crumbling of the surfacing material along the edges of the cracks takes place and the ingress of water may lead to loss of adhesion between surfacing and deck. Joints in the surfacing and at kerbs are especially vulnerable, even when sealing compound has been used. Deterioration of the surfacing, as it progresses, will lead to increasingly severe impact loading under traffic, particularly at joints.

2.6.39   Sliding of the surfacing material can take place because of a weakness in shear or loss of adhesion at the interface with the waterproofing system or between the waterproofing system and deck. Sliding may be caused by braking forces or forces due to gear changes of heavy vehicles (for example, at the start of inclines) and may lead to humps or depressions in the surfacing at kerbs and expansion joints.

2.6.40   The surfacing will, with the passage of time, lose its skid resistance due to abrasive polishing under traffic. However, a check on skid resistance will usually undertaken as part of the highway maintenance.

2.6.41   Where the carriageway is surfaced in concrete, the surfacing may become cracked, abraded, delaminated or even pot holed. Abrasion may be aggravated by scaling due to frost action (see paragraphs 3.4.43-3.4.46).

2.6.42   Kerbs may become damaged or dislodged, for example, by the action of a snowplough blade. This may damage the seal between it and the surfacing allowing the ingress of water.

Revetments

2.6.43   Revetments typically comprise an armour layer, to protect against erosion by weathering, water, currents or wave action, and an underlayer, to restrain subsoil movement and act as a drainage zone. Although revetments sometimes fail through disruption of the armour layer, the most frequent failures occur due to the failure of the underlayer.

2.6.44   Inspectors should record evidence of ground or soil movement, including significant depressions and bulges in the revetment or adjacent ground. A common failure of block systems on geotextiles is that of sliding or movement of the blocks. Other obvious defects such as loss or disruption of components of the armour layer should be recorded. The revetment should be checked for signs of undermining or scour at its edges, particularly where it is subject to flowing water or flooding.

Utilities

2.6.45 Utilities and their associated ducts and covers can represent a point at which water may find its way through to the bridge deck. Covers and frames may be cracked or dislodged due to accidental traffic loading, allowing water to enter the ducts and hence to travel away from the point of ingress. Waterproofing of utility ducts may be damaged by work on the services.

2.7 DEFECTS ENCOUNTERED ON OTHER HIGHWAY STRUCTURES

Culverts

2.7.1 Culverts are generally constructed in concrete, steel or masonry and specific material defects are described in Sections 3-5. They are also subject to many of the same defects as bridges given earlier. *BA 87 The management of corrugated steel buried structures* [12] and *BA 88 The management of buried concrete box structures* [13] (also see Section 4.9) describes typical defects found in these structures.

2.7.2 The general condition of trash screens (if fitted) and other ancillary items such as restraint systems or handrails should be noted. It may be appropriate to comment on the apparent effectiveness of the trash screens or even suggest improvements as these items require frequent maintenance and, when blocked, can cause flooding or scour at the headwalls and aprons. In some cases minor alterations to the layout and operation of the trash screen may improve effectiveness and reduce maintenance requirements. In fast flowing waters transported stones and cobbles may damage the culvert or any protective coating while in slow flowing waters silting may result in flooding and pose a danger when inspecting the structure.

2.7.3 It is also important to check that there is no evidence of settlement or other instability in the adjacent embankment or overlying carriageway and if any is detected to record the amount so it can be monitored at subsequent inspections. The condition of any joints should also be checked to ensure that there is no loss of fines from the backfill through the joint.

Retaining Walls

2.7.4   The principal causes of defects listed in paragraph 2.1.1 are equally applicable to retaining walls, however, the following supplement and expand on the causes that are of particular importance to their functionality and stability.

- *Movement* – Failures of retaining walls may often be predicted, and prevented, by studying the effects on the adjacent ground. Incipient wall movements may be indicated by the development of tension cracks in the ground at the rear of the wall and from changes in ground profile either to the rear or the front of the wall. Tension cracks may be very narrow and can be curved or straight in plan. Bulging in front of the toe of the wall could indicate forward movement of the toe. Inspectors should check for signs of movement by sighting along vertical and horizontal lines of the structure.

- *Ineffective drainage* – The effects of water, either by creating pressure behind a wall or weakening the backfill, can be a major cause of instability in retaining walls. Weep holes should be inspected to ensure that they are functioning correctly and are not blocked. Seepage of water at cracks or joints above the level of the weep holes can indicate a build-up of water due to blocked drainage. Attention should be paid to any areas showing signs of permanent seepage or dampness. Retaining walls can also be destabilized by problems with adjacent services. Water from defective mains or sewers can increase pressure behind the wall, or cause the ground in front of or under the wall to become waterlogged. Poor reinstatement of trenches behind or in front of walls may also disrupt drainage.

- *Changes in ground level or loading* – Retaining walls can be destabilized by excavations in front of the foundation or surcharges of fill behind the wall. For example, a service trench excavated in front of a wall could remove much of the restraint afforded by the ground, leading to serious movement or even collapse. Any evidence of any changes in ground level due to excavation or the building up of ground levels close to a retaining wall should be reported and ground levels should be compared with those recorded at previous inspections. If the wall is adjacent to water, the footings should be examined for scour.

Structural Defects

2.7.5   The following, also illustrated in Figure D.1, are the most common modes of structural failure of all types retaining walls:

- *Slip of the surrounding earth* – This is caused by a deep rotational failure of the earth mass below the retaining wall foundation. It is either the fault of design inadequately assessing soil strength or the removal of support at the toe of the wall. Strong soil underlain by weaker strata may fail by sliding on the top surface of the weaker strata. Also, if a wall is built on unstable sloping ground there is possibility that the whole slope may fail.

- *Overturning (Rotation about the base)* – The main cause of rotation or overturning is settlement at the toe of a wall. This may be due to increased pressure on the rear of the wall e.g. by application of additional surcharge loads, or an increase in hydraulic pressure; a decrease in the stabilising (passive) pressure on the face of the wall e.g. caused by excavation, scour, poor backfilling or compaction of trenches; and/or deterioration of the soil beneath the toe, e.g. bearing capacity failure or soil beneath the toe removed by piping (i.e. when a high water pressure gradient and a high seepage velocity combine to remove fines from beneath the structure). In anchored walls, fracture or slip of the anchor will cause the wall to fail by rotating about a point near the bottom of the wall.

- ***Sliding*** – Sliding will occur when the pressure on the rear of a wall is greater than the friction on the base and the passive soil reaction on the front of the wall. Increased pressure on the rear of the wall and reductions in passive pressures on the face of the wall, as described above for rotation failures, may also be the cause of sliding failure.

- ***Rotation about the top*** – Failure by rotation about a point near the top of the wall can occur in anchored walls due to a reduction in support in front of the wall. The causes of reduced passive pressure on the face of the wall are the same as those that can cause overturning (see above). The wall fails by the toe kicking outwards and rotating about the line of anchorage.

- ***Structural failure or deformation*** – Structural failure or substantial deformation may occur where excessive pressures are experienced, the structure has been inadequately designed or constructed, or deterioration of the construction materials has occurred.

- ***Settlement*** – Settlement of walls constructed on granular material is usually complete by the end of construction. However, walls constructed on saturated, cohesive material, may continue to settle for a considerable period of time. Total settlement is not generally a problem; only differential settlement between different parts of the wall, or between the wall and adjacent structures, is likely to cause damage. Differential settlement between the heel and toe may result in tilting of the wall. Differential movement is usually easier to detect than movement of a whole wall.

Slip

Overturning

Overturning due to
bearing capacity failure

Sliding

Rotation about tie

Rupture of tie

Overstressing of wall

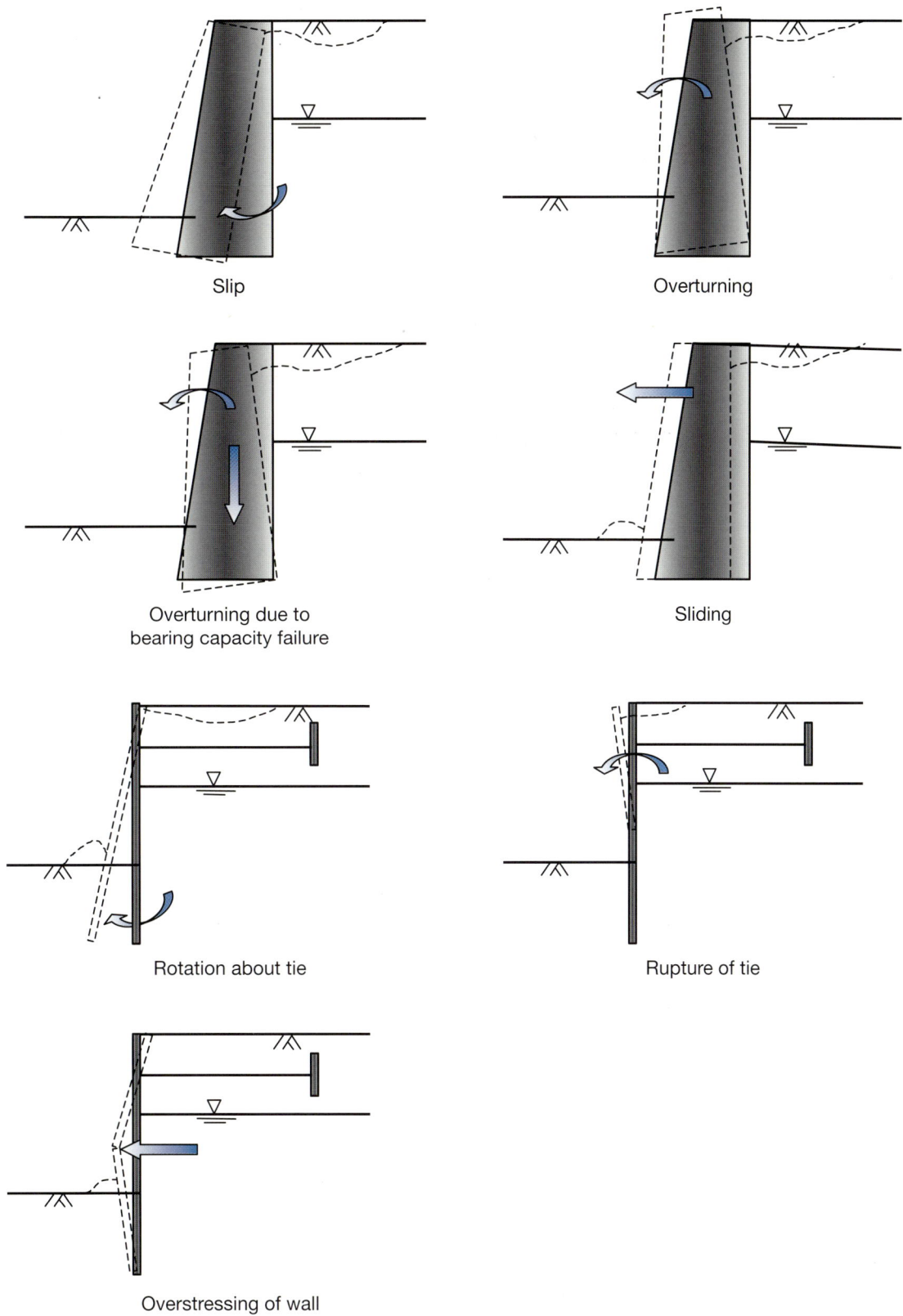

Figure D.1 – Typical modes of failure for retaining walls [14]

Defects Dependent on Material and Wall Type

2.7.6   In addition to the failure mechanisms described in paragraphs 2.7.4-2.7.5 retaining walls may also be subjected to the type specific problems summarised in Table D.3. Specific material defects are described in Sections 3-6. Retaining walls are particularly susceptible to damage from water and chemical attack due to their proximity to the ground. Inspectors should therefore pay particular attention to the condition and performance of drainage systems.

Table D.3 – Defects on Different Types of Retaining Walls	
Wall Type	**Issues to be aware of during inspections**
Gravity Walls	
Mass Concrete	• Mass concrete walls will normally have been constructed in low strength concrete, although walls adjacent to water may have used stronger mixes to improve durability.
	• Large pieces of stone or old concrete (plums) may have been cast into the centre of mass walls to reduce the amount of concrete required. These should not have been placed near the surface but may be visible if workmanship was poor.
	• The sealant of any expansion joints should be examined to confirm that it is continuing to prevent the ingress of stones, soil or grit and that it maintains its resilience and bond to the adjacent faces. Where joints have allowed the penetration of stones, subsequent expansion may have caused spalling of the concrete faces.
	• In walls subject to hydrostatic pressure, uplift may occur at horizontal joints that are not sealed or where seals have failed. Hydrostatic pressure may develop at unsealed joints where a drainage system is non-existent or ineffective. The buoyancy created in more or less horizontal fissures or joints may give rise to tension cracks in the wall. This can cause failure by sliding along the plane of the crack or by tilting of the section of wall, leading to crushing of the material at the face.
Unreinforced Masonry	• Unreinforced masonry walls may be constructed from bricks, stone masonry or precast concrete units. De-icing salts may cause deterioration of some types of masonry that appears as rapid weathering of the faces subject to splashing or spray from nearby carriageways.
	• Masonry walls may become saturated by rainfall, by upward movement of water from foundations, by downward movement from the top of the wall and, in the case of undrained walls, laterally from backfill. Waterproofing systems may have been installed to the rear of masonry walls. Evidence of failure of the system may be seen as percolation of water through the structure.
	• Joints constructed in masonry walls will be subject to the same defects as mass concrete walls described above. Similarly, the buoyancy created by hydrostatic pressure may lead to failure by sliding or tilting at courses in the masonry.
	• Masonry walls are particularly susceptible to bulging and cracking caused by localised loading and by tree or root growth.

Continued

Table D.3 – Defects on Different Types of Retaining Walls (continued)	
Wall Type	**Issues to be aware of during inspections**
Gravity Walls	
Gabions	• The permeable nature of gabions will usually arrest problems due to hydraulic pressure. However, this porosity can allow fines from the earth behind the wall to be leached out through the gabions. This may reduce porosity and increase the hydraulic pressure causing displacement of the gabions. Alternatively, voids may be formed behind the wall and settlement may occur.
	• Gabion boxes are susceptible to vandalism, erosion, corrosion and other durability concerns such as exposure to sunlight or harmful chemicals. Inspectors should check the integrity of individual boxes where these are accessible and ensure that these remain sealed and filled with stone. Erosion is a particular problem where boxes are situated in running water or tidal areas. Boxes in splash zones are most at risk from corrosion.
	• The life of a gabion wall is normally limited by the effective life of the individual boxes unless the stone filling remains stable after failure of the box. Stability may be aided by some vegetation growth, but trees will ultimately disrupt the structure.
Crib Walls	• Inspectors should check for cracks in crib elements and for movement of individual elements, such as headers, protruding from the face of the wall. This may be indicative of a shear failure of connections, allowing the headers to be pushed forward and out of the wall, or of foundation failure.
	• The consolidation of backfill during erection can cause deformation of crib walls, leading to bulging (vertical), bowing (horizontal) and steps, none of which are necessarily indications of failure. However records of bulging, bowing and steps should be made for comparison with previous and future records. Guidance on construction tolerances is given in *BA 68* [15]; the most relevant for post-construction inspections are +20mm in a 4.5m template for bulging and bowing and +5mm for steps at joints.
	• Timber cribs may be formed from whole logs or sawn timbers. Inspectors should inspect the condition of exposed timbers for signs of deterioration, such as rotting (see Section 6).
	• Reinforced concrete crib units are usually precast, however, workmanship problems such as poor compaction or low cover may still occur because of the complexity of the shapes being cast.
	• A feature of crib walls is that they are amenable to the planting of vegetation in their faces. This is usually facilitated by removing some of the fill and replacing it with sufficient topsoil to sustain plant growth. Planting should not be so dense as to prevent proper inspection of the cribs.

Continued

Table D.3 – Defects on Different Types of Retaining Walls (continued)	
Wall Type	**Issues to be aware of during inspections**
Gravity Walls	
Reinforced Soil	• The vast majority of reinforced soil walls are provided with hard facings to resist surface erosion. These may consist of reinforced concrete, galvanised steel, stainless steel or other materials. In some temporary walls the reinforcing mesh may be wrapped around to form the face. The reinforcing elements may take the form of sheets, grids, meshes, strips, tubes, etc. They may be made from steel, aluminium, plastics, geotextiles or other tensile products. A large number of proprietary systems are available for soil reinforcement, and details of particular products should be studied, by consulting the structure records, prior to inspection. • Stability of reinforced soil walls is dependent upon the integrity of the soil reinforcement or anchor ties. A reduction in the cross-sectional area of a metallic reinforcement, due to corrosion for example, could lead to deformation of the wall. In addition, shear failure can occur between the reinforcing element and the fill. • The form of construction of the retained face is typically such that no fill can be leached out from between the units. The condition of joints and sealants between units should be inspected where visible. • The fill behind rigid facing units may settle after construction is complete. Facing units and connections to the reinforcing elements should have been designed to accommodate predicted movements. However, where movements exceed those predicted, localised failure and disruption to the face of the wall may become apparent. Failure of joints between units may allow fill to leach out onto the face of the wall. • The anchor blocks of soil nails will be protected either by encapsulation in a grout filled anchor cap or in a sprayed concrete facing. The condition of any exposed caps and fixings should be examined.
Cantilever Walls	
Reinforced Concrete	• The condition of the movement joints sealant should be examined to confirm that it is continuing to exclude stones and other hard material, and that it maintains its resilience and bond to the concrete.
Reinforced Masonry	• Only the masonry elements will normally be visible; evidence of defects in the concrete and reinforcing elements may appear as rust staining on the face of the wall. The condition of drainage, joints and waterproofing systems should be inspected and recorded.

Continued

Table D.3 – Defects on Different Types of Retaining Walls (continued)

Wall Type	Issues to be aware of during inspections
Embedded Walls	
Steel Sheet Piles	• Steel piles are subject to corrosion; corrosion of piles subject to a typical UK atmospheric environment averages approximately 0.035mm per side per year [16]. Designers often specify a heavier section to allow for predicted corrosion loss, or make use of corrosion resistant steel. Alternatively, the exposed faces may be painted or encased in concrete. • The line and verticality of the wall, or sections of wall, may not have been true at the time of construction due to combinations of driving tolerances and ground conditions. Special tapered piles may have been used to restore verticality. This should not have affected the overall integrity of the wall and inspectors should look for evidence of movement subsequent to construction. In particular inspectors should record signs of bulging or bending of the wall. Such failure may, for example, be due to inadequate design or installation, an increase in water levels or deterioration of the piles or tie rods. Inspection of tie rods will usually only be possible by excavation, and should not be undertaken without justification of a particular concern.
Concrete Sheet Piles	• Precast concrete piles are not often used but, if they are, inspectors should pay particular attention to the joints between units as these are often not watertight and will allow seepage with possible loss of fines and deterioration of the units.
Timber Sheet Piles	• Timber piles should be checked for decay, especially in areas where they are alternately wet and dry. In order to do this, approximately 0.3m depth of earth has to be removed from around the pile and the timber probed or bored. Holes made for testing in the piles might promote decay and should, therefore, be filled with treated wooden plugs. Although piles may appear sound on the outer surface, some may contain interior decay. Creosoted piles, for example, may become decayed in the core area where the treatment has not penetrated, even though the outside surface shows no evidence of deterioration. Sounding with a hammer may reveal an unsound pile. • Timber piles in salt water should be checked for damage by marine organisms, which will attack timber in the inter-tidal area.
Insitu Concrete Bored Piles	• Both types of insitu concrete bored pile walls, i.e. Contiguous (close bored) and Secant, usually have a facing, which may consist of structural concrete, sprayed concrete, precast concrete panels or masonry. Inspectors should check by hammer tapping that structural or sprayed concrete is not delaminated, and that the precast panels and masonry remain satisfactorily connected to the piles and capping beam. • In the case of both pre-cast panels and masonry facing, behind-the-wall drainage may have been provided. Signs that this is not working may be indicated by efflorescence, staining or damp areas on the surface of the facing.

Continued

Table D.3 – Defects on Different Types of Retaining Walls (continued)	
Wall Type	**Issues to be aware of during inspections**
Embedded Walls	
Diaphragm	• Joints between panels are cast or pre-formed. They may achieve watertightness by keyed construction but this cannot always be guaranteed as the work proceeds blind below ground. Remedial work to mitigate this by grouting may have been attempted, but it will not always have been completely successful.
	• Insitu walls will have a rough surface finish since they are cast against excavated soil faces. Consequently they will usually be faced with insitu concrete or a semi independent cladding system, similar to those discussed for insitu bored pile walls above. This can make inspection of the structural wall difficult or even impossible.
	• Precast panels will have more even and clean surfaces, more accurately positioned reinforcement, better tolerances and improved watertightness. Hence these panels and the joints between them are usually left exposed and can be inspected more easily.
Soldier Piles	• The piles may be of reinforced concrete or steel sections, whilst the sheeting can be steel, precast concrete or timber. Sheeting planks can be separated by up to 50mm to facilitate drainage of ground water.
	• Particular attention should be paid to the sheeting, the ends of which may only have small areas of bearing.

Sign/Signal Gantries and Masts

2.7.7   In addition to the principal causes of defects listed in paragraph 2.1.1, which are equally applicable to sign/signal gantries and masts, the functionality of the following elements is of particular importance to their structural integrity.

Foundations and Anchorages

2.7.8   Integrity of the foundations is essential to the stability of sign/signal gantries and masts. Foundation failure may be due to a number of factors, including ground movement and impact loading. The foundations of many sign/signal gantries and masts are located in the highway verge or on embankments, where settlement or slip is more likely. Foundations may also be affected by erosion or softening of the ground due to watercourses or defective drainage.

2.7.9   Sign/signal gantries and masts columns should be checked for vertical alignment. Lack of verticality in any direction may indicate incipient failure. Excessive vibration during the passage of a heavy vehicle may also indicate that the structure is defective.

2.7.10   Anchorages, fixings and the ground surrounding the base should be checked for signs of movement. All bolts and nuts should be tight and welds should be in good condition.

2.7.11   Anchorages for the ends of catenary cables should be inspected for signs of movement, of both the anchorage block relative to the ground, or slip of the cable relative to the anchorage.

Attachments and Fixings

2.7.12   Attachments, brackets and fixings for any equipment supported by sign/signal gantries and masts should be checked for movement, deterioration and the functioning of any moving or operational parts. In particular, bolts or studs should be tight, of the correct length and with nuts properly engaged. Welds should be in good condition. All equipment should be safely retained by fixings.

2.7.13   Sign/signal gantries and masts often have many locations where water may pond or protective coatings incur damage, creating opportunities for rapid corrosion or loss of section. For example, fixing brackets and straps for signs and electrical conduit can damage the protective coating and impede drainage, increasing the potential for corrosion of the structural member and/or the fixing. Particular attention should be paid to such components as fixings, which may be of a relatively small cross-section and hence are particularly sensitive to corrosion.

2.7.14   Dissimilar metals in contact or close proximity can result in galvanic corrosion. Some types of fixing, which penetrate hollow structural members, such as blind rivets, may have been used. These can create a path for water to enter the member. Signs of rust emanating from the low end of a hollow member indicate that corrosion is occurring.

2.7.15   Inspectors should note any alterations, which have been made to sign/signal gantries or their attachments, since the last inspection. For example, new signs may have been installed or cabling replaced. This may have caused damage to the structure or its protective coating.

2.7.16   Access ladders, platforms, walkways, handrails and similar components should be included in the inspection, together with lightning conductors and any drainage.

Catenary Cables

2.7.17   Catenary cables should be inspected for signs of slip at the end anchorages and at the saddles at the top of each column. The fixings clamping individual light fittings to the cable should be checked for signs of slip.

2.7.18   The protective coating around the cable will eventually deteriorate. Any sign of breakdown or corrosion should be noted.

2.7.19   All damage and deterioration of the cable could have serious consequences, and should be recorded. An example could be fractured wires, which can be evident as a partial unravelling of the cable. Other damage could include impact damage adjacent to the ground level anchorages, or wear at the column saddle or at light fixings.

Hoists and Winches

2.7.20   Hoists and winches associated with sign/signal gantries and masts should be inspected in accordance with the requirements set out in the maintenance manual for the equipment. In general, all elements should be inspected to check for corrosion, distortion, wear, lack of lubrication, loose or missing parts, etc.

3   Concrete Defects

3.1   OVERVIEW

3.1.1   This section describes typical defects that occur in concrete structures and in the concrete elements of structures built mainly from other materials. The different types of defects are described with particular emphasis on identification and likely causes.

3.2   DEFECTS CAUSED BY STRUCTURAL DISTRESS

Structural Cracking

3.2.1   Structural cracking may occur due to overload, structural deficiency, vehicle impact or unanticipated structural action. Vertically aligned crack patterns in the vicinity of the mid-span of a beam and diagonal cracking at the ends can be indicators that cracking is of a structural nature. However, the presence of very fine examples of these crack patterns may be due to the normal loading experienced during construction, which has influenced the pattern of shrinkage cracks. It is therefore difficult in many cases to be certain whether a structure is marginally under-designed and it may require long-term monitoring (see Volume 1: Part E: Section 4.3) of such cracks before a definitive diagnosis can be made.

3.2.2   The three types and patterns of structural cracking which may occur in a simply supported beam are illustrated in Figure D.2 and include:

- **Flexural Cracks** - Evidence of flexural (bending) overstress is usually found within the middle third of the span of a simply supported beam or slab. However, in continuous construction, the cracks may not be within this zone and the Supervising Engineer will need to establish the structural behaviour of the element in order to ascertain whether the cracks are due to bending stresses. Flexural cracking usually appears as a series of fine, parallel cracks at right angles to the direction of the span. In two-way spanning slabs, however, different patterns may be evident and these should be analysed by the Bridge Engineer. Cracking may be confined to a localised area of a bridge if it has been formed from concrete that would not have met the minimum material specification. Examples of severe cracking have been recorded in parts of a bridge deck due to a single truckload of poor quality concrete.

- **Shear Cracks** – Shear cracks are usually only visible in beams. The typical pattern for such cracks is that they propagate at an angle of approximately 45° away from the junction of the beam with its support. However, the crack direction can be influenced by the reinforcement. The cracks tend to be widest at the support and may not continue across the entire element. There are rarely more than a few shear cracks in a simple connection between a beam and its support. Monitoring of shear cracks should record the crack width in at least three locations evenly spaced along the length of each crack, but ignoring the last 10% of any cracks that do not continue across the entire element. The location of the end of the crack should be recorded at the point where it is no longer visible under good light conditions.

- **_Torsion Cracks_** – Torsion cracks tend to be similar in form to shear cracks except that they run from the top of the beam or slab above the support, dropping down at an angle of 45° to below the centreline of the beam or slab. Monitoring of the cracks should be carried out in a similar manner to shear cracks.

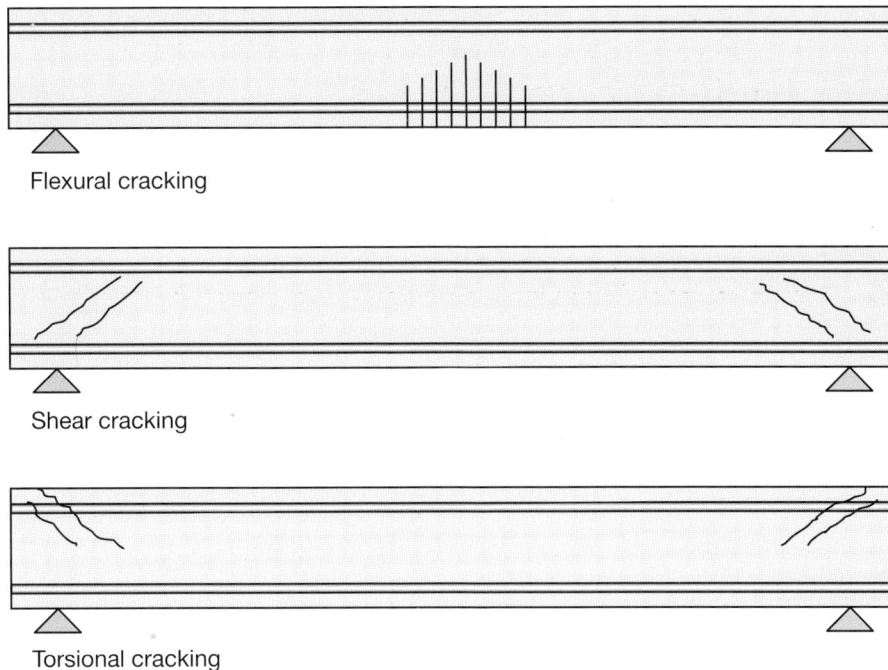

Flexural cracking

Shear cracking

Torsional cracking

Figure D.2 – Typical structural cracks in beams

3.3 DEFECTS ARISING DUE TO THE MATERIAL NATURE

Non-Structural Cracking

3.3.1   Cracking may be non-structural or superficial and affect only the appearance of a structure or it may be indicative of a more serious underlying problem, e.g. formed due to structural loading (see paragraphs 3.2.1-3.2.2). Inspection reports should describe the main features of cracking, including length, width, variation of width along its length, location, distribution, macro pattern and, in some cases, depth. Cracks may be categorized as:

- **_Passive cracks_** – do not open or close in response to cyclic loads or temperature changes, e.g. plastic settlement cracks formed at the time of concrete casting.

- **_Active cracks_** – will either follow a hysteresis cycle as occurs with a normal structural crack or progressively widen and lengthen, dependant on the initial cause of cracking.

3.3.2   Much debate exists as to the significance of cracking in concrete elements. Current opinion is that non-structural crack widths should not exceed structural crack widths permitted in reinforced concrete, i.e. they should not exceed 0.3mm wide. Cracks above this width are termed significant, as the durability of the structure may be affected. Non-structural cracks are not necessarily passive as they can penetrate through the full thickness of an element, and may then open and close in response to thermal cycles or live loading. In such instances non-structural cracks may require monitoring to ensure they do not reach significant widths.

3.3.3 To determine whether cracks are growing they should be mapped and compared to the previous inspection records. There are several methods for detecting crack widening such as tell-tales, Demec studs and automatic remote monitoring systems and these are described in detail in Volume 1: Part E: Section 4.3. Further, the Concrete Bridge Development Group publication *Guide to Testing and Monitoring the Durability of Concrete Structures* [17] provides a structured method for describing crack widths and spacing.

3.3.4 Cracking may occur at any time throughout the life of concrete both in its plastic, i.e. before hardening, and hardened state as a result of the material nature and the nature of its constituent materials. Cracks before hardening can be due to plastic shrinkage, plastic settlement, and formwork movement during casting or poor workmanship (e.g. insufficient vibration leading to poor compaction and planes of weakness). In hardened concrete, cracking can occur due to physical effects such as long term drying shrinkage or crazing; chemical reactions such as reinforcement corrosion, alkali-silica reaction (ASR), carbonation or delayed etringite formation; or due to thermal action such as freeze/thaw cycles or early thermal contraction. Crack patterns can either be regular, as with linear cracks that follow the layout of the reinforcement below, or irregular, as with map cracking or crazing.

3.3.5 The different types of non-structural cracking are summarised in Figure D.3 and Table D.4. Their causes and significance are described in the following sections, which are arranged in the approximate chronological order that cracks may form in a typical concrete structure.

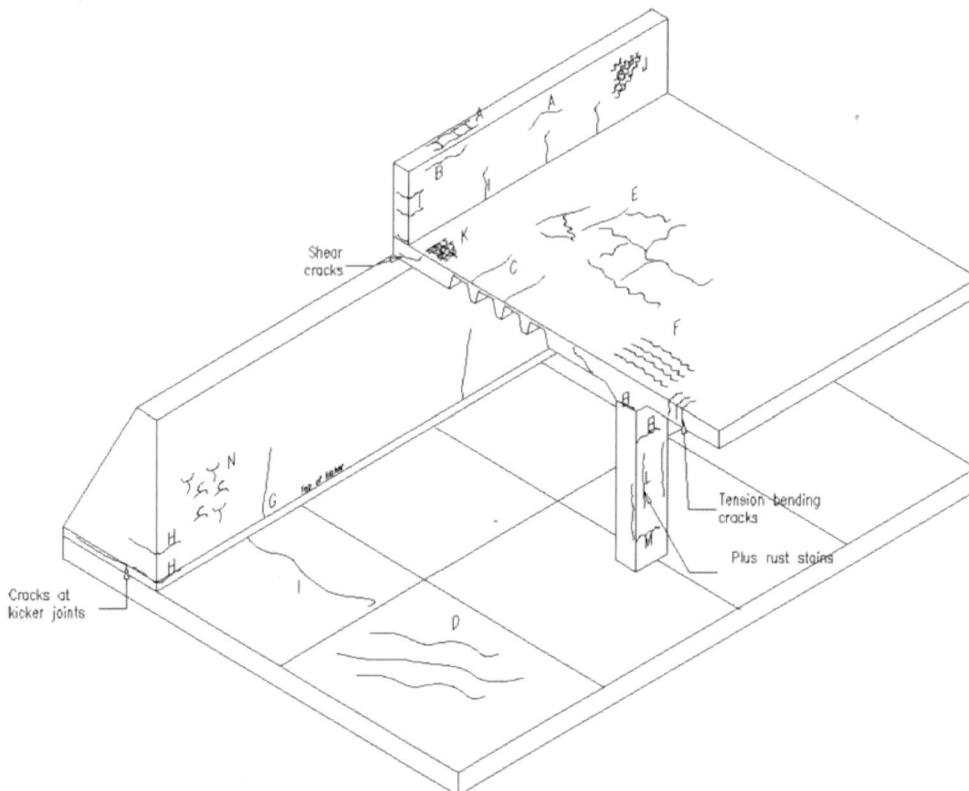

For legend, see Table D.4

Figure D.3 – Examples of non-structural cracks in a hypothetical concrete structure [18]

Table D.4 – Types of Non-Structural Cracks [18]

Cracking type	Legend (Figure D.3)	Features	Most common location	Primary cause	Secondary cause	Paragraphs
Plastic settlement	A	Over reinforcement	Deep sections	Excess bleeding	Rapid early drying conditions	3.3.6-3.3.11
	B	Arching	Top of columns			
	C	Change of depth	Trough and waffle slabs			
Plastic shrinkage	D	Diagonal	Roads and slabs	Rapid early drying	Low rate of bleedin	3.3.12-3.3.14
	E	Random	Reinforced concrete slabs			
	F	Over reinforcement	Reinforced concrete slabs	Ditto plus steel near surface		
Early thermal contraction	G	External restraint	Thick walls	Excess heat generation	Rapid cooling	3.3.15-3.3.16
	H	Internal restraint	Thick slabs	Excess temperature gradients		
Long-term drying shrinkage	I		Thin slabs and walls	Inefficient joints	Excess shrinkage Inefficient curing	3.3.17-3.3.20
Crazing	J	Against formwork	'Fair faced' concrete	Impermeable formwork	Rich mixes Poor curing	3.3.21-3.3.22
	K	Floated concrete	Slabs	Over-trowelling		
Reinforcement Corrosion	L		Areas subject to leakage or spray from road run-off	Chloride ingress		3.3.23-3.3.25 3.4.8-3.4.13
	M		Sheltered areas	Carbonation	Porous concrete Low cover	3.3.23-3.3.25 3.4.3-3.4.7
Alkali-silica reaction	N		Damp locations	Reactive aggregate plus high alkali cement		3.3.26-3.3.31

Plastic Settlement Cracking

3.3.6   Plastic settlement cracks occur on the top surface of slabs, beams or walls, particularly with deep pours, or on the sides of walls and columns. These cracks are formed as the direct result of restraint between the hardening concrete and either the reinforcement or the formwork, but may also be caused by changes in the section depth.

3.3.7   Plastic settlement of concrete is another term for the natural process of sedimentation, where the solid materials consolidate to the bottom and surplus or bleed water, as is commonly termed, is displaced to the top of the concrete mix. A concrete mix may have a high rate of bleed, as a result of the relative proportions of the materials used in the mix design or a long duration of bleed, due to delayed setting, e.g. retarded mixes, cold weather, etc. In both cases restraint to the settlement of solids would cause cracking.

3.3.8   When settlement is obstructed by reinforcement at the surface of a slab, beam or wall, the concrete above the bar will 'tear' forming a crack that leads directly to the bar. In severe cases, a void may form underneath the bar, but this would only be visible if exposed in a cored section. Surface cracking will mirror the position of the reinforcement, with long runs of cracks over the top layer of steel. Cracks may also form over the lower perpendicular layer of top steel, i.e. secondary or distribution steel. This may lead to a grillage crack pattern. All surface cracks are significant, as there is the potential to provide a direct water path to the reinforcement. The resulting concrete surface will have a corrugated or 'waffle' appearance, being slightly higher over the bar positions than between them. The effect will be enhanced if a straightedge is placed on the concrete surface.

3.3.9   Horizontal cracks may form in columns and walls, spaced at the same centres as the link bars. The cause is the settling concrete tearing as it passes between the bar and the formwork. The tear will therefore be approximately the same width as the reinforcement cage, often tapering downwards at the ends. 'Sand runs' may also feature; these appear as irregular vertical channels, a few millimetres in width and depth, where bleed water has formed an escape path to the surface, and washed out the cement to leave a sandy channel. The pattern of regular-spaced cracks may be accompanied by much finer horizontal tears, only a few centimetres in length, caused by formwork restraint to the downwards movement of the concrete.

3.3.10   Settlement cracks may also form if there is a change in depth. An example would be a thin slab with downstand beams or column heads cast integrally. At the change in depth, the deeper section will settle more than the surrounding, producing a tear at the surface.

3.3.11   Plastic settlement may result in very wide cracks, which often extend to the reinforcement and may reduce the durability of the concrete. However, this type of cracks is usually sealed during the construction phase. Particular care is needed when looking for plastic settlement cracks on the top surface of elements, as they may have been filled with dust.

Plastic Shrinkage Cracking

3.3.12   Plastic shrinkage cracking is caused by rapid drying of the surface of freshly placed concrete. This type of cracking occurs before the concrete has stiffened due to cement hydration; typically in the first few hours after mixing. During this time, the surface is highly sensitive to moisture loss arising from high temperatures or drying winds. As water is lost, tensile strains are set up in the top surface and the partially set concrete 'tears' to, potentially, substantial depths.

3.3.13   As cracking occurs in the wet or 'plastic' state, the reinforcement plays no part in controlling the size or spacing of cracks so the pattern may be random polygonal, i.e. a large scale map pattern, or essentially parallel in nature; the spacing may be as much as 2m apart. The resulting crack pattern may be governed by factors such as wind direction and the pattern of cracks rarely follows the position of the reinforcement. Cracks may initially be minor but can subsequently be enlarged by other mechanisms. Plastic shrinkage cracks vary in their length from a few centimetres to several metres and are typically wider towards the middle than the ends.

3.3.14 The width and depth of cracks may be substantial as they are formed by a tearing process. Surface crack widths of 1mm are not uncommon and crack depths can extend through the full thickness of slabs as opposed to stopping at the reinforcement (see paragraphs 3.3.6-3.3.11 on plastic settlement cracking). The size and width of these cracks may cause spalling due to freeze/thaw cycles (see paragraphs 3.4.43-3.4.46) or lead to reinforcement corrosion (see paragraphs 3.4.1-3.4.13). However, such cracks are usually sealed at the construction stage. Particular care is needed when looking for plastic shrinkage cracks on the top surface of elements, as they can easily be filled with superficial dust.

Early Thermal Cracking

3.3.15 Early thermal cracking is normally found in thicker walls and deck slabs and forms within a few days of the concrete being cast. The cracking is a consequence of the heat generated during the hydration cycle of the cement used in the mix. In thicker sections, the concrete becomes partially self-insulating, resulting in substantial increases in temperature in the core of the element. Concrete cracking may then occur due to:

- A differential temperature between the centre and edges of the concrete element, as the cooling surface concrete contracts but is restrained by the warmer core area. The cracks are widest at the surface and typically affect the outer region of the element and appear to radiate from the warmer central core.

- Restraint from contact with previously constructed concrete elements or other materials, such as compacted fill. In this case, the hardened concrete cools and contracts in its hardened state and is restrained by friction and/or mechanical pinning of reinforcement that passes through the joint. The contraction is a consequence of the thermal expansion caused by hydration temperature rise. In contrast to cracks formed by differential temperatures, restrained thermal cracks are long, often diagonal, and can penetrate several metres into the element. Typically, these cracks will be found where a thick slab is cast onto the walls of a portal, e.g. a subway, or where a wall is cast onto a solid base. The cracks will be widest close to the point of restraint.

3.3.16 Uncontrolled early thermal cracks can be of significant width. However, bridges are usually provided with sufficient reinforcement to control the problem and it is likely that any significant cracks will have been treated during the construction stage. *BA 24* [19] contains further discussion on the causes of early thermal cracking.

Drying Shrinkage Cracking

3.3.17   Long-term drying shrinkage is a natural process as concrete loses moisture and contracts over its life. Cracking can only occur where the concrete is restrained, by reinforcement or friction, as discussed in paragraph 3.3.15, with the cracks being widest at the point of restraint.

3.3.18   Shrinkage of hardened concrete may continue over a significant period during the life of a structure, even if the moisture content of the concrete is in equilibrium with its environment, as it is influenced by a variety of factors relating to its constituent materials, such as:

- The type and shape of aggregates, e.g. aggregates originating from some geographical areas, or some manufactured lightweight aggregates, with low elastic modulus, are particularly susceptible to high shrinkage.

- The mix water/cement ratio, e.g. mixes with a high ratio have a tendency to higher shrinkage.

3.3.19   Drying shrinkage cracks are most likely to occur in thin slabs or long walls. They typically propagate along the line of existing cracks, such as those caused by early thermal contraction. The cracks are unlikely to be of significant size, as the design should have included controlling measures, but, their size may be influenced by other effects and the cracks may widen with time.

3.3.20   There is little benefit to be derived from detailed monitoring of shrinkage crack widths, as they are usually very fine and of little consequence to structural integrity or durability.

Crazing

3.3.21   Crazing is characterised by an irregular close spaced map pattern of very fine cracks, often hexagonal and typically between 6mm and 75mm apart. Crazing is rarely more than a few millimetres deep. It can occur in either vertical faces, in formed surfaces, or on the upper surface of a slab. The small scale of the map pattern makes crazing readily identifiable. Over a long period of time the cracking is likely to become accentuated by the deposition of dirt into the cracks.

3.3.22   Crazing is caused by shrinkage of the surface relative to the mass of the concrete, commonly due to differential moisture movement during curing. Crazing is only significant where appearance is important. Generally, the very fine cracks do not affect the structural integrity of the concrete and are unlikely to cause deterioration of the surface.

Cracking Due To Reinforcement Corrosion

3.3.23   The early signs of reinforcement corrosion (see paragraphs 3.4.1-3.4.13) appear as cracking on the surface, followed by spalling (see paragraphs 3.3.33-3.3.38) of the cover concrete. This type of cracking is indicative that concrete quality and cover to the reinforcement are inadequate for the exposure conditions, either local to the area of cracking or generally within the structure.

3.3.24   The expansive corrosion product, i.e. rust, formed due to reinforcement corrosion, generates bursting stresses that cause cracking and spalling of the concrete cover zone. Even light corrosion would cause expansion sufficient to produce a crack that will propagate to the surface. The surface cracking will appear parallel with, and often be directly over, the line of the reinforcement, often with accompanying staining as the rust leaches to the surface.

3.3.25 Reinforcement corrosion will usually be accompanied by hollowness, i.e. delamination (see paragraphs 3.3.33-3.3.38), even in the absence of rust staining. The first signs of cracking are unlikely to pose an immediate threat to the serviceability of the structure, but the formation of the crack may accelerate the corrosion process and this will be greater where the defect is in an exposed location and in the presence of chloride contaminated water from de-icing salts or in a sheltered location where the surrounding concrete is carbonated (see paragraph 3.4.3-3.4.7). If cracks and corrosion are neglected or where the affected area is extensive, deterioration may eventually compromise the structural integrity.

Map Cracking and Alkali-Aggregate Reaction

3.3.26 Map cracking is a characteristic of internal expansion in the concrete, which is caused by a reaction, generally termed alkali-aggregate reaction (AAR), between minerals in the aggregate and sodium and potassium alkalis in the concrete pore solution. There are three types of AAR reactions that occur in concrete, and these may be defined as follows:

- *Alkali-Silica Reaction (ASR)* – A reaction between alkali pore fluid and siliceous minerals in some aggregates forming a calcium silicate gel which is expansive in the presence of water.

- *Alkali-Silicate Reaction* – A reaction between alkali pore fluid and layer silicate minerals. This reaction is not widely accepted to be significantly different from ASR [20].

- *Alkali-Carbonate Reaction (ACR)* – A reaction between alkali pore fluid and certain argillaceous (i.e. clay rich) dolomitic limestones. ACR is rare, but nevertheless there have been a few reported cases United Kingdom.

3.3.27 Cracks appear as a regular network of inter-connected cracks on the macroscopic scale. On a smaller scale each individual crack follows an irregular, apparently random path. The early stages of map cracking may appear as individual 'three-legged' star patterns, typically 200-500mm apart. As the cracking progresses, these patterns join, initially in groups of two or three, and eventually link to form the pattern known as map cracking.

3.3.28 Bands of light coloured concrete bordering the cracks often accompany this form of cracking, although the lines of the fine cracks may be marked by dark discolouration with the appearance of persistent dampness. A hygroscopic gel forms in the aggregate or at its margins and subsequent absorption of pore solution, causes it to swell and exert pressure. This gel may exude from the cracks and cause staining on vertical surfaces if it is washed down by rain. It can stand proud on horizontal surfaces, but it is more likely that it will have carbonated and dried out, and assumed the appearance of efflorescence. In its more extreme form AAR can cause cracking, in which case it is possible that hygroscopic gel will be seen beneath the spall.

Atkins

3.3.29   If AAR is suspected it can only be confirmed by laboratory testing, such as petrographic analysis (see Volume 1: Part E: Section 5.4: Paragraphs 5.4.22-5.4.23) and/or the core expansion test (see Volume 1: Part E: Section 5.5: Paragraphs 5.5.3-5.5.5).

3.3.30   AAR is a process that takes time to develop and the crack pattern is highly distinctive. In structures less than five years old, AAR cracking is unlikely and other causes should be considered. It should be noted that cracks associated with AAR initiate from within the structure of the concrete and propagate to the surface, where they will be at their widest. Where the concrete is restrained there will be a tendency for the cracking to be orientated along the line of reinforcement.

3.3.31   Where possible, inspections should be carried out in good daylight, in drying conditions following rain (because faint cracks and other defects are highlighted by differential drying). This may also help highlight which areas of a structure are susceptible to dampness. Any consideration of AAR will be assisted by knowledge of the source and type of cement and aggregates used in the construction. Once the presence of AAR is suspected a Special Inspection should be carried out in accordance with *BA35* [21]. The confirmed presence of AAR is not always a cause of immediate concern and it may be possible to maintain a structure in service through the adoption of an appropriate management strategy, although expert advice should be sought.

Other Cracking

3.3.32   Cracks can occur for a variety of other reasons, including accidental impact damage where cracks may be associated with loss of concrete and an early sign of frost damage, where a pattern of fine, largely parallel cracks appear near the edges of the upper portion of the vertical face of exposed concrete. Cracking may also occur due to local overstress, caused, for example, by seizure of bearings. Details of all cracks should be recorded as they may indicate underlying, but not immediately apparent, problems.

Delamination and Spalling

3.3.33   Delamination, spalling and incipient spalling usually present at the concrete surface are caused by expansive forces within the concrete mass due to reinforcement corrosion (see paragraphs 3.4.1-3.4.13), AAR (see paragraphs 3.3.26-3.3.31) or local overstressing. It can also be caused in a previous repair where the repair material has not effectively bonded with the concrete.

3.3.34   Delamination is the detachment, or partial detachment of a region of concrete from the parent mass of concrete. The region of delamination may still remain attached and there may be little visual evidence, except of a slight lifting in the surface, although it may be quite extensive. An area which has almost broken free from the parent concrete, but is still partially attached, is described as an 'incipient spall'. Eventually, the piece of concrete may break away completely, forming a 'spall'.

3.3.35   An area of incipient spalling may first be indicated by a crack, with the concrete to one side slightly raised from the other side of the crack. There may also be signs of corrosion products (rust) leaching from the crack. In certain cases, there may be no outward signs of distress, with delamination occurring within the concrete, e.g. corrosion in a slab causing separation in the plane of the reinforcement but with the corrosion product expansion temporarily insufficient to cause the crack to migrate to the surface. Incipient spalling and delamination can be revealed by a hammer test, as the affected areas give a hollow sound when tapped. It has been found that a ½ kg geologist's hammer is a most effective tool for this purpose.

3.3.36   Spalling is apparent where the delaminated piece of concrete has become completely detached from the parent mass, exposing the underlying reinforcement. The exposed reinforcement may show varying degrees of corrosion, from minor damage to severe corrosion and significant loss of cross sectional area.

3.3.37   Delamination and spalling can adversely affect the visual appearance of the structure and accelerate the deterioration of the concrete element. The consequential loss of steel cross-sectional area due to corrosion can affect the structural integrity of the element. It is not normally possible to determine whether the structural integrity is affected from visual inspection, and therefore it is important to properly record the damage for evaluation by the Supervising Engineer.

3.3.38   A piece of loose concrete poses a significant safety hazard, particularly if there is public access beneath the structure. All such pieces should be removed, if they are accessible during inspections. Where they are not accessible, for example during a General Inspection, they should immediately be reported to the Supervising Engineer, so that action can be initiated for their removal.

3.4   DEFECTS INSTIGATED BY EXTERNAL AGENTS

Reinforcement Corrosion

3.4.1   Corrosion of reinforcement occurs when there is a breakdown of the protective oxide film that forms naturally on the surface of reinforcing bars when they are cast into concrete. When the film breaks down, the steel surface oxidises and an expansive corrosion product, i.e. rust, generates bursting stresses beneath the unreinforced cover zone, leading to cracking (see paragraphs 3.3.23-3.3.25). The two main causes of reinforcement corrosion are carbonation of the concrete and penetration of chloride ions, with the time to corrosion influenced by the quality of the concrete (i.e. the cement content, water/cement ratio, compaction and curing), the cover to the reinforcement and the exposure conditions. Other coexisting defects, like cracking or honeycombing (see paragraphs 3.6.1-3.6.3), are likely to increase the risk of corrosion.

3.4.2   Some structures often contain areas where the cover to the reinforcement is less than that required by current design standards. In older structures this may be due to the use of less onerous requirements at the time of construction, but often it is due to construction errors or poor workmanship. The reduction in cover makes the reinforcement particularly susceptible to corrosion. Low cover can be detected by a covermeter (see Volume 1: Part E: Section 5.2: Paragraphs 5.2.2-5.2.3).

Carbonation

3.4.3   Carbonation is caused by atmospheric carbon dioxide reacting with the alkalis present in hydrated cement to reduce the alkalinity of the matrix. Initially, only the outer part of the cover zone will be affected, with the effect of carbonation seen as a sharp front of reduced alkalinity. However, given sufficient time, the process will continue to the depth of the steel reinforcement. When the carbonation front reaches the steel reinforcement, the protective oxide film will be destroyed and the steel will corrode if a sufficient supply of moisture and oxygen is present.

3.4.4   The rate at which concrete will carbonate depends on many factors including cement content, original water-cement ratio and ambient humidity. Where a concrete surface is continually damp, the rate of carbonation will be very slow and very difficult to predict. For areas that remain relatively dry however, the rate of carbonation will increase in proportion to the square root of the time of exposure and this would be dependant on the age of the structure.

3.4.5   The presence of carbonation, although not visually apparent, may be detected using phenolphthalein indicator, which turns pink in the uncarbonated zone but remains clear where carbonation has taken place (see Volume 1: Part E: Section 5.7: Paragraphs 5.7.1-5.7.2). For concrete complying with the *Specification for Highway Works* [22] the depth of carbonation is unlikely to penetrate more than 30-40mm over the design life of the structure.

3.4.6   Carbonation is therefore only likely to be a problem with older structures; or where the concrete has become degraded for other reasons, e.g. due to early cracking or honeycombing; where the concrete quality is poor, e.g. due to poor curing; where the concrete is excessively porous; or where there is insufficient cover to the steel reinforcement.

3.4.7   The resulting general corrosion, arising from a uniform advance of the carbonation front, may cause cracking and spalling over substantial areas of a structure. This will further accelerate the rate of deterioration.

Chloride Ion Penetration and Pitting Corrosion

3.4.8   The most common cause of deterioration in concrete highway structures in the United Kingdom is corrosion of the reinforcement due to the presence of free chloride ions. These chlorides come mainly from de-icing salt, although problems can occur at coastal sites from wind-borne salt spray.

3.4.9   Water containing chloride ions from road salt reaches the surface of the concrete through spray from passing traffic and leakage or seepage from expansion joints, blocked drainage or defective waterproof membranes [23]. The chloride ions on the surface then penetrate the concrete by migration through the concrete pore solution.

3.4.10   When the concentration of chloride ions adjacent to the reinforcement exceeds a threshold level, the protective oxide layer on the reinforcement is destroyed. This creates alternate anodic and cathodic zones along the length of the reinforcing bar, resulting in localised corrosion of the anodic areas. The size and severity of the anode reaction depends on environmental and other factors, but in general the rate of reaction due to chloride-induced corrosion can be several times that for general corrosion caused by carbonation of the concrete.

3.4.11   Under certain circumstances, particularly in very damp conditions, deep pits can form in the bar at the anode sites due to rapid localised corrosion and in severe cases the whole cross section of the bar may corrode. The corrosion product is mostly contained within the pit formed and the expansive pressure exerted on the concrete mass will be less than that of general corrosion. Therefore, it is less likely to give rise to cracking and, may be undetectable on the surface of the concrete. Therefore, it will only be possible to visually confirm pitting corrosion by removing the cover concrete. Rarely, the corrosion product may be absorbed into the concrete and appear as brown rust staining on the surface (see paragraph 3.4.40). Brown stains should always be thoroughly investigated, although they can have other causes, such as iron pyrites inclusions, because pitting corrosion can quickly reduce the capacity of a structural element.

Mott MacDonald

3.4.12 Chloride ion penetration of the concrete often causes more extensive (or general) corrosion of the reinforcement. In such cases the corrosion products will exert expansive pressure, leading to cracking and eventually spalling. In some instances corrosion may start as localised pitting corrosion but develop into more extensive corrosion once the concrete has cracked.

3.4.13 Where chloride-induced corrosion is suspected, specialist investigation techniques should be considered [23], including half-cell potential mapping of the concrete surface (see Volume 1: Part E: Section 5.7: Paragraphs 5.7.3-5.7.6).

Corrosion of Prestressing Tendons

3.4.14 Prestressed structures are normally designed to halt the formation of cracks transverse to the direction of prestress in the concrete, so the development of any cracking of this nature is indicative of potentially serious implications. The inspector should be aware that prestressed concrete bridges can suffer considerable loss of strength through hidden deterioration without showing visible signs, i.e. these structures are less likely to exhibit warning sags or cracks than reinforced concrete bridges [24].

3.4.15 Most prestressed concrete bridges contain reinforced concrete decks acting compositely with prestressed concrete beams. Considerable stresses can build up at the reinforced/prestressed interface due to both loading and shrinkage stains. It is important, therefore, to check for cracks or any other indication that the joint between the reinforced and prestressed elements is failing.

Pretensioned Wires

3.4.16 In pretensioned concrete beams the prestress is generally provided by a large number of small high-tensile wires. These are mainly concentrated in the bottom flange of the beam but there are frequently wires in the webs or top flange too. In most beams the wires are straight, but some of the larger beams may have deflected wires. The wires are cast directly into the concrete and can therefore suffer from corrosion due to carbonation or the ingress of chlorides in the same way as normal reinforcement. Cracks following the line of the reinforcement are usually indicative of corrosion, or even failure, of the pretensioned wire. The prestressing forces generally suppress transverse cracks.

3.4.17   Many bridges with pretensioned concrete beams are constructed with only nominal gaps between the bottom flanges of the beams. With some of these the spaces between the beam webs are filled with concrete to form a solid slab, while with others voids are left between the webs. Such voids are difficult to inspect and it is normally necessary to insert an endoscope (see Volume 1: Part E: Section 4.2: Paragraphs 4.2.12-4.2.13) either into the gap between the beams or through a drilled hole. In some cases the void is used as a service bay for utilities which may allow the ingress of water from the ducts or openings through the curtain walls. Inspection may be carried out by the removal of the cover slabs where they have been installed.

Corrosion of Post-tensioned Tendons in Grouted Ducts

3.4.18   In post-tensioned concrete the stressing force is applied by wires or tendons tensioned between anchorages cast into the structure. The tendons are usually grouted into ducts inside the concrete slab, webs or flanges of the superstructure, in order to provide the tendons with a protective alkaline environment. The tendons may follow curved profiles, particularly in webs, and may either extend for the full length of the bridge or be stopped off at intermediate anchorages

3.4.19   Post-tensioned concrete bridges with grouted ducts are particularly vulnerable to corrosion and severe deterioration of the tendons if the ducts have not been filled completely with good quality grout. If the grouting is incomplete, moist air, water or de-icing salts can enter the voids, and initiate corrosion both general and pitting corrosion. Many post-tensioned bridges are of segmental construction, whereby precast concrete units are stressed together to form the bridge superstructure. The joints between units are usually either narrow (about 100mm) unreinforced concrete or much thinner glued joints. Joints can form a plane of weakness, creating paths for water and contaminants to enter tendon ducts. However, water and salts may also enter ducts at other construction joints or anchorages.

3.4.20   Cracking will not normally be evident due to the expansion space of the void and prestressing forces. Longitudinal cracking along the line of the tendons can indicate severe corrosion or that the tendons have severed and re-anchored; any such cracks should be investigated.

3.4.21   Post-tensioned ducts may, on occasion, become displaced during the concrete pour. If this is suspected it should be investigated (see Volume 1: Part E: Section 5.8).

External Post-Tensioned Tendons

3.4.22   In some bridges the prestressing tendons lie outside the structural concrete. The tendons are usually positioned alongside the webs of the bridge beams between anchorages. External tendons have also been used for strengthening existing bridges. The tendons are protected from corrosion by galvanising, by casting concrete around them after tensioning or by sheathing them in a protective medium.

3.4.23 Tendons encased in concrete are liable to corrosion due to carbonation or the ingress of chlorides in the same way as normal reinforcement. If the concrete casing is placed after stressing (usual case), it may contain shrinkage cracks and these will provide paths for chloride ingress.

3.4.24 Tendons sheathed in grease, plastic or some other medium are liable to corrosion if there is any flaw in the sheathing. Corrosion or broken wires may be evident from a careful inspection of the sheathing. Shining a powerful torch along the strand can reveal a ripple on the surface of the sheathing if a broken wire has unwound from the strand. Dusting the sheath with a uniform coating of talcum powder can assist this process [25].

3.4.25 Corrosion of external tendons can occur at end anchorages or where the tendons pass through a diaphragm. It is often more difficult to provide corrosion protection in these areas making them more vulnerable to attack. They are also more difficult to inspect.

3.4.26 Inspection of unbonded prestressing tendons should be carried out with care, since any failure of a wire or strand could seriously injure inspectors.

Chemical Attack

3.4.27 In certain aggressive environments, concrete may be subject to severe chemical attack. This is usually observed as a softening and disintegration of the surface layers of the concrete.

3.4.28 Acidic ground water may weaken hardened cement in foundations and the buried part of substructures; the attack begins at the surface and advances into the concrete, reducing the alkalinity of the outer layers, in much the same way as carbonation (see paragraphs 3.4.3-3.4.7). Where water flows over the concrete, the rate of attack will be increased. This will be further accelerated, where the water flow is turbulent, as damaged concrete will also be eroded (see paragraphs 2.3.17-2.3.18). White efflorescence (see paragraphs 3.4.38-3.4.42) is often a sign that the concrete is being attacked in this way.

3.4.29 Limestone aggregate can be weakened or even dissolved if it is exposed to acidic ground water. Miscellaneous spillages onto concrete structures should be recorded and their effects evaluated. Concrete may be attacked by a wide variety of compounds, some of which may superficially appear to be innocuous.

Sulfate Attack

3.4.30 Concrete may be attacked by naturally occurring sulfates in ground water or other sources of sulfate contamination, e.g. sulphuric acid from pollution. Sulfate ions penetrate deeply into concrete, reacting with the tricalcium aluminate phase of Portland cement. This reaction produces expansive forces, resulting in microcracking and weakening of the cement matrix. While concrete mixes are normally designed for the prevailing exposure conditions and, where necessary, sulfate resisting concrete mixes are used, local exposure conditions and concentrations can exceed the design limits, e.g. in situations where evaporation concentrates the sulfate content or the sulfate is present in other materials such as de-icing salts.

Thaumasite Form of Sulfate Attack

3.4.31 Thaumasite is a form of sulfate attack (TSA) that has been observed to cause damage in some buried concrete elements. An unusual combination of factors is required before a structure is liable to TSA, however the primary risk factors include:

- Presence of sulfate and sulfides in the ground;

- Presence of mobile groundwater;

- Presence of carbonate in coarse and/or fine concrete aggregates; and

- Low temperatures, since thaumasite formation is most active below 15°C.

3.4.32 During thaumasite formation, the calcium silicate binder in the cement matrix is broken down resulting in a reduction in strength of the concrete until it eventually deteriorates into a soft mass. This transformation starts at the soil-concrete interface and proceeds inwards towards the centre of the concrete. Advanced TSA in buried concrete is therefore visually very distinctive as it typically exhibits a while pulpy mass at its surface.

3.4.33 Earlier stages of thaumasite formation, however, may not be so visually apparent, as attack will be limited to thaumasite within voids and will exhibit no softening of the matrix. Thaumasite formation and earlier stages of TSA can be identified by laboratory analysis (see Volume 1: Part E: Section 5.5: Paragraphs 5.5.6-5.5.7).

3.4.34 On a newly exposed concrete surface, TSA generally appears as a softening of the concrete, which forms a white friable layer. This may be restricted to arrises or form as larger patches or bands on plane surfaces. The formation of this white friable surface layer, can, but not always, be accompanied by general volumetric expansion or surface 'blistering'. Removal of the soft surface is likely to reveal concrete with a network of fine cracks parallel to the surface and haloes of white (thaumasite) material around aggregate particles. The softened surface of TSA quickly dries out on exposure, making detection more difficult.

Highways Agency

3.4.35   Whilst the appearance of concrete affected by TSA is visually distinctive, the exact nature of the deterioration requires further confirmation.

3.4.36   This may be provided by black or green surface staining on reinforcement corrosion, when it is first exposed.

3.4.37   Further information on TSA is contained in the report of the Thaumasite Expert Group, *The Thaumasite Form of Sulfate Attack: Risks, Diagnosis, Remedial Works and Guidance on New Construction* [26]. Specialist advice should be sought on any structure believed to be affected, or liable to TSA.   A Special Inspection will probably be necessary.

Leaching and Staining

3.4.38   Deterioration due to leaching and staining can take a variety of forms. However, the cause of leaching and its significance can conveniently be classified by the three common colours and textures of the exudates.

3.4.39   White or cream coloured efflorescence is a deposition of salts on the surface of concrete, which have leached from the concrete. Leaching occurs when water penetrates through cracks or porous areas in the concrete, such as parapets that have drying shrinkage cracks with ponded water behind. Leakage may also occur through failed expansion joints or in concrete decks where waterproofing has failed. The water will penetrate most readily where sections are thin with high porosity and at construction joints, particularly if the concrete is honeycombed (see paragraphs 3.6.1-3.6.3). The white efflorescence indicates that water is taking calcium hydroxide from the concrete and into solution (see paragraphs 3.4.27-3.4.29). On the surface, the solution is neutralised by atmospheric carbon dioxide and forms calcium carbonate deposits. In severe cases, the leaching of salts may lead to the formation of stalactites, which are the accretion of salts produced by evaporation of dripping water. Also, because alkali is being leached from the concrete, corrosion of the reinforcement may eventually occur.

Atkins

3.4.40   The most serious form of staining is the brown efflorescence or streaks typical of reinforcement corrosion. As with white efflorescence, the staining will often be seeping from cracks or through the surface of apparently sound concrete. Such staining may also be seen emanating from expansion joints, bearing areas and cold joints (see paragraph 3.6.4) in the concrete. All staining of this type should be investigated for evidence of association with corrosion of structural reinforcement (also see Volume 1: Part E: Sections 5.6-5.7). Brown staining may result from other causes, such as aggregates, surface embedded metal and leaching of bituminous waterproofing materials, but the main concern is reinforcement corrosion, as it can weaken the structure.

3.4.41   A grey-white deposit at the surface adjacent to cracks may be due to drying out of exuded gel, formed as a result of ASR (see paragraphs 3.3.26-3.3.31). However, other causes should be considered and discounted, before this is diagnosed as ASR.

3.4.42   Leaching, staining and algal growth are readily identifiable indications of either current or historic water leakage. The inspector should note the location of staining in order to determine the source of the water leak. Where secondary consequential defects have occurred, such as reinforcement corrosion, further testing may be appropriate (see Volume 1: Part E: Sections 5.6-5.7). It is important to establish the source of leakage and to determine whether this is still active.

Scaling and Frost Action

3.4.43   Scaling is associated with a loss of surface concrete in the form of flakes. It is more usually observed in the exposed faces of structures. It is caused by freezing whilst wet (i.e. repeated freeze/thaw cycles, particularly in combination with direct exposure to de-icing salts). When de-icing salts are applied to the surface of frozen concrete, the action of rapid thawing cools the sub surface region. This causes expansion and can result in micro cracking [27].

3.4.44   Air-entrained concrete is used to resist frost action, but it is not unknown for even this concrete to suffer scaling problems. These are often traced to inadequate entrainment. Specifications also permit higher strength concretes (Grade C50 and above) as an alternative to entrained air mixes. These are also not totally immune to frost action. In particular, aggregate 'pop-outs' have been reported, caused by the different thermal expansion characteristics of the aggregate and cement matrix.

3.4.45   Scaling is only an aesthetic problem, affecting the first few millimetres of the concrete, but it can exacerbate and initiate other forms of deterioration. The roughened surface of the structure will drain less easily and retain quantities of de-icing salts. The former will add to freeze/thaw effects, whilst the latter will facilitate chloride ion migration and accelerated reinforcement corrosion. It has also been shown that removing the surface layer of concrete can increase the rate of carbonation.

3.4.46   Severe scaling will often occur in areas of cracking, because the cracks form water traps and will be forced apart when the water freezes. As a result deep spalling may occur on either side of the cracks.

3.5   DEFECTS CAUSED BY ACCIDENTAL OR DELIBERATE DAMAGE

Fire Damage

3.5.1   Concrete is reasonably resistant to the effects of heat and it is unusual that a structure will be sufficiently affected that major remedial works are required. The effects of heat can cause micro cracking and decomposition of the hardened cement paste matrix, affecting its durability and its ability to prevent corrosion of the reinforcement. In moderate cases, the heat expansion of the concrete causes aggregate particles to 'explode' from the surface of the concrete. In more severe cases, a degree of spalling may result due to the large differential temperatures created within the concrete. Such spalling can occur at relatively low temperatures and the remaining concrete is likely to be unaffected. Where spalling occurs, exposing the reinforcement directly to the fire, the effects of heat can permanently reduce the yield strength of the steel. In severe fires, the concrete strength can be severely impaired if it reaches temperatures in the region of 600°C. In such cases someone with specialist knowledge should inspect the damage.

3.5.2   Most structural concrete changes colour to pink or red at temperatures of between 300°C and 600°C, then grey between 600°C and 900°C and buff at higher temperatures. Pink or red concrete is likely to have reduced strength, and may have to be replaced, whilst grey and buff concrete is likely to be porous and friable and would almost certainly need to be replaced.

3.5.3   Further information on the evaluation of the effects of fire is provided in *Appraisal of Existing Structures*, [28], *Assessment and Repair of Fire-damaged Concrete Structures* [29] and *Spalling of Concrete in Fires* [30].

Impact Damage

3.5.4   Impact damage on concrete structures can range from scoring or gouging of the surface, to spalling and fracture of primary structural members. Damage is most likely on bridge soffits but walls or parapets close to traffic lanes are also frequently impacted.

3.5.5   Minor scoring or gouging is unlikely to be of immediate concern, but it will reduce the cover to reinforcing bars, and in time lead to corrosion. Spalling is a frequent consequence of impact, exposing reinforcement or prestressing tendons. If left untreated, corrosion will occur leading to a loss of structural capacity. Severe impacts may also damage reinforcement or tendons sufficient to reduce strength immediately.

3.6   DEFECTS ARISING DUE TO CONSTRUCTION ERRORS

Honeycombing

3.6.1   Honeycombing of concrete is symptomatic of poor detailing or workmanship and is usually found in construction with dense reinforcement or inaccessible corners of formwork. It is caused by inadequate compaction, and is recognized as open structured coarse aggregate without a dense matrix of cement mortar between them. In itself this would generally only affect the appearance, but honeycombed concrete offers virtually no resistance to the ingress of water and air. It therefore provides little protection to the reinforcement from corrosion or fire.

3.6.2   Cracks and honeycombing may also occur due to localized poor compaction. At the interface between the areas of different compaction, 'tear' cracks may form, often accompanied by honeycombing. This is often located in the vertical faces of deep sections or at confined corners of formwork. These cracks are not necessarily associated with the location of the reinforcement and may be combined with other effects such as plastic settlement if the uncompacted concrete settles at a different rate.

3.6.3   The cracking and associated honeycombing may be significant, particularly in exposed areas. The extent of such cracking should be carefully investigated, as small areas of honeycombing visible on the surface may conceal large areas hidden beneath a skin of cement paste.

Cold Joints

3.6.4   A cold joint is a discontinuity in the concrete resulting from a delay in the pouring sequence, where earlier batches of concrete have stiffened before the remainder can be placed. Vibro-compaction across the interface therefore becomes impossible, resulting in a plane of weakness with poor compaction during the placement of a pour. It can vary in severity from a visible cold joint line in the finished face of the concrete, with good quality concrete either side, to severe segregation of material and honeycombing (see paragraphs 3.6.1-3.6.3) adjacent to the cold joint line.

Formwork Movement

3.6.5   Where the formwork has moved before stiffening of the concrete, a ridgeline or step is likely to form on the surface. The area immediately surrounding the ridgeline is unlikely to be fully compacted and may show signs of grout leakage and honeycombing (see paragraphs 3.6.1-3.6.3). In severe cases, where there has been significant movement of the formwork, a wide 'tear' crack may form in the concrete along the line of the ridge.

3.6.6   Cracking can also occur if supporting falsework fails or is removed before the concrete has gained adequate strength. The pattern of this cracking will be unpredictable as it depends on the particular circumstances and the remaining support. In severe cases, this may significantly reduce the structural capacity.

Peeling

3.6.7   Peeling has the appearance of thin sheets or flakes of mortar, typically more than 50mm across, that have peeled away from the concrete surface like skin shedding. Peeling should not be confused with frost damage (see paragraphs 3.4.43-3.4.46), as it forms at the construction stage and should not deteriorate further. It is caused by the surface mortar adhering to the formwork as the forms are removed.

3.6.8   Peeling primarily affects the appearance and the surface integrity of the concrete. It may have received treatment during construction to restore the smooth profile. If left untreated, the rough surface may facilitate the acceleration of other forms of deterioration (see paragraph 3.4.45).

Blow Holes

3.6.9   Blow holes are small regular or irregular cavities in the surface of the concrete. They are typically less than 25mm in diameter and result from the entrapment of air bubbles during the placement of the concrete. Large blow holes will reduce the effective cover to the reinforcement and may affect resistance to carbonation and chloride ion ingress. They also form areas for water and salt collection, and may accelerate the rate of freeze/thaw damage and chloride ion penetration.

3.7 DEFECTS ASSOCIATED WITH PROTECTIVE COATINGS AND REPAIR SYSTEMS

Coatings and Renders

3.7.1 Some exposed concrete surfaces are coated with protective or decorative paints or other coatings. For example, acrylic paints are used both to reduce the rate of carbonation and to improve appearance. The service life of a coating is generally much less than that of the concrete structure. The coating will eventually exhibit defects such as cracking, flaking, chalking and peeling. Flaking, or other forms of separation from the substrate, may occur over large areas. The cause may be incorrect choice of coating, inappropriate surface preparation, loss of moisture from within the concrete or exposure to severe conditions such as water leakage.

3.7.2 Concrete surfaces exposed to chloride contamination have, on some bridges, been impregnated with silane or other hydrophobic pore-lining impregnants in accordance with *BD 43* [31]. Impregnation does not change the appearance of the concrete so it is difficult to tell whether impregnation has been undertaken. It may be possible in wet weather to distinguish a silane treated concrete from an untreated concrete due to the lack of surface moisture which would be retained on the silane treated surface. Also, there may be more algal or fungal growth on the untreated surfaces. However, these are not definitive tell-tale signs. Even destructive and laboratory testing will not provide conclusive evidence of the presence of silane treated surfaces. A record drawing showing the areas treated, date of treatment and material used would aid the inspection, if available.

3.7.3 Areas of concrete below ground level or not on general view may be coated with bituminous or epoxy coatings to waterproof or otherwise protect the concrete. Such coatings may be visible at ground level or when inspecting areas such as bearing shelves.

3.7.4 Exposed concrete may also have been rendered with a thin layer of cementitious mortar or concrete, either trowel applied or sprayed. Typical failures of renders include cracking, delamination or spalling. Shrinkage cracking of the render may occur soon after application, leading to relatively rapid deterioration. In some instances deterioration of the structural concrete behind the render may continue; where this leads to expansive forces, such as those due to corrosion of the reinforcement, the render may delaminate.

3.7.5 The inspector should endeavour to determine the likely cause of any deterioration in the coating or render and also distinguish defects in the coating or render from those due to deterioration of the underlying concrete. Where coatings or renders have been applied for protection, defects will remove that protection, allowing the structural concrete to deteriorate.

Repairs and Protection Systems

3.7.6 Repairs and corrosion protection systems applied after construction may eventually become defective. These include passive systems such as patch repair skim coats, mortars and concrete, as well as active systems such as cathodic protection.

3.7.7 All repaired areas of the structure should be carefully inspected for signs of failure of the repair product or system. Typical failures can include cracking or delamination of the repair product, continued corrosion within the repair, and incipient anode corrosion where an area of rust staining or spalling has formed around the perimeter of the existing patch repair. The last case will occur where the extent of the repair was inadequate and chloride-contaminated concrete remains at the boundaries.

3.7.8   Active corrosion protection systems, such as cathodic protection, should be subject to a continuous operation and maintenance monitoring [32]. However, some surface defects may still occur to paint and anode systems and these should be recorded. Common problems include vandalism of the system, delamination of sprayed concrete overlays to cathodic protection mesh anodes and delamination of paint anodes. Water leakage onto cathodic protection anode systems can be a major cause of premature deterioration, especially in areas of ponding on horizontal faces of elements, such as crosshead beams, where paints will rapidly degrade.

3.7.9   Cathodic protection systems may produce surface defects where small pieces of steel present in the cover zone are not electrically connected into the cathode, e.g. construction debris in the soffits of slabs and beams such as tie wires and nails. The results are seen as rust staining and spalling of the concrete cover, and should not be confused with spalling caused by corrosion of the main reinforcement.

3.8   MINOR DEFECTS

3.8.1   There is a variety of minor defects which in themselves do not generally affect more than the visual appearance of the concrete, such as:

- *Rust stains* – may appear on the soffit of a structure where, for example, steel tie wires have dropped onto the bottom formwork during erection of the reinforcement and with time will corrode. Not all rust coloured stains, however, indicate active corrosion. It is common to see staining of concrete surfaces as a result of corrosion of the exposed reinforcement during construction. Additionally, some aggregates contain iron-rich particles, e.g. iron pyrites, which can result in rust stains if exposed on the surface. Stains may also appear on finished concrete surfaces from contaminated formwork caused by such things as corrosion products from the exposed reinforcement during construction.

- *Sand streaks and sand pockets* – may appear over the life of structure, as weathering erodes the surface skin of cement on the concrete surface.

These defects are due to excessive bleed of the concrete mix and where the concrete was not fully mixed during the concrete pour respectively.

- **Stratification** – may occur where the concrete has segregated during the concrete pour and this appears as a roughly horizontal banding, with a higher proportion of fines tending towards the upper portion of the zone and coarser aggregate occupying the lower section. It is possible, where severe, that such a region would be more susceptible to carbonation.

- **Severe dusting of the surface** – can be due to several causes. In slab construction, escape of bleed water from the concrete can produce a weak laitance layer on the surface. Normally, this would be removed as part of preparation for applying waterproof bridge deck membrane. Similar laitance defects can be seen on the top of all surfaces that are left with an untrowelled finish and the dusting will continue until the weak layer has been removed, though this is normally less than 1mm in thickness. Dusting can also be caused by lack of curing of the concrete, leading to a weak outer layer prone to abrasion and wear. Concrete mixes made using cement replacement materials are particularly prone to dusting and weakness, if the surface is not properly cured.

- **Grout runs** – may occur at the face of the concrete if formwork was not adequately sealed against previously cast concrete. Apart from their poor appearance, grout runs are not normally significant. However, a severe loss of grout may have led to honeycombing of the upper section of concrete (also see paragraphs 3.6.1-3.6.3).

- **Corrosion of formwork fixings** – formwork is often retained in place by tie bars through the concrete pour. Cones or similar fixings at each end should be removed when the formwork is struck and the resulting holes made good with mortar. However, sometimes pieces of fixing or even formwork are left in position, the holes are not filled or are poorly filled, or the tie bars are left protruding or with inadequate cover. Such defects will usually result in corrosion of the exposed or near-surface steelwork.

- **Removal of formwork** – may also cause damage to the concrete. This is most likely to occur where there are protruding ribs or features, but also occurs adjacent to formwork ties.

4 Steel Defects

4.1 OVERVIEW

4.1.1 This section describes typical steel defects, which occur in steel and steel/concrete composite structures and in the steel elements of structures built mainly from other materials. The different types of defects are described with particular emphasis on identification and likely causes.

4.2 DEFECTS CAUSED BY STRUCTURAL DISTRESS

Deformation and Distortion

4.2.1 Distortions may be present in steel members or plates for a number of reasons. They could be due to initial distortions, residual stresses, lack of fit, initial out-of-flatness or out-of-straightness of the component before fabrication, inadequate design, excessive loading, buckling under compression loading, external impact or fire damage. Distortions out of plane in the forms of waves, kinks or warping can considerably reduce the resistance to compressive forces. Any increase in distortion is significant and may reduce the load-carrying capacity of the structure. Some deformations of members may have been assumed in the design, such as vertical sag on long span girders.

4.2.2 Thermal stresses and strains may exceed design limits if bearings are incorrectly installed or cease to function as intended, e.g. by freezing or ceasing to slide due to corrosion or faulty alignment. Similarly, if expansion joints are incorrectly installed or cease to move, then additional stresses may be transmitted to the structure. Increased stress may also result from failure or yielding of adjacent components or from section loss due to corrosion of the member itself (see paragraphs 4.4.1-4.4.5). Stresses and strains may be set up within the member by exposure to temperature extremes during fabrication or subsequent repair works. This may occur during hot-dip galvanising, particularly when double dipping is required due to the size of the member, or more commonly during welding or weld repairs. Other errors that could lead to overstress are site substitution of the wrong grade of steel or incorrectly designed bracing.

4.2.3 The location and extent of the deformation or distortion should be measured and recorded for structural members and plates. The loading on the structure at the time of the measurement should be recorded. A datum measurement taken when there is no live loading on the structure should be taken to enable future comparisons to be made. If deflections are measurable under live loading then these should also be recorded.

Fatigue Damage

4.2.4 Fatigue is the process by which a structural member or element eventually fails after repeated applications of cyclic stress. Failure may occur even though the maximum stress in any one cycle is considerably less than the fracture stress of the material. Characteristically, a fatigue fractured surface displays two distinct zones: a smooth portion indicating stages in the growth of the fatigue crack, and a rough surface, which represents the final ductile tearing or cleaving. Typically, fatigue failures do not exhibit any significant ductile 'necking'.

4.2.5   Fatigue failure is the most common cause of cracking and fracture of steelwork structures (see paragraphs 4.3.3-4.3.6). Fatigue crack growth under the action of repeated traffic loading is a major concern for steel highway bridges. The risk of fatigue-induced failure in the existing bridge stock is unknown, however, potential problems may exist in bridges:

- which have not been designed for fatigue;

- which have been designed to inadequate fatigue criteria;

- where materials and manufacturing controls may not have been adequate;

- where structural changes may have occurred, which may include the addition of new fixtures or repair of damage using, for example, welded cleats or brackets, flame cut holes or strengthening plates;

- where operational changes have occurred, such as alterations to carriageway layouts; or

- where there is evidence of resonance occurring in any of the structural members.

4.2.6   Guidance on fatigue susceptible details can be obtained from *BS 5400: Part 10: Code of Practice for Fatigue* [33].

Wear

4.2.7   Evidence of wear may be found in moving parts such as pins in trusses and joints. Wear may occur if the joint or structure is experiencing excessive movement or vibration, and may be exacerbated by a lack of maintenance or by the specification or installation of an unsuitable material for the part.

4.3   DEFECTS ARISING DUE TO THE MATERIAL NATURE

Delamination

4.3.1   Delamination may be defined as separation into layers within the thickness of the steel in a direction parallel to the surface. Delamination may occur as a result of a plane of weakness being formed in the steel section during the manufacturing process. These planar defects originate from entrapped non-metallic matter or from shrinkage cavities known as pipes formed in the ingot. If the pipe is exposed by, for example, trimming off the head of the ingot, then its surface will oxidise and this will prevent the cavity from welding up during re-heating and rolling.

4.3.2   Where a lamination is wholly within the body of a plate or section and is not excessively large it will not impede the load-bearing capacity in terms of stresses which are wholly parallel to the surface of the member. However, laminations can cause problems if they are at or close to a welded connection.

Cracking and Fracture

4.3.3   Cracks are potential causes of complete fracture and usually occur at connections and changes in section. The most common causes are fatigue and poor detailing practices that produce high stress concentrations. Elements that have been modified since initial construction are also potential problem areas. Fracture of any member, bolt, rivet or weld is obviously serious and can have important structural implications.

4.3.4   Fatigue starts with fabrication flaws or at sites with high surface stress concentrations such as weld toes, irregular cut edges and flame cut edges (see paragraphs 4.2.4-4.2.6). It then proceeds by the growth of these flaws until a final failure mode, such as brittle fracture or buckling, occurs. The initial fabrication flaws may be large or small but in many cases are too small to be detected by eye. Cracks may range in width from hairline to sufficient to transmit light through the members. The variations in length and depth are of a similar magnitude.

4.3.5   Cracks may also be present in welds because of poor welding techniques or the use of steel with poor weldability. If cracks are detected it is likely that they will be repeated in similar details within the structure.

4.3.6   The location, extent and width of cracks and fractures should be recorded. In addition, deep pits, nicks or other defects that may cause stress concentrations should be measured and recorded.

4.4   DEFECTS INSTIGATED BY EXTERNAL AGENTS

Corrosion Defects

4.4.1   All ferrous alloys are susceptible to corrosion deterioration. Unprotected steel, in the presence of oxygen and moisture and in the absence of contaminants (clean atmospheric conditions) corrodes at approximately 0.02 mm/year. The rate of deterioration is therefore very slow and it will be many years before the integrity of a structure is compromised. However, corrosion is accelerated by continuous (or even intermittent) wet conditions or by exposure to aggressive ions, such as chlorides in de-icing salts or in a marine environment, and other atmospheric industrial contaminants. In these conditions, steel becomes vulnerable to both pitting and general corrosion. Pitting corrosion is local large reduction in parent metal and can cause a serious reduction in load-carrying capacity. It can also lead to local high stresses, which may increase the risk of fatigue failure. Other causes of steel corrosion are animal waste, the spillage of farm fertilisers, the flux used in welding (if not neutralised), and direct contact with dissimilar metals, but, with the exception of bi-metal contact, these are less likely to cause significant structural deterioration.

Highways Agency

4.4.2　　Structural steelwork is normally protected against corrosion by a paint coating system. Weathering steel, which does not require a protective coating, is the only exception. Corrosion is usually associated with the breakdown or inadequacy of the protective system (see Section 4.7).

Loss of Section

4.4.3　　It is important to assess the magnitude, location and form of corrosion and to identify its cause. Particularly vulnerable locations are areas that experience water leakage and those where water may collect, e.g. horizontal surfaces and joints. All loss of effective structural section should be assessed.

4.4.4　　Where a build-up of rust scale is present, visual observation is usually inadequate to evaluate section loss. Where access is available, the rust scale should be removed to base metal and this should be measured using callipers, ultrasonic thickness meters, or other appropriate methods (see Volume 1: Part E: Section 6.2). Steel corrosion products, are expansive and occupy a greater volume than the parent metal, e.g. rust occupies some 4 to 8 times the volume of the iron from which it was formed, so that even an apparently heavy layer of rust may, when removed, reveal only a small loss of section. Also there is a tendency to distort bolted and riveted connexions, when the corrosion occurs between the faying surfaces.

4.4.5　　Inspectors should be alert to any evidence of the corrosion of steel encased in concrete or masonry. Particular attention should be paid to junctions of steelwork with concrete, masonry and other structural materials. Mating and rubbing surfaces are particularly susceptible to corrosion because of the exposed conditions in which they operate.

Bimetallic Corrosion

4.4.6　　Bimetallic corrosion can occur where uncoated dissimilar metals are in contact in damp or wet conditions, i.e. in the presence of an electrolyte. Cast and wrought iron in contact with carbon steel are not prone to such corrosion as they all have similar electro-chemical potentials. However, problems can arise when galvanised or stainless steel or non-ferrous metals are in contact with iron or carbon steel.

Stress Corrosion Cracking

4.4.7　　Stress corrosion cracking is the unexpected sudden failure of normally ductile metals subjected to a constant tensile stress in a corrosive environment, especially at elevated temperature. This type of corrosion can happen 'unexpectedly' and rapidly after a period of satisfactory service leading to catastrophic failure of structures. The stresses can be the result of directly applied loads, or can be caused by the type of assembly or residual stresses from fabrication.

4.4.8 Certain stainless steels and aluminium alloys crack in the presence of chlorides, mild steel cracks in the present of alkali and high-tensile structural steels crack in an unexpectedly brittle manner in a whole variety of aqueous environments, e.g. where there is a presence of chlorides and/or hydrogen.

Weathering Steel

4.4.9 Weathering steel owes its corrosion protection to the formation of a stable protective oxide film, which seals the surface against further corrosion. The film is dark brown with a lightly textured 'rusty' surface. Unlike other types of structural steel, weathering steel has a carefully controlled non-ferrous content to ensure that the oxide film adheres tightly to the substrate. Nevertheless, the surface oxide is slowly worn away, and replaced by a new film, causing a very slow loss of section over the life of the structure. The rate of loss depends on the alloy content, air quality and the frequency with which the surface is wetted by dew and rainfall and dried by the wind and sun. Designs using weathering steels typically allow for a 1mm to 2mm sacrificial loss to sections and fasteners [34].

4.4.10 For most applications of weathering steel the weathered appearance of the finished structure is of prime importance. It is critical that any welds, which are exposed to public view, weather in the same manner as the adjacent material. Procedures and consumables have therefore been developed which produce welds that have a composition closely matching weathering steel. Economics sometimes dictate that multi-run welds are sometimes used in which only the top most run is of a weathering consumable i.e. the weld is 'capped'.

4.4.11 When steel structures are painted, the paint seals bolted connections from the ingress of water. This is not possible with a weathering steel structure, and water is likely to find its way to the interior joint surfaces. This may initiate corrosion, especially if the bolts have a different electro-chemical potential. Joints in weathering steel should therefore be closely inspected.

4.4.12 Weathering steel bridges should have built-in access for maintenance and close examination of critical areas, particularly those around movement joints [34].

4.5 DEFECTS CAUSED BY ACCIDENTAL OR DELIBERATE DAMAGE

Fire Damage

4.5.1 Steel progressively weakens with increasing temperature, e.g. the yield strength at room temperature is reduced by about 50% at 550°C, and to about 10% at 1000°C. There is therefore a risk that steel members may fail by buckling or deflecting if they are inadvertently heated during a traffic accident fire. The extent of failure will depend on the loading that the member is carrying, its support conditions, and the temperature gradient through the cross section.

4.5.2 Secondary effect damage can occur in bearings, movement joints and other structural members if they are unable to accommodate the large expansions that may occur in a fire. It is unlikely that this will have been allowed for in design.

4.5.3　In a severe fire unprotected steelwork will lose practically all its load bearing capacity, deform and distort and will not be suitable for reuse. In a less severe fire, damage may be limited and it may be possible to retain members after checking for straightness and distortion and the mechanical properties. Bolted connections often fail through shear or tensile failure or thread stripping. Any section yielding could have caused severe weakening of connections and it is important that these are properly inspected.

4.5.4　Fire will cause blistering and flaking of paintwork. Specialist personnel should be commissioned to inspect any serious damage and provide advice on its repair.

4.5.5　Further information on the assessment of the effects of fire is given in *Appraisal of Existing Structures [28]* and *Appraisal of Existing Iron and Steel Structures [35]*.

Impact Damage

4.5.6　Impact damage to a steel structure is usually obvious and may vary from scoring of the paint to deformation of an element. In severe cases the damage may render a steel member incapable of carrying any load.

4.5.7　Cracking and fracture may occur as a result of impact damage. The inspector should be aware that these defects may occur several metres away from the impact site.

4.6　DEFECTS ARISING DUE TO FABRICATION ERRORS

Quality of Welded Connections

4.6.1　Welding is widely used in the jointing of structural steel components and there are two main forms:

- ***Fillet welds*** – are external to the thickness of the jointed members. They are often used for non-fatigue sensitive joints between plates inclined to each other.

- ***Butt welds*** – require chamfering either one or both edges of the jointed elements to form a V-groove and the weld metal is deposited within the V. Butt joints may join the ends of two members or the end of one to the surface of the other, i.e. T-butt weld. In cases where welding access is only available from one side, a backing strip may be provided. This can be a fatigue sensitive detail under fluctuating loads.

Missing Welds and Welds of Poor Quality

4.6.2　Defects such as missing welds and welds of poor quality should have been identified and corrected during fabrication, but this cannot always be presumed. Protective coatings will also mask defects.

4.6.3　The leg length of fillet welds should be recorded using a fillet weld gauge where there is evidence of cracking or other defect. Where the leg lengths are unequal both figures should be recorded. Note should be made of whether the weld is convex or concave and for concave welds the throat thickness should be measured (see Figure D.4).

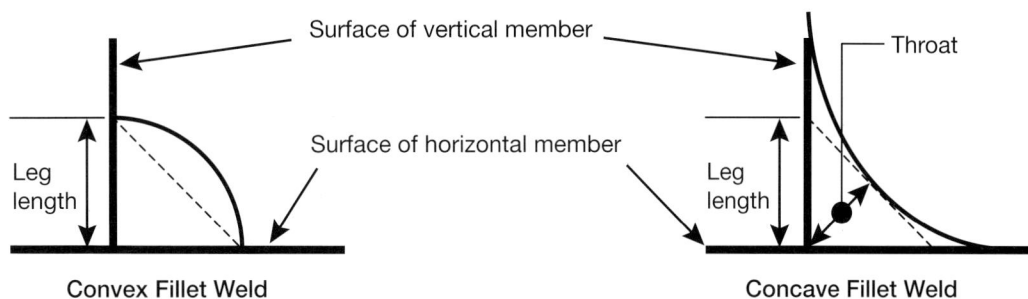

Figure D.4 – Equal leg fillet weld

4.6.4 Evidence of undercutting or porosity should be recorded for all types of welds. The main causes and features of these defects include:

- **Undercut** – is formed when a groove is melted into the parent material by the arc action and is not subsequently filled in by the weld metal. It is readily visible when not covered by a protective coating (see Figure D.5). For welds subject to fatigue stress, undercut at the toe of a weld is a serious fatigue defect, more so for fillet than butt welds.

- **Porosity** – is caused by gas entrapment leading to cavities in the weld metal. The cavities are generally spherical but can be elongated to produce piping. Porosity often occurs as air entrapment at the start of the weld and this is commonly known as 'start porosity' (see Figure D.6). Porosity does not significantly impair the structural strength but, if extensive, it may indicate other problems associated with the weld and could mask or prevent detection of more serious defects when using certain examination techniques, such as radiography, ultrasonics, etc.

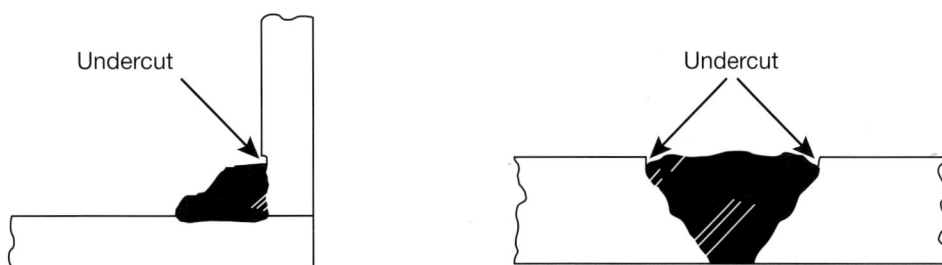

Figure D.5 – Undercut weld [36]

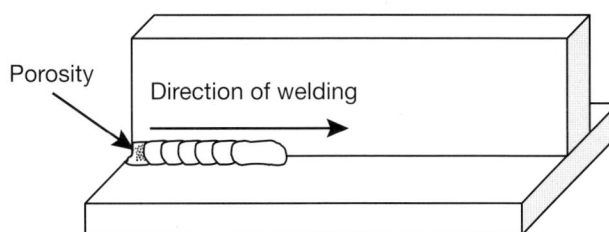

Figure D.6 – Start porosity in a fillet weld [36]

Surface Cracks in Welds or Adjacent Areas

4.6.5   Cracks visible at the surface are potentially the most serious form of weld defect. In welded members cracks may originate within the weld and extend into the adjacent parent metal. Incipient cracks may escape detection at time of welding, but later extend under service loading. It should be noted that visual inspection of welds is extremely limited and specialist non-destructive testing is likely to be necessary. Use of an illuminated magnifying glass will aid the examination of welds and adjacent areas for hairline cracks.

4.6.6   Welding is prone to cracking from fatigue, induced by a large number of fluctuations of stress, at levels that can be well below those which would cause failure if applied as single loading. Weld terminations and returns are the areas most susceptible to fatigue cracking. If problems are suspected they can be further investigated using microscopic or non-destructive testing. If cracks are detected, it is likely that similar details within the structure will also be affected. Defects in welds increase the fatigue risk (see paragraphs 4.2.4-4.2.6).

4.6.7   Weld cracking usually occurs following welding due to one of the following reasons:

- *Weld metal solidification cracking* – Weld metal solidification cracking is widely known as hot cracking and occurs during cooling and solidification of the weld. Figure D.7 shows a typical longitudinal crack in a fillet weld. Such cracks may be due to material composition or weld restraint and bead shape.

- *Heat affected zone cracking* – The heat affected zone (HAZ) due to a weld is the area of parent metal immediately adjacent to the weld bead. It is affected by the heat input during welding and the cooling immediately afterwards. Within the HAZ the microstructure of the steel will have been affected; this may lead, in some conditions, to the steel becoming brittle and susceptible to cracking. A typical HAZ crack is illustrated in Figure D.8.

- *Lamellar tearing* – Lamellar tearing is caused by the presence of manganese sulphides and silicate inclusions that occur in steel making. When the billet is formed and rolled into plates these are extended into thin planar inclusions. Lamellar tearing can result when a large weld is made with the boundary of the weld running parallel to an inclusion. The tear occurs due to the considerable shrinkage stresses, which can occur across the thickness of the plate on cooling. Restraint due to the joint geometry and plate thickness can affect the level of stress. The tears are usually close to the HAZ of the weld and are usually stepped, as shown in Figure D.9. Tearing is generally completely below the surface and not visually detectable.

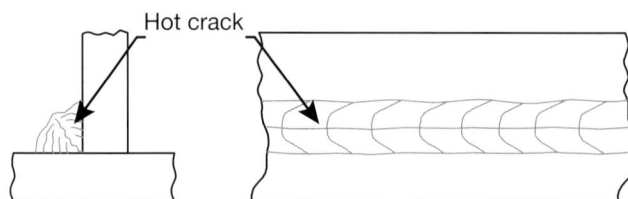

Hot crack

Figure D.7 – Hot crack in fillet weld [36]

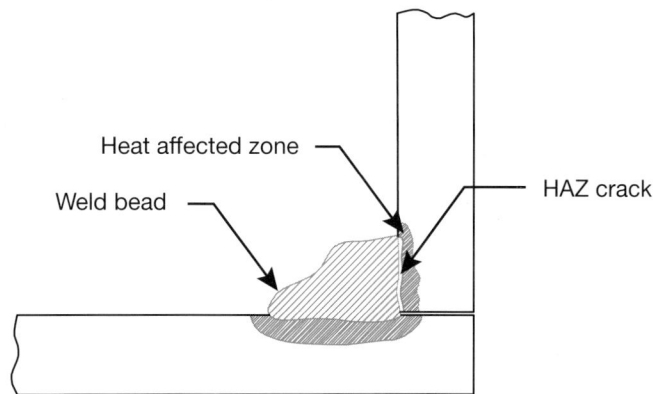

Figure D.8 – Heat affected zone crack [36]

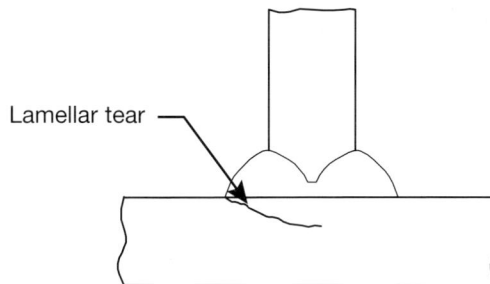

Figure D.9 – Lamellar tear [36]

Joint Slip and Tearing

4.6.8　Failure of connections may also occur if welds are subject to overloading or impact, or if the connection has been incorrectly designed or poorly fabricated. A close visual inspection should reveal if there is any evidence of slip. This can sometimes be seen by the development of cracks in the coating film or by signs of rubbing.

Other Defects in Welded Connections

4.6.9　Other weld defects, which are unlikely to be visible but which may be detected by specialist non-destructive testing methods, include:

- Incomplete weld penetration, where the weld metal has failed to penetrate and fill the root of the weld;

- Lack of side wall fusion, where the weld metal has not fused to the parent metal over parts of the side of the weld; and

- Slag inclusions, where slag from the welding flux has become incorporated into the weld.

4.6.10　Further information on welding defects is given in *Introduction to the Welding of Structural Steelwork* [36].

Quality of Bolted and Riveted Connections

4.6.11　In older steel bridges rivets were used for joining plates together, but, with a few exceptions, their use is nowadays limited. The simplest form of riveted joint is the lap joint where the rivets hold together two overlapping plates. The lines of forces in adjoining plates are not coincident, so there is a tendency to cause the joint to rotate. A more satisfactory joint is one in which the plates butt against each other, with cover plates fixed across the joint on both sides of the plates.

4.6.12   Bolts with nuts may be used in the same way as rivets. In order to achieve the highest strength from a bolted joint the threaded part of the bolt shank should stop at the surface of the plate, as the weakest section of a bolt is at the root of the thread where not engaged by the nut. Nuts should be sufficiently tight to ensure that bolts will not work loose, but excessive tightening is detrimental.

Joint Slip, Open Holes and Loose Rivets, Bolts and Nuts

4.6.13   A close visual examination should be made to find out whether there is any evidence of slip or movement at cover plates, washers or bolt heads. This is sometimes indicated by the development of cracks in the paint film. The connection should also be inspected for signs of rubbing or rusting. Slippage is particularly important in bolted connexions, as it may indicate a defective joint even though the bolts appear to be tight.

4.6.14   Riveted and bolted connections in shear should be checked for condition and for loose elements. Section loss to the heads of rivets should be recorded. Incorrect or misaligned seating or engagement of nuts on threads should also be recorded.

Transport Scotland

4.6.15   An indication of the integrity of suspect bolts can be gained by tapping them with a hammer. Tight bolts when hit will produce a ringing sound which penetrates the whole member, whereas loose bolts will give a dull thud. Any movement will obviously indicate a faulty bolt. Missing or unsound rivets or bolts should be reported.

4.6.16   Since virtually no riveting is done today, restoration projects sometimes use cup-headed bolts to give the impression of riveting. An alternative procedure is to use dome-head caps over ordinary bolts. The inspector should also be aware of the historic fraudulent practice of 'dummy rivets', i.e. of moulded putty with a covering of paint, used by less scrupulous erectors.

4.6.17   Where joint slip has occurred it may be necessary to check the types and dimensions of connectors. With bolts, it should usually be possible to reach the bolt shank projecting beyond the nut and measure this diameter. Earlier bolts and nuts were not standardised, and it may be necessary to examine thread depths. More recent bolts can often be identified from head markings.

4.6.18   If it is necessary to remove a bolt for examination, it should be replaced with a new bolt, nut and washers. Bolts should be removed and replaced one at a time. Where a bolt is removed, the exposed surfaces should be examined for signs of corrosion.

4.6.19   Rivets are not readily removed or separated in order to measure their shank diameter. It may be possible to strike or grind off a rivet head to check the diameter, but this will not always be safe or practicable. *The Appraisal of Existing Iron and Steel Structures* [35] contains useful guidance relating shank diameter to head and tail dimensions of the rivet.

Cracks, Tears and Distortion Adjacent to Holes

4.6.20   Fatigue can cause both cracking and fracture of bolts, rivets and connections. Fatigue behaviour of connections is improved by detailing the structure so that local restraints do not impose secondary deformations and stresses. Performance can be adversely affected by concentrations of stress at holes, openings and re-entrant corners. Cracks are most likely to propagate from the rivet or bolt holes.

4.6.21   Where practicable, fatigue may be confirmed by microscopic examination of a small sample; fatigue failure does not exhibit significant 'necking' (except at the stage of final rupture when the remaining intact steel is insufficient to sustain the load).

4.6.22   Punching holes in steel causes work-hardening and a consequent loss of ductility. Riveted plates are vulnerable to cracking and fracture, particularly in early steels with higher levels of sulphur and phosphorous.

4.6.23   Overloading of the connection may cause tearing or distortion adjacent to the bolts or rivets. Overloading may also cause deformation of the plates forming the connection. This may be due to faulty design, increased loading or use of the incorrect grade of steel. Shearing forces may also distort bolts or rivets

Quality of Bonded Connections

4.6.24   Recent developments have produced structural connections that utilise resin adhesives, e.g. strengthening by bonding steel plates or carbon fibre sheets to beam and deck soffits. The connexion is primarily achieved by chemical bonding. The process differs from welding as the jointing material is different in type and composition from the elements to be joined and in that it hardens at normal temperatures without fusion of the bonding materials. Resin adhesives are slow to harden at low temperatures and the bond can be impaired by vibration and dampness until it has set.

4.6.25   A number of bridges in the United Kingdom have been strengthened by bonding steel plates to beam and deck soffits with resin adhesives. The method is usually used for strengthening steel or cast iron members, but it has also been used to strengthen concrete beams and slabs. Carbon fibre advanced composites are beginning to be used as the bonded element in preference to steel (see Section 6.7).

Quality of Steel/Concrete Composite Connections

4.6.26   Steel/concrete composite bridges are common form of construction; most composite bridges have a reinforced or prestressed concrete deck slab with steel girders. Composite bridges rely on interaction between the structural steel and the concrete.

4.6.27 Shear connectors are provided to transfer horizontal shear (caused by loading, shrinkage and temperature differentials) between the steel beams and the concrete slab. They also anchor the slab to the beam against vertical separation. Connectors may be headed studs, bars with hoops, channels or friction grip bolts. They are protected from corrosion by the surrounding concrete.

4.6.28 Failure of shear connectors may be indicated by evidence of separation between the top flange of the steel member and the concrete. If separation has occurred the unprotected top surface of the steel and/or the shear connectors may be corroding. The steel/concrete interface should be examined for evidence of separation and rust staining. Close visual inspection may also reveal signs of movement or rubbing between the steel and the concrete, indicating that composite action has broken down.

4.7 DEFECTS ASSOCIATED WITH PROTECTIVE SYSTEMS

4.7.1 Paint systems suffer from various forms of deterioration such as cracking, flaking, chalking and peeling. The life of a paint system is normally much less than that of the steel member it is protecting. Early detection of breakdown is beneficial because it will substantially reduce the amount of preparation that is involved in repair and reapplication. Delays to the maintenance of paint systems can result in rapidly accelerating increased costs.

4.7.2 Maintenance of galvanised-only steelwork becomes extremely difficult if remedial action is delayed until rusting is well established. There is no reliable estimate for time to first maintenance of galvanising. The appropriate point at which to initiate painting relies on informed inspection reporting. This also applies to systems incorporating aluminium or zinc metal spray with sealer.

CSS Bridges Group

4.7.3 The cause of any white deposits on the surface of paint over zinc metal spray should be investigated; these may be the first signs of breakdown of the zinc. If left untreated corrosion of the zinc will become extensive. Aluminium metal spray is less easily attacked, breakdown usually occurs because the aluminium has been badly applied. Breakdown of paint over galvanising is often due to the poor adhesion of a wrongly selected paint system.

4.7.4   A general assessment of the condition of the protective system should be made. If this is indicating that repainting is necessary, specialist personnel should undertake a more detailed pre-specification paint survey, i.e. Special Inspection. Details are given in *BD 87* [37].

4.7.5   Local failure of systems or coatings should be remedied as soon as possible unless the failure can be dealt with during planned maintenance. Common types of failure are:

- ***Blistering*** – Blistering of coatings is generally caused either by solvents which are trapped within or under the paint film, or, by water which is drawn through the paint film by the osmotic forces exerted by hygroscopic or water soluble salts at the paint/substrate interface. The gas or the liquid then exerts a pressure stronger than the adhesion of the paint.

- ***Corrosion blistering*** – Coatings generally fail by disruption of the paint film by expansive corrosion products at the coating/metal interface. General failure can result from inadequate paint film thickness. However, local or general deterioration can occur when corrosion is due to water and aggressive ions being drawn through the film by the osmotic action of soluble iron corrosion products, as the attack will start from corrosion pits.

- ***Flaking*** – Flaking or loss of adhesion is generally visible as paint lifting from the underlying surface in the form of flakes or scales. If the adhesive strength of the film is strong then the coating may form large shallow blisters. Causes include:

  i.   Loose, friable or powdery materials on the surface before painting;

  ii.   Contamination preventing the paint from 'wetting' the surface, i.e. oil, grease, etc;

  iii. Surface too smooth to provide mechanical bonding;

  iv.   Application of materials in excess of their pot life.

- ***Chalking*** – Chalking is the formation of a friable, powdery coating on the surface of a paint film caused by disintegration of the binder due to the effect of weathering, particularly exposure to sunlight and condensation. This is generally considered the most acceptable form of failure since maintenance surface preparation consists only of removing loose powdery material and it is usually unnecessary to blast clean to substrate.

- ***Cracking*** – Cracking may be visible in increasing extent, ranging from fine cracks in the top-coat to deeper and broader cracks.

- ***Pinholes*** – Pinholes, or holidays are minute holes formed in a paint film during application and drying. They are caused by air or gas bubbles (perhaps from a porous substrate such as metal spray coatings or zinc silicates) which burst, forming small craters in the wet paint film which fail to flow out before the paint has set.

4.8 SPECIAL: CLOSED MEMBERS

Structural Hollow Sections

4.8.1 Steel bridge components may be fabricated from structural hollow sections (SHS) - rolled steel sections of circular or rectangular cross-section. The internal surfaces of SHS are not normally corrosion protected as the ends are generally sealed with welded end plates, and connectors.

4.8.2 Flaws in the sealing welds may allow the penetration of moisture and contaminants, leading to corrosion. Such corrosion may become visible as it emerges from cracks or flaws at the ends of the member. This is more likely to be observed at the low end of inclined members. Accumulated water may freeze and expand causing bulging and even splitting of small cross-section members.

Closed Steel Members

4.8.3 Closed steel members in the form of fabricated box beams or columns may either be sealed or ventilated.

4.8.4 Seals are provided at access covers in some structures to prevent the ingress of air and water. Even if the seals are not fully airtight, they will restrict the entry of pollutants, birds, rodents and other animals. It is not uncommon to find an accumulation of water in closed members together with bird/animal excrement, and mould and fungus. All these can damage the surface protection. Additionally accumulated water may freeze and expand causing plate distortion or even splitting in small cross-section members.

Enclosures

4.8.5 Enclosures provided in accordance with *BD 67* and *BA 67 Enclosure of Bridges* [38, 39] allow a relaxation in the specification of the paint system. This is because they encase the below-deck steelwork, which has been shown to reduce the rate of breakdown of the protective coating.

4.8.6 Enclosure systems and materials may vary from structure to structure. Guidance on inspection periods, details and items of particular concern should therefore be included in the maintenance manual for each structure.

4.9 SPECIAL: CORRUGATED STEEL BURIED STRUCTURES

4.9.1 Defects in corrugated steel buried structures (CSBS) are generally concerned with:

- the structural condition: alignment, cross-sectional shape and the integrity of the joints and seams;

- the material condition: thickness and soundness of protective coating, residual thickness of steel and condition of invert paving; and

- the general condition of ancillary structures: headwalls, aprons, adjacent earthworks, restraint systems, etc.

4.9.2 Typical defects affecting the structural or material condition of the CSBS are outlined in the following sections. Further information on the deterioration of such structures is given in *The Durability of Corrugated Steel Buried Structures* [40].

Transport Scotland

4.9.3   Reinforced concrete headwalls or ring beams, structural steel collars, ties or ground anchorages are methods of supporting the ends of CSBS. Culverts also often have concrete aprons and inverts (see paragraphs 2.7.1-2.7.3 for a discussion of general problems with culverts and Section 3 for concrete defects).

Structural Defects

4.9.4   Corrugated steel buried structures are flexible structures, and are therefore liable to deformation of the cross-sectional profile if subjected to excessive or inappropriate loading. Their strength and stability is derived from soil/structure interaction.

4.9.5   Local deflections in the upper part of a CSBS can often be traced to construction plant, particularly heavy earth moving equipment, during and immediately following backfilling. This may have lead to overstressing of the steel shell, with deformations not becoming apparent until the structure is loaded by covering fill. Gross distortions in cross-section may be the result of either the backfill or foundation failing to provide adequate support. Closed invert multi-radii structures are most susceptible to deformation from inadequate foundation support. Transverse distortion of a structure is more likely when:

•   the depth or properties of the subsoil varies;

•   poor construction practice has occurred, such as uneven loading due to compaction;

•   excavation has taken place close to one side of the structure;

•   there is a difference in the level of the embankment on either side of the structure; or

•   the applied dynamic load from the overlying carriageway varies across the structure.

4.9.6   Transverse distortion should be relatively small and should not increase much in service. Continuing movement is a sign of inherent instability and should be investigated.

4.9.7   CSBS are often installed through embankments constructed on poor ground, as their flexibility allows them to accommodate some longitudinal settlement. These settlements are a consequence of the consolidation of the underlying soils, and it may be several years before they are complete. The key to acceptable behaviour is the selection and placement of the backfill, and provided that this has been properly controlled, the structure should be able to accommodate significant movements.

4.9.8   However, excessive deformation will reduce the load capacity of the structure and can cause failure of the joints and seams between the individual plates of the CSBS. This will affect the integrity of the structure, and may permit water to pass into the backfill, leading to washout and progressive collapse.

Material Defects

4.9.9   All surfaces of corrugated steel buried structures are normally galvanised with a secondary protective coating of bitumen. Inaccessible surfaces are provided with a sacrificial thickness of steel that, together with the galvanising, should achieve the required service life without maintenance [41]. Exposed surfaces, which can be maintained, are usually coated with an additional protective coating over the galvanising; high build epoxy coatings are frequently used.

4.9.10   The most common problem is the breakdown of the protective coating. Initially, the secondary bituminous or epoxy coating may be affected, and if this is allowed to continue the galvanising will be exposed which will then deteriorate permitting corrosion of the structural steel. Deterioration of coatings can result from:

- ***Corrosive water seeping into or flowing through a structure*** – Contaminated water flowing through the structure can damage the invert and sidewalls. The presence of chloride or sulfate ions or naturally occurring acidity (or some combination of these) will accelerate the rate of attack. Aggressive chemicals may also find their way into watercourses from manufacturing or agricultural sources.

- ***Erosion of the invert and sidewalls of a structure*** – The protective coating on the invert and sidewalls can be eroded by mechanical action of water and hydraulic sediments flowing through culverts. Culverts on steep gradients and carrying intermittent flows are particularly susceptible to this form of damage. Protective grills, gates or trash screens can be installed to prevent the entry of larger cobbles and boulders, but they require regular inspection and maintenance if blockage is to be avoided. Livestock passing through CSBS farm accommodation structures may also wear away the bitumen coatings as they brush against the sides.

- ***Cycles of wetting and drying*** – Cycles of wetting and drying can damage exposed bituminous coatings. This is most likely to occur along the wet/dry line of a culvert sidewall.

- ***Weathering at the exposed ends of a structure*** – This is more likely where the ends are bevelled (cut) to the slope of the embankment.

- ***Corrosion due to corrosive water seeping through from the backfill*** – Chloride or sulfate contaminated water seeping from the backfill can enter through bolted seams and joints, and this may lead to severe local deterioration of the buried surfaces of a structure and to corrosion at the bolt holes. The corrosive water could come either from aggressive materials in the surrounding fill, from the natural ground or from de-icing salts applied to the road surface above.

- ***Corrosion of exposed surfaces due to the presence of contaminants in the air*** – It is comparatively rare for structures to be damaged by atmospheric contamination, but cases have been observed at coastal locations where galvanizing has not been provided with secondary protection.

- ***Corrosion generated by stray electric currents*** – Stray currents (originating from direct-current distribution lines, substations, or street railway systems, etc.) flowing into a structure through damp soil are a serious problem because they enter and leave metallic structures carrying some of the metal into solution in the ground until eventually the metal is dissolved. The exact rate at which this happens will depend on the type of metal and the magnitude of the prevailing currents (which in turn depend on many other variable factors), but it is capable of causing considerable damage if allowed to continue for a long period of time.

5 Masonry Defects

5.1 OVERVIEW

5.1.1 This section describes typical defects that occur in masonry structures and in the masonry elements of structures built mainly of other materials. The different defects are described with particular emphasis on identification and likely causes.

5.2 DEFECTS CAUSED BY STRUCTURAL DISTRESS

Excessive Loading

5.2.1 Excessive loading, particularly when applied as a point loading, may cause localised crushing of masonry or even displacement of individual masonry units.

5.2.2 The lateral pressure of earth behind abutments, wing walls and retaining walls is generally constant so the retaining structure remains stable. An increase in this pressure may cause forward movement or tilting. Such movement of an abutment will distort the shape of an arch bridge and may cause transverse cracking of the arch barrel. Cracking may have occurred as the earth pressures stabilised, shortly after construction. Recent cracks would indicate that movement is occurring.

5.2.3 Increased imposed loading, either live or dead loads, a change in ground levels, or a rise in the water level behind abutments, wing walls and retaining walls will raise the earth pressure and may cause distress. Dead loads may have increased by adding fill or by widening the structure. Live loads are generally small in relation to dead loads on masonry structures and therefore less likely to be the cause of structural distress. However changing traffic patterns may lead to significantly increased live loads; a typical example is an arch bridge on a minor road that has become the access to a quarry.

CSS Bridges Group

5.2.4 Lateral forces or pressures in the fill material, especially if it becomes saturated and freezes, or those due to traffic loading may destabilize spandrel walls on arch bridges; vertical traffic loading increases lateral forces as it is distributed through the depth of fill. Centrifugal forces can also be generated on curved bridges by turning and braking traffic, and these will be transmitted into the structure through the fill. Spandrel walls are more vulnerable to damage or displacement if no footway exists to prevent vehicles passing close to the side of the bridge. Without footpaths, vehicular impact is more likely and the effects of the lateral loading generated by the vehicle through the fill will be more acute. The consequence of this loading may be for the wall to tilt by rotating outward from the arch barrel, to deform by bulging, to slide on the arch barrel or to displace outwards detaching part of the arch ring, as indicated in Figure D.10.

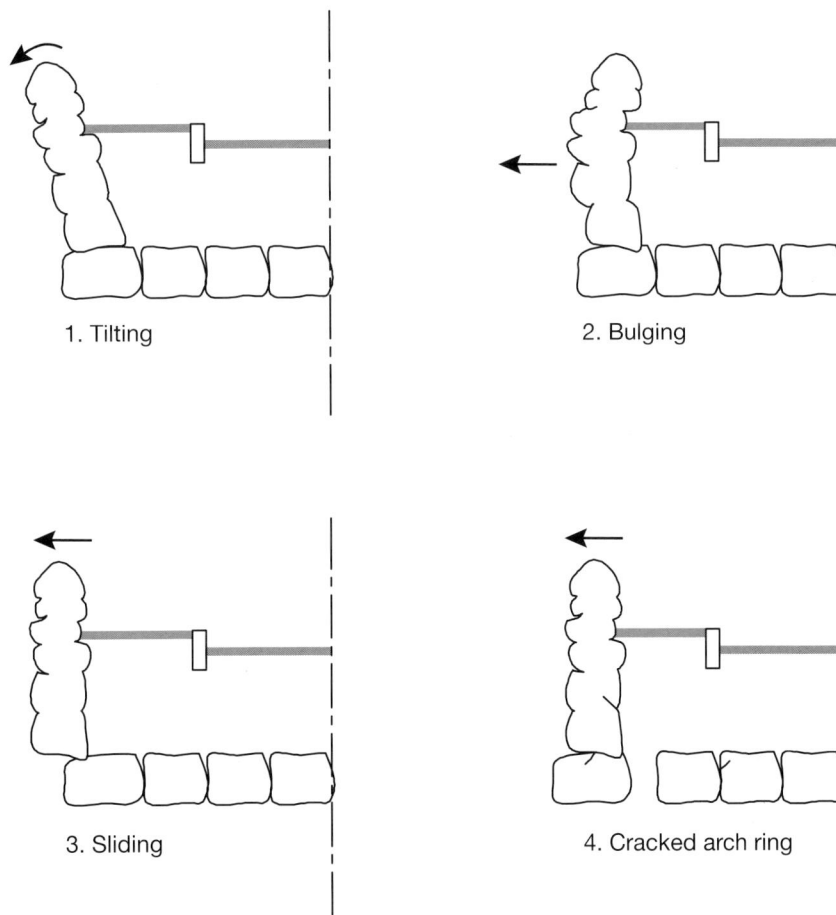

1. Tilting

2. Bulging

3. Sliding

4. Cracked arch ring

Figure D.10 – Spandrel wall failures

Arch Shape Deformation

5.2.5   The thickness and shape of the arch is important in determining the load carrying capacity of masonry arches. A detailed dimensional survey may be required if a load assessment is to be carried out, particularly if the arch is severely deformed. For other inspections a visual check of the shape or a simple dimension survey may be adequate. The maximum rise of an arch above springing level at midspan, can be obtained by stretching a line between springing points and measuring to the ring, either by dropping a line down from on top of the bridge or by use of an extending measuring rod.

5.2.6   Flattening of the arch may be a sign of outward movement of the abutments. Movements may be more easily identified by evidence of dips in the courses of the spandrel walls or the parapets above the arch. Sighting from a distance may aid checking of the arch profile and identification of deformations.

5.2.7   Distortion may have occurred at the time of construction, or shortly afterwards. To determine if the distortion is stable or progressive measurements should be taken from fixed survey points and the data compared with that obtained from subsequent surveys.

Structural Cracking

5.2.8   Cracks in masonry construction may arise from a number of causes, including unbalanced thrusts from arches and vaults, foundation movements, and thermal and moisture movements of the masonry itself or of other structural elements. Movement of cracks may be seasonal, because of climatic fluctuations, or diurnal and not progressive. They may only affect the appearance but can also be indicative of a more serious underlying defect. The inspector should observe many aspects of the cracking, including length, width, variation of width along its length, location, distribution, and, in some cases, depth. The displacements forwards, backwards and sideways of the masonry on each side of a crack should also be recorded. The current extent of the displacement should, if possible, be marked and dated on the surface of the structure to assist future inspections; a tell tale will produce a more precise record.

5.2.9   Cracks or deformations, which may have occurred soon after the structure was built, are not usually as serious as those of recent origin, identified by clean faces and possibly loose fragments of masonry. A further important consideration is whether the cracking is still 'live' and developing. Where this is suspected, more frequent inspections may be necessary. Cracks may on occasion form in the mortar only, and it is important that cracking and joint deficiencies should not be confused.

5.2.10   Accurate sketches should be drawn of all cracks, with sufficient detail to allow subsequent surveys to determine whether there has been developing deterioration. Crack movements may be detected by using tell-tales or more precise measuring devices such as Demec gauges. For continuous measurement of crack movement, automatic remote monitoring systems are also available (see Volume 1: Part E: Section 4.3).

5.2.11   The most serious form of cracking is that caused by structural inadequacy or overloading. The four types of cracking which can occur in different parts of masonry arch structures, their causes and significance are described below.

- *Longitudinal cracks* – Differential settlement or movement across the width of an abutment or pier will produce longitudinal cracks in the arch barrel, as the structure splits apart, dividing the barrel into independent sections. Closer crack spacings are more serious, as these may suggest that the barrel is subsequently splitting into many more independent sections. If accessible, the depth of the cracks should be probed to reveal whether or not the whole thickness of the arch barrel has been cracked. Confirmation, if required, can be achieved by digging a trial-pit above the crack location.

- *Transverse (lateral) cracks* –These may be accompanied by permanent deformations of the arch shape and are caused by partial load failure of the arch or by movement at the abutments. Transverse cracking can result from forward, backward, rotational or differential movement between abutments. Transverse cracking may also occur due to the settlement of a pier relative to abutments or other piers.

- ***Diagonal cracks*** –These normally start near the sides of the arch at the springings and spread up towards the centre of the barrel at the crown. They are generally due to subsidence at the sides of the abutment or pier and are caused by the resultant twisting of the arch. Extensive diagonal cracks indicate that the barrel is in a dangerous state. Diagonal cracks at the spandrel walls and ring separation from the spandrels, though uncommon, would indicate that the arch is flexing and that the bridge could be in a serious condition.

- ***Longitudinal cracks near the spandrel walls*** – Longitudinal cracks near the edge of the arch barrel may be a sign that the spandrel wall has been forced outward and, instead of the spandrel wall sliding on the extrados (i.e. the exterior curve) of the arch, the arch ring itself has cracked. An alternative cause of cracks in the arch ring near the spandrel walls may be due to the deformation under loading of the arch. These deformations will cause stress/strain concentrations at the junction between the portion of the arch barrel stiffened by the spandrel wall directly on top of it and the more flexible remainder of the arch barrel. Since deformations are greatest near the crown of the arch cracks will tend to develop near the crown and progress outwards towards, but not necessarily as far as, the springings.

5.2.12 The first three types of cracking will have an adverse affect on the load carrying capacity of the arch; the fourth, whilst also undesirable, is unlikely to affect the carrying capacity directly but may affect the stability of the road surface and the durability of the structure.

5.3 DEFECTS ARISING DUE TO THE MATERIAL NATURE

Arch Ring Separation

5.3.1 Many arch barrels are constructed with multiple rings of brick, with the rings bonded together only by mortar. This mortar layer may deteriorate due either to chemical attack or loading, or a combination of these factors. The load capacity can be significantly reduced if ring separation has occurred. Separation within the barrel of an arch may be detected by hammer tapping to detect 'drumminess' (a dull hollow sound) as opposed to a 'ring' if fully bonded.

Defective Mortar and Pointing

5.3.2 The load carrying capacity of a masonry arch is dependent upon the thickness of the arch ring. If the mortar is missing, loose, or friable, then that depth of the ring affected is unable to transmit load and contribute to the strength of the arch. The depth from the face and the extent of the area affected by mortar loss or deterioration should be recorded. Defective mortar will also reduce the strength of masonry abutments and walls.

5.3.3 The profile and finish of mortar joints are important factors in minimising the amount of rainwater that will penetrate masonry. A recessed joint, particularly if left textured from raking, rather than being 'tooled', will tend to increase rain penetration, especially when exposed to heavy wind-driven rain. Recessed joints can therefore increase the risk of frost attack as the arrises of the masonry units become saturated, which increases the susceptibility to freeze / thaw cycling.

Highways Agency

5.3.4   Flush joints formed by simply cutting off the surplus mortar with an upward stroke of the trowel are more rain resistant but are often left with small cracks between the mortar and the masonry unit, which encourage water penetration. The joint finish, which is most resistant to rain penetration, is the flush joint, which is then 'tooled' to a concave 'bucket handle' finish or trowelled to give a 'struck and weathered' joint. This tooling presses the mortar into intimate contact with the brick to 'seal' the joint and the smooth compact surface is more water resistant.

Displaced or Missing Stones or Bricks

5.3.5   Deterioration of mortar, localised loading or large structure movements may result in masonry units becoming loose or displaced. Sometimes individual voussoir stones that have dropped to the extent that they protrude from the soffit of the arch will move back when struck with a hammer. There is a danger of loose units falling out, particularly if they are in a soffit. For safety reasons, the inspector should effect a temporary repair by installing small wedges to hold the units in position.

5.3.6   The displacement of individual masonry units should be noted; particular emphasis should be made to those at the crown of arches with small depths of cover. Displacement may be due to uneven masonry, projecting above the extrados of the barrel, attracting concentrated loads. Another reason could be a hard spot directly above the unit, such as a pipe flange bearing directly upon the arch. It is unfortunately not uncommon for sections of the crown to be cut away to provide room for service pipes to cross arch bridges, and the bedding of these pipes at the crown may cause displacement of stones. Reinstatement of trenches with foamed concrete may also provide a concentration of wheel loads.

5.4   DEFECTS INSTIGATED BY EXTERNAL AGENTS

5.4.1   The age of most masonry structures means that they will have been subjected to prolonged exposure to wind, rain, frost attack and large variations in temperature and humidity. Deterioration under these conditions may occur due to one or a combination of two or more of the following factors:

•   Erosion by water and wind and water borne particles, by frost attack and by vegetation root growth;

•   Chemical/biological attack due to acids, sulfates and chemicals either water-borne or released by water, or from air-borne pollution;

- Efflorescence staining;

- Moisture and thermal movement of bricks and blocks.

5.4.2 These are discussed in more detail below. Further information is given in *The Maintenance of Brick and Stone Masonry Structures* [42], in *A Guide to Repair and Strengthening of Masonry Arch Highway Bridges* [43] and in *Masonry Arch Bridges: Condition Appraisal and Remedial Treatment* [44].

Water Attack

5.4.3 The durability of masonry will be seriously affected if it remains saturated for long periods of time. Masonry may become saturated by rainfall, by upward movement of water from foundations, by downward movement from the top of walls and the extrados of non-waterproofed arches and, in the case of undrained walls, laterally from the backfill. Deterioration due to water entering or passing through masonry may occur because of all or some of the actions described below. Water is probably responsible for more damage to masonry structures than any other single cause. Preventing masonry from becoming saturated and maintaining the waterproofing and drainage systems is an important consideration. The inspector should look for evidence that drainage systems, such as weep pipes and drain pipes are functioning. A problem particular to some multi-span viaducts is that there may be a lack of drainage installed to the arch valley above intermediate piers.

5.4.4 Water percolating through the arch of a bridge or any other masonry structure is an obvious sign that either the waterproofing and/or drainage has failed or was never installed. This may wash the fines out from the spandrel fill, creating voids and causing local damage or sinking to the carriageway surfacing. Waterproofing to the arch barrel is often damaged or removed during the installation of utilities or service ducts; it may be possible to match leakage on the soffit with the location of a trench reinstatement in the surfacing.

5.4.5 Inspectors should consider the recent incidents of rainfall, as this may indicate whether the source is natural or is a leaking water pipe or sewer. This may require testing of the water seeping through the masonry to confirm the source.

Erosion

5.4.6 Erosion by wind and water borne particles will cause loss of surface texture and colour changes with a greater effect on softer than harder materials.

CSS Bridges Group

5.4.7   Sedimentary rock such as sandstones and limestones will be more susceptible to erosion than igneous and metamorphic rock and bricks. However, porous sedimentary stones will absorb water penetrating through defects, which will reduce the effect on the mortar. Granite, slate and other impermeable stones will absorb little or no water and the amount available to affect the mortar will be greater. Also, the bond between mortar and non-porous stone will be lower than for porous stone.

Frost Attack

5.4.8   Masonry is susceptible to freeze thaw damage when saturated. Damage is caused by the 9% volume expansion on freezing of water in the pore system of porous materials. It is not necessarily the coldest or wettest winters that lead to frost damage, but rather recurring freeze/thaw cycles of the saturated masonry. Freeze/thaw cycling of masonry, which is wet, but not saturated, causes little damage. Frost will cause spalling of the face of saturated porous brick or stone. Wet mortar will soften, crumble, and spall when subjected to frost attack.

5.4.9   Some types of bricks, especially those not intended for external use, can suffer serious degradation due to frost attack. Parapets are the elements most commonly affected.

Salt Crystallisation

5.4.10 Water will carry dissolved salts through the pore structure of brick, stone and mortar. The source of the salts may be external, e.g. de-icing salts, or they may be released from the masonry itself. The salts will crystallise within the pores just below the surface if the surface is subject to rapid drying. The stresses created by this can cause local fragmentation and spalling of the surface, similar in appearance to the damage caused by freeze/thaw action, but typically this will be accompanied by efflorescence staining.

Chemical Attack

5.4.11 Mortars are less stable materials than brick or stone and are therefore generally at greater risk from chemical attack.

5.4.12 Rainwater is slightly acidic, containing dissolved carbon dioxide, which dissolves calcium carbonate. Calcium carbonate is the main binding agent in lime-based mortars and hence these are at most risk from water percolating through masonry. Mortars will become loose, sandy or friable, and will eventually be displaced from the joints. When this occurs it can lead to the loss of bricks or blocks. Staining or efflorescence due to re-precipitation of the dissolved salts may be visible.

5.4.13 Rainwater running off one type of stone onto another may cause damage. Runoff from limestone or any lime based product such as concrete will damage sandstone and even some granites.

5.4.14 Groundwater can be mildly acidic, particularly where it originates from peat moors or densely forested areas. Mortar subjected to this will become soft and friable and hence more susceptible to other forms of attack such as erosion and frost. Groundwater can also be contaminated with domestic, farm or industrial pollutants, which can severely affect mortars and masonry.

5.4.15   Masonry adjacent to carriageways may be splashed with contaminated run-off by the passage of vehicles. Significant erosion may occur, especially with some sandstones and limestones, reducing the thickness of the structural element.

5.4.16   Steel members cast into masonry may corrode, creating expansive forces that can crush, crack or deform adjacent masonry.

Sulfate Attack

5.4.17   Water-soluble sulfates can cause deterioration of mortar by softening and expansion. The expansive forces can cause spalling of adjacent brick or stone. However, this will only occur in wet or saturated conditions. The most common sulfates are the freely soluble potassium, sodium and magnesium salts and the less freely soluble calcium sulfate. These may be present in groundwater and can affect masonry below ground level and any masonry in contact with the ground. Some bricks contain sulfates that may leach out into the mortar; in particular old solid bricks with unoxidised centres (blackhearts), and some Scottish bricks often have significant amounts of soluble sulfate.

5.4.18   Expansion of unrestrained masonry, or stress induced cracking where restrained, is evidence of sulfate attack. Greater expansion will occur in more permanently wet areas; hence the centre of a wall beneath the drier surface layer may be at greater risk. Evidence on the face of a wall may therefore be limited, but fine cracking in the centre of mortar joints could be an indication of sub-surface expansion.

Atmospheric Pollution

5.4.19   Atmospheric pollutants such as carbon dioxide and sulphurous gases may increase the deterioration of masonry. Limestones are particularly vulnerable and may experience crumbling or delamination of the surface.

Vegetation

5.4.20   Lichens and mosses are indicators of wet conditions. They can generate acids that not only damage limestones, sandstones and mortar, but will also etch granite.

5.4.21   Plants and trees may often gain a foothold in masonry, growing from joints or cracks. Once established, they can cause rapid deterioration. The invasion of plant and tree roots into mortar beds is one of the common sources of joint failure and cracking.

Staining

5.4.22   Staining is normally only a visual problem and does not in itself pose a danger to masonry. However, it may be indicative of other problems and of current or historic water leakage. The four common chemical staining processes are:

- *Silica staining* – indicated by white surface deposits of silica fines which are insoluble in most acids;

- *Lime staining* – due to leaching of calcium carbonates and evident as patches of white carbonates, which are soluble in acids;

- **Efflorescence** – due to the crystallisation of soluble salts, as evidenced by white areas on the surface that are soluble in water;

- **Leaching of coloured compounds** – typically iron impurities in mortars appearing as brown stains.

5.4.23 The inspector should note the location of staining in relation to the structure as a whole in order to determine the source of the water leak.

Moisture and Thermal Movements

5.4.24 Changes in temperature or moisture content will cause masonry to expand or contract. The ability of masonry with a softer mortar to flex and accommodate movement will prevent cracking; conversely, a stronger but inflexible mortar may result in cracking of the mortar joints or even the bricks or stones.

5.4.25 Where other materials form part of the structure, such as a steelwork deck built onto masonry abutments, or where bearings have become seized, the forces generated due to thermal and moisture movements may disrupt and crack the masonry, or cause tilting, rotation or sliding of the whole element.

5.5 DEFECTS CAUSED BY ACCIDENTAL OR DELIBERATE DAMAGE

Fire Damage

5.5.1 Masonry performs relatively well when subjected to fire. Fired bricks in particular are quite stable. Sandstones tend to weaken and fail beyond temperatures of 570°C. Similarly mortar tends to become weak and friable above 500°C. Generally if brickwork walls are otherwise satisfactory following a fire, their load bearing capacity can often be assumed to be satisfactory. Cracked and deformed structures are however unlikely to have a satisfactory load bearing capacity. Where any doubts exist specialist advice should be sought.

Impact Damage

5.5.2 The age and the arch form of construction means that there are many existing masonry bridges where parts of the structure project into the minimum standard carriageway headroom. This increases the risk that they will be struck by high vehicles. However, in contrast to other forms of construction, the damage is often only superficial, usually resulting in shallow scores and scratches. Nevertheless, repeated contacts can produce a chase in the intrados (i.e. the interior curve of an arch) and this can affect the integrity of the arch. However, the greatest effect can occur at the exit side of the bridge, where a single vehicle impact can result in the spandrel being 'hooked' off.

5.5.3   Masonry parapets or retaining walls with rough faces may be damaged by stones being plucked out if the wall is hit by a vehicle. Dry-stone walls are particularly susceptible to such damage.

5.6   SPECIAL: DEFECTS ARISING DUE TO ALTERATIONS TO MASONRY STRUCTURES

5.6.1   The age of masonry structures, combined with the changes in traffic requirements, has resulted in many of them being subject to a range of maintenance, strengthening and improvement works. Some of these works will be visible, while others will not and bridge records should be consulted before undertaking inspections. The most common techniques are discussed briefly below, with particular emphasis to the issues that are peculiar to the inspection of masonry structures.

Underpinning

5.6.2   Underpinning by installing small diameter reticular piles is a technique often used to strengthen foundations subject to settlement or to support additional load. Piles are typically bored through and cast into the existing abutment, foundation, pier or retaining wall. Their location in the structure will determine whether there is evidence of their installation, e.g. ends of piles visible in top of foundation.

Ground Anchors

5.6.3   Ground anchors may be used to restrain abutments, wing walls or retaining walls from sliding and/or rotation. A steel tendon is installed in a cavity drilled through the structure, with one end grout anchored into a load-bearing stratum whilst the other is attached to the structure by a mechanical anchor; the anchor head is normally visible as a bearing plate bedded against the structure with the tendon and locking arrangement extending from it. Anchors are usually inclined, prestressed and permanent. Since anchors may be highly prestressed a loss of section due to corrosion could have serious consequences.

5.6.4   Adequate corrosion protection should be maintained for the whole length of the tendon including the anchor head, which may comprise a bearing plate, locking nuts, friction grips and the bare tendon. It will only be possible to view the anchor head, but the inspector should check the condition of any existing corrosion protection and record any loss of section or significant corrosion.

Concrete Saddle

5.6.5   The most common means of strengthening an arch barrel is to cover it with a mass or reinforced concrete saddle or provide a relieving arch. The advantages of these methods are that they not only strengthen the arch but also improve load distribution and tie together any cracked sections. A saddle also leaves the external appearance of the bridge unchanged. Care needs to be taken, when designing or installing such strengthening, to ensure that the thrust is transmitted to the abutment and that the abutment is capable of carrying the additional load. It is usual to cast the saddle directly onto the existing extrados, thus creating composite action, but sometimes a smooth debonding layer is introduced to minimise the live load carried by the existing arch.

5.6.6 The saddle will be buried and hence its presence or condition will not be apparent to the inspector unless trial pits are excavated. It is usual to waterproof the reinforced concrete saddle and therefore there should be no signs of water leaking through the arch in these circumstances.

Relieving Arches

5.6.7 Where there is a large depth of fill, where there are many utilities or where the headroom beneath the bridge is not critical and appearance is not important, it is often economic to place a relieving arch underneath. This may be formed of curved steel beams, sprayed concrete, or by placing a corrugated metal or glass reinforced liner under the arch and pumping concrete into the space between the liner and the existing intrados.

Steel beams

5.6.8 Steel beams (colliery arches) rolled to the shape of the arch intrados may be used as a permanent strengthening. The beams are set into chases cut in the abutments or fixed to their face. Gaps between the beams and the soffit may be filled with grout or packed with wedges. Inspection of steel components should give due consideration to the deterioration mechanisms described in Section 4. Special attention should be paid to joints and to the steel/packing/masonry interfaces to check that reliable contact is being achieved.

Sprayed Concrete

5.6.9 Sprayed concrete is commonly used to improve load capacity by increasing the thickness of the arch ring. If the existing arch is particularly flexible or inadequately waterproofed or drained then the sprayed concrete will also deteriorate. The inspector should record crack patterns, water seepage, leaching of salts and any other defects, particularly evidence of reinforcement corrosion (see Section 3). It is possible for the concrete to separate from the masonry intrados; therefore, when access permits, the inspector should carry out a hammer sounding survey to test for debonding (see Volume 1: Part E: Section 4.2: Paragraphs 4.2.2-4.2.4).

Prefabricated Liners

5.6.10 Liners of corrugated or plain steel sheets, glass reinforced cement, or plastic may be attached to the intrados as permanent formwork; the gap is then filled with grout or pumped concrete. The lining will usually accommodate movement without cracking but bulging or deformation of the lining may occur. Corrosion of permanent steel components may also be a problem (see Section 4).

Concrete Slab

5.6.11 If the abutments cannot accept extra thrust then a concrete slab may be built above the crown of the arch, spanning onto the abutments. The slab will be remote from the arch, supported on compacted fill or weak concrete, and its upper surface may form the carriageway running surface. When this occurs it should be included in the inspection report; defects in concrete elements are discussed in detail in Section 3.

Strengthening by Incorporating Stainless Steel Bars

5.6.12   Stitching is used to tie masonry together. It is usually carried out by grouting deformed stainless steel bars into holes drilled through layers of masonry. Holes may be up to 40mm diameter and visible on the surface, located on each side of major cracks.

5.6.13   Various other methods have been developed for incorporating steel reinforcement into an arch to strengthen it. One such method involves encasing stainless steel bars in longitudinal and transverse rebates cut into the arch with a non-shrink, flexible adhesive. The appearance is that of additional mastic or mortar joints.

5.6.14   Increasing their size or adding stabilizing constructions can strengthen abutments, wing walls and retaining walls. An example would be the addition of buttress walls. Walls may have been rebuilt in reinforced concrete with masonry facing.

Tie Bars, Pattress Plates and Spreader Beams

5.6.15   The traditional way to restrain outward movement of spandrel walls is to install systems of tie bars and spreader beams. Many systems will have been installed early in the life of arch bridges and hence may have deteriorated significantly.

5.6.16   Tie bars may pass completely through a structure with a pattress plate at each end, or may be anchored within the structure with only a single visible pattress plate at the exposed end. Alternatively, a system of beams held in place by tie bars may replace individual pattress plates. Tie bars are installed hand tight as their function is only to restrain further movement and the pattress plates should have been bedded on mortar.

5.6.17   The tie rods are either installed by drilling through the arch or laid in a trench excavated from the road surface. Different methods of providing corrosion protection are available and currently the tie bar would be either stainless steel or have been grouted into a plastic sleeve, wrapped in waterproof tape, or galvanised. The pattress plates and other exposed steel would be painted.

5.6.18   Due to the age of many installations it is unlikely that the tie bars are stainless steel and their corrosion protection may be deficient. Therefore the tie bars may have suffered some corrosion whilst the external steelwork will often have suffered some corrosion (see Section 4).   Inspectors should note any evidence of continued movement of the spandrel walls that could indicate that the tie rods have failed. Failure may be due to corrosion or physical damage, which could have occurred during excavation of service trenches.

Widening of Arch Bridges

5.6.19   Old arch bridges will often have been widened to accommodate growth in traffic volumes. Many older arch bridges may have been widened using similar materials and to the same profile as the existing structure. However, as reinforced concrete is much cheaper than masonry it has been common practice to widen the barrel in concrete and to reserve the use of masonry for the spandrels and outer ring of voussoirs. Extensions consisting of steel or concrete beams with spans equal to those of the existing bridge are also common. An alternative method of widening is to provide a concrete slab across the bridge which cantilevers on both sides of the existing arch.

5.6.20   Joints to accommodate movement or settlement often separate the old bridge and extensions. These should be inspected for evidence of excessive movement. Evidence of damage or deterioration of masonry could indicate that the new structure is placing additional excessive stresses or strains on the old structure. The increase in the dead weight or an eccentric dead or live load pattern may adversely affect foundations that had been performing satisfactory for many years previous to the widening. Defects in concrete and steel elements are discussed in Sections 3 and 4.

6 Defects in Miscellaneous Materials

6.1 OVERVIEW

6.1.1 This section describes typical defects that occur in structures and structural elements fabricated of materials other than concrete, steel and masonry. It examines the most commonly used alternative materials, namely stay cables, cast iron, wrought iron, aluminium, timber and advanced fibre reinforced composites (glass and carbon fibre). The different defects are described with particular emphasis being placed on identification and likely causes.

6.2 STAY CABLE SYSTEMS

6.2.1 With the popularity and rapid growth in the use of cable-stayed bridges in the United Kingdom and worldwide, the need to reliably ascertain the condition of stay cables is taking on added significance. Defects that can occur on the cable members of cable-supported bridges include:

- Failure of the protective paint system

- Pitting corrosion

- Surface rust

- Section loss

- Fatigue cracking

- Impact damage

- Fire damage

6.2.2 The causes of these defects are similar to those affecting steel that are discussed in more detail in Section 4. Further guidance is contained in *Synthesis 353: Inspection and Maintenance of Bridge Stay Cable Systems – A Synthesis of Highway Practice* [45] and in the *Bridge Inspector's Reference Manual* [46].

6.3 CAST IRON

6.3.1 There are several types of cast iron, but that usually found in structures is known as grey, or flake graphite cast iron, from the dull grey appearance of a freshly fractured surface. In addition to main structural members, cast iron was often used in handrails and balustrades, and in trusses and lattices. It was rarely used in ties, as it is brittle and relatively weak in tension.

Atkins

6.3.2   Many of the defects which cast iron exhibits are similar to those of steel (see Section 4), although it should be recognised that the homogeneity and purity of cast iron is inferior to that of present day steels. It is important to distinguish between cast iron, wrought iron and steel as they have different properties. The structure records, appearance or structural forms provide useful indications to the type pf material, but the only definitive method of distinguishing between cast iron, wrought iron and steel is chemical analysis and metallographic examination. Further, there are a number of characteristics, which may assist in identifying the material and these are listed in Volume 1: Part F: Appendix F and described in the *Appraisal of Existing Iron and Steel Structures* [35].

Distortion

6.3.3   Being brittle, cast iron is likely to break rather than distort. Any significant distortions in cast iron members have usually been caused during the casting process.

Cracking and Fracture

6.3.4   Cracks are common defects in cast iron. They may be caused by a number of mechanisms:

- cooling of the metal after casting;

- restrained shrinkage, especially at re-entrant corners;

- 'cold spots' where an earlier splash of molten iron cooled and solidified without being absorbed by further molten iron;

- blowholes;

- water accumulating in hollow members and causing cracking when it freezes; and

- overloading, particularly in tension.

6.3.5   The first four mechanisms listed above derive from defects in the detailing or casting of the member. In many cases, therefore, the flaw will have been present from new. However, it may have developed further during the life of the structure. Some flaws will have been deemed acceptable at the time of construction, while others may have been concealed.

6.3.6   Blowholes are common defects in cast iron. They are caused by internally trapped air reaching the surface as the iron solidifies. These may act as points of initiation for subsequent cracking. Large castings may contain hidden voids.

6.3.7   Since cast iron is brittle and relatively weak, it is liable to crack or fracture when subjected to significant tensile loading.

6.3.8   Cracked or fractured cast iron is sometimes repaired using a proprietary cold stitching technique, which involves drilling holes across and along the line of the crack and driving in metal inserts, which locks the parts together.

Corrosion

6.3.9   During the casting process, silica in the moulding sand fuses and coats the surface of the casting, forming a barrier to oxygen. The cast surfaces of cast iron are therefore highly resistant to corrosion, but corrosion may still become significant because of the age of the structure. Cut surfaces, however, rust quickly in moist air.

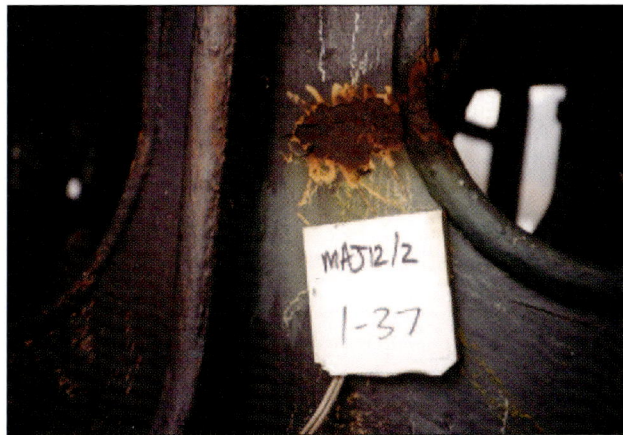

6.3.10   Cast iron is particularly susceptible to chloride-assisted corrosion from salt in sea water or road de-icing salt. Areas within splash zones, where the cast iron is alternately wetted and exposed to air, are particular cause for concern.

6.3.11   Cast iron can be subject to bi-metallic corrosion where it is in contact with dissimilar metals. This includes contact with galvanised (zinc-coated) steel and stainless steel (which is a nickel and chrome alloy).

Graphitisation

6.3.12   Corrosion of cast iron may occur by a process known as graphitisation. In this form of corrosion the iron is replaced by graphite with no significant change in volume but with a considerable loss of strength. Graphitisation occurs in salt or acidic water, or in ground bearing such water, and requires little oxygen. It therefore typically occurs below ground or water level.

6.3.13   The effect of graphitisation is to weaken the cast iron significantly, although this may not become evident until it is struck or loaded abruptly. Graphitisation can be recognised by a soft, black and blistered surface that can be easily broken away with a knife or other hard and sharp instrument.

Impact and Fire Damage

6.3.14 Cast iron is an inherently brittle material, with relatively poor resistance to impact. It is therefore likely to shatter under substantial dynamic or shock loadings. By contrast, it has a relatively high resistance to fatigue.

6.3.15 Cast iron does not melt at the temperatures normally encountered in a fire and temperatures up to at least 400°C do not adversely affect its basic strength. However its brittleness means that it tends to shatter when subjected to thermal shock or restraint against thermal movement. Thermal shock may occur when cold water is applied during or following a fire. As this is the normal technique used in extinguishing fires, it is a major cause for concern with this material.

6.4 WROUGHT IRON

6.4.1 Wrought iron is found in older bridges, with larger elements built up from relatively small components, connected together using wrought iron rivets and bolts. Wrought iron was also commonly used for cables and forged links. Other applications include trusses and lattices, handrails and balustrades.

6.4.2 The features that may assist identification of cast iron, wrought iron and steel are summarised in Volume 1: Part F: Appendix F and described in the *Appraisal of Existing Iron and Steel Structures* [35]. Not all of the features listed are necessarily found together.

6.4.3 Many of the defects which wrought iron exhibits are similar to those of steel (see Section 4).

Distortion

6.4.4 Wrought iron behaves elastically up to its yield point; further extension or deflection will cause permanent plastic deformation. Typical causes of visible distortions include:

- Overloading causing a permanent set or buckling (elastic or inelastic);

- Impact or fire damage;

- Deliberate shaping or profiling; and

- Fabrication defects, such as out-of-true rolling.

6.4.5 Section 4 describes causes of deformation and distortion of steel members and these are also applicable to wrought iron.

Cracking

6.4.6 Like steel, wrought iron is a ductile material that can undergo substantial deformation before fracture, which results in characteristic 'necking' at the fracture zone. Cracks are potential causes of complete fracture and usually occur at connections and changes in section. The most common causes are fatigue and poor detailing practices that create high stress concentrations. Elements that have been modified since initial construction are also potential problem areas. Fracture of any member, bolt, rivet or weld is serious and can have important structural implications.

6.4.7   Fatigue is likely to occur in highly stressed or reversibly stressed components. However, wrought iron has a relatively high resistance to fatigue. Fatigue cracks will begin at fabrication flaws or at sites with high surface stress concentrations such as weld toes, irregular cut edges and flame cut edges. Once initiated there is the potential for them to grow until a final failure occurs by brittle fracture or buckling. The initial fabrication flaws may not be significantly large to be detected by eye. Cracks may range in width from hairline to sufficient to transmit light through the members. The variations in length and depth are of a similar magnitude.

6.4.8   Cracks may also be present in welds because of poor welding techniques. They are often associated with separation of the welded members.

6.4.9   Punching of holes through wrought iron members can cause local work hardening with consequent loss of ductility. This may lead to the development of cracks radiating from the holes.

6.4.10   If cracks are detected it is likely that they will be repeated in similar details within the structure. Cracking and fracture may occur as a result of impact damage. The inspector should be aware that these defects may occur several metres away from the impact site. The location, extent and width of cracks and fractures should be recorded. In addition, deep pits, nicks or other defects that may cause stress concentrations should be measured and recorded.

Wear

6.4.11   Evidence of wear may be found in moving parts such as pins in trusses and joints. Wear may occur if the joint or structure is experiencing excessive movement or vibration, and may be exacerbated by a lack of maintenance or by the specification or installation of an unsuitable material for the part.

Corrosion

6.4.12   Although corrosion of wrought iron is relatively slow, it may reach significant proportions because of the age of the structure. In general the corrosion products of wrought iron cause expansion and can be readily detected.

6.4.13   Wrought iron is susceptible to chloride-assisted corrosion, typically from salt (sea) water or road de-icing salt. It is especially vulnerable to this form of attack within the splash zone, where the metal is alternately wetted and exposed to air.

6.4.14   Wrought iron may also be subject to bi-metallic corrosion where it is in contact with dissimilar metals. This includes contact with galvanised (zinc-coated) steel and stainless steel (a nickel and chrome alloy).

Delamination

6.4.15   Delamination is the separation of the material into layers within the thickness of the member in a direction parallel to the surface.

6.4.16   Delamination of wrought iron is caused by corrosion occurring along lines of slag inclusions, which, due to the rolling process, run parallel to the longitudinal axis of the element. The slag is unaffected by corrosion, so the expansive forces generated cause the rust to detach in flaky sheets. Corrosion leading to delamination occurs within the element and therefore deterioration may be greater than is apparent at the surface. An experienced inspector can obtain useful qualitative information by tapping with a hammer.

6.4.17   This same phenomenon causes wrought iron rods to rust into 'cake wedges', often with further delamination on the perimeter.

Impact and Fire Damage

6.4.18   Wrought iron, being a ductile material, is likely to deform or distort when struck. Impact damage to a wrought iron structure is usually obvious; it can vary from scoring of the paint or metal surface to deformation of an element. In severe cases the damage may render the member incapable of carrying load.

6.4.19   Like steel, wrought iron begins to lose strength above 200°C. The effect of heating wrought iron is to anneal it, reducing its strength to that of the source metal. Therefore a wrought iron member subjected to a fire may initially fail by its ability to sustain load being impaired. Initial modes of failure are likely to be column buckling or excessive deflection for beams. The temperature at which failure occurs will depend on the loading that the member is carrying, its support conditions, and the temperature gradient through the cross section.

6.4.20   Wrought iron has a high coefficient of expansion and will expand considerably during a fire. The expansion will normally be far greater than that allowed for in the design and, may cause damage either close to the fire or at locations considerably remote from the heat source.

6.4.21   In a severe fire unprotected wrought iron members will distort and will not be suitable for reuse. In a less severe fire only a few members may be affected and the straightness, distortion and mechanical properties of these should be checked. Bolted or riveted connections often fail during a fire, through shear or tensile failure or thread stripping. If there has been yielding during the fire, severe weakening of connections and sections is a possibility that will need to be considered.

6.4.22   Protective coating systems will generally be damaged by fire, with blistering and flaking likely. Where damage appears to be serious, specialist personnel, who can advise on remedial treatments, should undertake the inspection.

6.4.23   Further information on the assessment of the effects of fire is given in *Appraisal of Existing Structures* [28] and *Appraisal of Existing Iron and Steel Structures* [35].

6.5   ALUMINIUM

6.5.1   A wide range of aluminium alloys has been developed with the individual properties adjusted for a specific purpose. However a number of characteristics are common to all these metals:

- high strength/density ratio, making them ideally suited to applications such as long-span structures where it is important to save weight;

- good general corrosion resistance, with associated low maintenance costs;

- readily formed and extruded into complex shapes.

Atkins

6.5.2   Aluminium elements may be joined by welding, bolting or riveting. The most frequent use of aluminium is for vehicle and pedestrian restraint systems, but it may occasionally be encountered or used for other structural members, such as sign gantry access walkways.

Cracking

6.5.3   Higher strength aluminium alloys are susceptible to stress-corrosion cracking, which can occur at stresses well below the yield stress. Such cracking gives the impression that the material is brittle, since it propagates without attendant plastic deformation. However, testing will continue to confirm that the alloy is ductile. Stress-corrosion cracking can occur as a result of residual stresses arising from manufacturing processes, which include quenching followed by machining [47].

Corrosion

6.5.4   Exposure of aluminium to road salts produces a thin oxide film on the surface, which forms a protective barrier against further corrosion. However rusting may be observed occasionally, from oxidisation of iron impurities within the alloy. Some aluminium alloys are susceptible to deterioration when exposed to high concentrations of atmospheric and other environmental pollutants. These cause a change in the chemical composition, leading to corrosion or pitting.

6.5.5   Aluminium alloys in contact with other metals such as steel, can cause localised corrosion due to galvanic action. Direct contact with concrete, cement mortars or other alkalis should be avoided, and separation films such as bitumen paint should be provided. Aluminium restraint system posts may react with the base plate mortar causing the base plate to crack and in some cases produce reaction gas within the posts causing these to bulge.

Fire damage

6.5.6   Aluminium melts between 600°C and 660°C, and suffers a significant loss of strength above 200°C. However it has high thermal conductivity, which enables it to dissipate heat to other elements quickly and therefore its temperature increases relatively slowly.

6.6   TIMBER

6.6.1   Timber is no longer used for the main members of vehicular structures in the United Kingdom. However it is used for the decking of some movable bridges, and there are also many timber footbridges. The two main problems with timber is decay and insect attack.

Decay

6.6.2   The commonest form of deterioration of timber is decay or rot. The principal forms of decay are generally termed dry rot and wet rot and both are caused by fungi. In the presence of moisture, the fungi grow by consuming the constituents of timber, leaving it soft and lacking in strength.

6.6.3   Dry rot, so called because affected timber looks dry, needs a source of moisture in order to become established but can spread into dry wood. Timber attacked by dry rot looks dry and brittle, developing deep cracks across the grain and breaking into brick-shaped pieces.

6.6.4   Wet rot can only attack wood with high moisture content: it will not spread into dry wood. Affected wood becomes soft, pulpy and wet, with the structure of the wood progressively breaking down.

6.6.5   Dampness, rust stains and vegetation growing from crevices are all signs that the timber may be decaying. Decay is primarily characterised by softness of the timber. Areas that are particularly susceptible to decay are those that are in contact with both water and air. For example:

- Parts in contact with the ground;

- Places where dirt, debris and water collect and vegetation grows (for example, in corners, between boards, close to kerbs or running strips, near deck ends and in truss joints);

- Around fixings. Water can sometimes penetrate timber sections through holes for fixings. This type of decay can be difficult to detect;

- Around splits in the timber. Splits are common in timber but they will only lead to decay if water can accumulate in them, for example, in splits in the tops of horizontal members.

6.6.6   Chemical treatment to prevent decay will not penetrate to the middle of the timber so even if the outside is sound, decay may still be occurring below the surface. Indications of hidden decay are:

- Water stains on the timber;

- Soft areas on the surface;

- Soft areas that are cracked into small blocks, which can be a sign of severe decay;

- A flat fungus growing on the surface of the timber, which is a sign of severe decay deep inside the timber.

Insect Attack

6.6.7 Insect attack can occur anywhere on a structure and can seriously weaken it. Insect holes usually have dust in them or near them. A few small holes (less than 5mm diameter) are not usually serious but if there are much larger holes, the problem is serious.

6.6.8 Many insects attack timber. The most damaging are the furniture beetles or 'woodworm', which bore holes of 2mm diameter, and forest longhorn beetles, which make larger holes up to 10mm diameter. In warmer countries termites are serious pests. These subterranean insects live in large colonies or nests and attack timber by boring into it where it is contact with the ground. Fortunately termites are very rare in the United Kingdom.

6.6.9 In salt water, teredo worms may attack any timber below the high tide level. These molluscs make large holes, up to 25mm diameter, and can cause significant damage. All timber structures at risk from this form of attack should, therefore, be checked.

6.6.10 Gribbles, a form of crustacean like a wood louse, also attack timber in contact with salt water. They bore myriads of small holes in the surface of the timber, up to 50mm deep. This activity weakens the surface, which is then eroded by abrasion and wave action, which in turn encourages the gribble to bore deeper. There is therefore a progressive reduction in timber section.

Splitting or Cracking

6.6.11 Splitting commonly occurs in timber as it dries out, and does not necessarily seriously affect the structure. However the following types of split should be treated with concern:

- Splits across the grain of the wood. These may be caused by overloading and will reduce the load-bearing capacity of the timber element;

- Splits that are orientated so that water can accumulate in them. These will lead to decay;

- Splits around connections such as bolt holes;

- Splits that are increasing in size.

6.6.12 Overstressing of a timber element may cause splintering, cracking or even shattering of the timber. This is often associated with sagging, buckling or other deformity.

6.6.13 Exposed timber surfaces are affected by weathering, the surface becoming rough and slightly corrugated. In advanced stages of weathering deep cracks appear in the timber. These can allow moisture into the heart of the element, leading to decay.

Defective Joints or Laminations

6.6.14 Loose or damaged joints can seriously affect the strength of the structure, and in some cases can also cause serious accidents.

6.6.15   Bolted joints can work loose as a result of shrinkage in timber or vibration from traffic. Steel connection components, such as plates, bolts, pins and cables, may also be subject to corrosion, particularly in saline environments. Additionally, oak when wet produces acids that can corrode ferrous connectors.

6.6.16   Some timber truss bridges have 'nail laminated' decks, in which the timbers are all nailed together with no gaps. The strength of these bridges relies largely on the connection between the deck and the truss. This connection must be checked from below, to see that the deck is tightly nailed to the top member of the truss.

6.6.17   In glued-laminated timber elements, separation of the laminations may occur due to degradation of the adhesive. Delamination may be seen at the edges of the timber, where the edges of laminations are exposed, or on top or bottom surfaces as blistering.

Fire Damage

6.6.18   Timber is naturally combustible and therefore will normally only survive where fires are either localised or rapidly extinguished, unless sections are of significant mass. The response of timber to heat is summarised in Table D.5.

6.6.19   Any charred timber (i.e. 'Blackened' in Table D.5) must be assumed to have lost all it's strength, but timber beneath the charred layer may be assumed to have suffered no significant loss of strength.

Table D.5 – Effects of High Temperatures on Timber	
Temperature	**Response**
120°C - 150°C	'Browns'
200°C - 250°C	'Blackens'
About 300°C	Evolves combustible vapours
400°C - 450°C (300°C if a flame is present)	Surface ignition and charring

6.7   ADVANCED COMPOSITES

6.7.1   Advanced composites are now used as primary load-carrying elements of structures, which has permitted the construction of footbridges and bridge enclosure systems, or as composite ropes and cables. Some early advanced composites were purely used in non structural roles, e.g. fascia panels. Advanced composites are now also used in repair applications, where thin plates, usually of carbon fibre, are bonded onto existing structures to provide extra strength (see paragraphs 4.6.24-4.6.25).

6.7.2   The principal reasons driving the development of advanced composites are their anticipated long-term corrosion and fatigue resistance and their very high strength/weight ratio.

Surface Erosion

6.7.3   Surface erosion may be due to a number of causes. Some types of fibres may be susceptible to hydrolytic attack by water or some chemicals. The alkaline environment of concrete will attack bare fibres of glass or aramid. Some types of advanced composite materials may be susceptible to attack by ultraviolet or infrared radiation. Depending on the type of material, this may cause crazing, resin-fibre interface failure, colour fading or loss of structural properties. However the protection provided by resins or sheathing materials is generally sufficient to prevent direct contact and subsequent attack. It is therefore important that the protective layer remains intact and is not damaged in the course of the inspection process.

6.7.4   A visual inspection of advanced composite elements should be carried out to check for any surface erosion or loss of section. The internal fibres and the resin act compositely to carry loads, therefore any loss will adversely affect the capacity of the section.

Delamination

6.7.5   All surfaces and edges should be inspected for signs of delamination. This may be near the composite surface or within the body of the material, and is generally denoted by a localised change in colour or opacity. A 'pat' test (see Volume 1: Part E: Section 9) may also be used, which produces a change in sound in delaminated areas. This is analogous to the hammer test for delaminated concrete.

6.7.6   The presence of moss growing on an advanced composite element may indicate a high moisture content in the material. This is likely to signify a loss of strength and possible delamination.

Physical Damage

6.7.7   Sheets and thin elements of advanced composite materials may be vulnerable to impact, particularly from concentrated point loads. Such impact could result in plastic deformation or complete rupture.

6.7.8   Bolted joints should be inspected for evidence of failure around points of stress concentration. Adhesive joints should be inspected for evidence of peeling or shearing. Ropes should be inspected for evidence of broken fibres at or near the end fittings.

Fire Damage

6.7.9   Glass and carbon fibres are resistant to quite high temperatures, but aramid starts to lose strength at about 200°C. However resins generally lose strength at much lower temperatures. Advanced composite elements, which show signs of having been subjected to fire or other intense heat source, should be assumed to have suffered significant loss of strength.

7 References for Part D

1. *BD 63 Inspection of Highway Structures*, DMRB 3.1.4, TSO.

2. *BD 21 The Assessment of Highway Bridges and Structures*, DMRB 3.4.3, TSO.

3. *BA 16 The Assessment of Highway Bridges and Structures*, DMRB 3.4.4, TSO.

4. *Management of Highway Structures: Code of Practice*, TSO, 2005.

5. *BA 79 The Management of Sub-standard Highway Structures*, DMRB 3.4.18, TSO.

6. *Road Geometry*, DMRB Volume 6, TSO.

7. *TD 27 Cross-Sections and Headrooms*, DMRB 6.1.2, TSO.

8. *Manual on Scour at Bridges and Other Hydraulic Structures*, May RWP, Ackers J & Kirby A, CIRIA C551, 2002.

9. *BA 74 Assessment of Scour at Highway Bridges*, DMRB 3.4.21, TSO.

10. *Bridge Scour*, Bruce WM & Stephen EC, Water Resources Publications, 1999.

11. *Application Guide 29: Practical Guide to the Use of Bridge Expansion Joints*, Barnard CP & Cuninghame JR, Transport Research Laboratory, 1997.

12. *BA 87 Management of Corrugated Steel Buried Structures*, DMRB 3.3.4, TSO.

13. *BA 88 Management of Buried Concrete Box Structures*, DMRB 3.3.5, TSO.

14. *Seawall Design*, Thomas RS & Hall B, Butterworths, London, 1992.

15. *BA 68 Crib Retaining Walls*, DMRB 2.1.4, TSO.

16. *Piling Handbook*, 8th Edition, Arcelor Group, Luxemburg, 2005.

17. *Technical Guide 2: Guide to Testing and Monitoring the Durability of Concrete*, Concrete Bridge Development Group, Concrete Society, Slough, 2002.

18. *Technical Report 22: Non-Structural Cracks in Concrete*, 3rd Edition, Concrete Society, Slough, 1992.

19. *BA 24 Early Thermal Cracking in Concrete*, DMRB 1.3.14, TSO.

20. *Handbook of Analytical Techniques in Concrete Science and Technology*, Ramachandran VS & Beaudoin JJ, Noyes Publications, New Jersey, USA, 2001.

21. *BA 35 Inspection and Repair of Concrete Highway Structures*, DMRB 3.3.2, TSO.

22. *The Manual of Contract Documents for Highway Works: Volume 1: Specification for Highway Works*, TSO.

23. *The Performance of Concrete in Bridges: A Survey of 200 Highway Bridges*, Wallbank EJ, HMSO, London, 1989.

24. *Collapse of Ynys-y-Gwas Bridge, Glamorgan*, Woodward RJ & Williams FW, Proc. Inst. Civ. Eng., Part 1 August 1988, pp. 635-689.

25. *The Strengthening of an Externally Post-tensioned Structure*, Robson A, Craig JM & Taylor AJ, in Bridge Modification 2: Stronger and Safer Bridges, Pritchard B (Ed.), T Telford, London, 1997.

26. *The Thaumasite Form of Sulfate Attack: Risks, Diagnosis, Remedial Works and Guidance on New Construction*, Thaumasite Expert Group, Department of the Environment, Transport and the Regions, London, 1999.

27. *LR117 - Frost Scaling on Concrete Roads*, Franklin RE, Transport Research Laboratory, 1967.

28. *Appraisal of Existing Structures*, 3rd Edition, Institution of Structural Engineers, London, 2007.

29. *Technical Report 33: Assessment and Repair of Fire-damaged Concrete Structures*, Concrete Society, Slough, 1990.

30. *TN118 - Spalling of Concrete in Fires*, Malhotra HL, CIRIA, London, 1984.

31. *BD 43 The Impregnation of Reinforced and Pre-Stressed Concrete Highway Structures using Hydrophobic Pore-Lining Impregnants*, DMRB 2.4.2, TSO.

32. *BA 83 Cathodic Protection for Use in Reinforced Concrete Highway Structures*, DMRB 3.3.3, TSO.

33. *BS 5400: Part 10 Steel, Concrete and Composite Bridges, Code of Practice for Fatigue*, British Standards Instirution, 1980.

34. *BD 7 Weathering Steel for Highway Structures*, DMRB 2.3.8, TSO.

35. *Appraisal of Existing Iron and Steel Structures*, Steel Construction Institute, Bussell M, Ascot, 1997.

36. *Introduction to the Welding of Structural Steelwork*, 3rd Edition, Pratt JL, Steel Construction Institute, 1989.

37. *BD 87 Maintenance Painting of Steelwork*, DRMB 3.2.2, TSO.

38. *BD 67 Enclosure of Bridges*, DMRB 2.2.7, TSO.

39. *BA 67 Enclosure of Bridges*, DMRB 2.2.8, TSO.

40. *PR1: The Durability of Corrugated Steel Buried Structures*, Brady KC & McMahon W, Transport Research Laboratory, 1993.

41. *BD 12 Design of Corrugated Steel Buried Structures with Spans Greater than 0.9 Metres and up to 8.0 Metres*, DMRB 2.2.6, TSO.

42. *The Maintenance of Brick and Stone Masonry Structures*, Sowden AM (Ed.), E & F N Spon, London, 1990.

43. *Report 204: A Guide to Repair and Strengthening of Masonry Arch Highway Bridges*, Page J, Transport Research Laboratory, 1996.

44. *Masonry Arch Bridges: Condition Appraisal and Remedial Treatment*, CIRIA C656, 2006.

45. *Synthesis 353: Inspection and Maintenance of Bridge Stay Cable Systems – A synthesis of Highway Practice*, National Cooperative Highway Research Programme, Transportation Research Board, Washington D.C., 2005.

46. *Bridge Inspector's Reference Manual*, Publication No. FHWA NHI 03-001 United States Department of Transportation, Washington, D.C., 2002.

47. *Corrosion: Volume 1 (Metal/Environment Reactions) and Volume 2 (Corrosion Control)*, 3rd Edition, Shreir LL, Sharman R & Burstein T, Butterworth-Heinemann, London, 1994.

Part E
Investigation and Testing

This Part of the Inspection Manual summarises a wide range of testing techniques that are currently available for highway structures. This Part is intended to raise awareness of the range of testing techniques available, and, for each technique, to summarise its purpose, what it involves, the outputs and any relevant considerations. Additional information sources are referenced and these should be consulted when developing a testing programme. However, it is important to remember that testing techniques are continuously being developed and, whilst some techniques will be superseded, other new tests are likely to be introduced.

1 Introduction

1.1 OVERVIEW

1.1.1 This Part of the Manual summarises the wide range of testing techniques that are currently available for highway structures. It is intended to raise awareness of and provide high-level guidance on the testing techniques available and, for each technique, to summarise its purpose, what it involves, the information provided and any relevant considerations. Where possible, documents that contain more comprehensive information on the testing technique have been referenced and these should be consulted when developing a testing programme.

1.1.2 In using this Part of the Manual it is important to bear in mind that testing techniques are continuously being developed and, whilst some techniques will be superseded, other new tests are likely to be introduced. Every effort has been made to check that the guidance provided here is relevant at the time of publication.

1.2 PURPOSE OF TESTING

1.2.1 The overall purpose of testing is to complement inspections by providing more in-depth and targeted information, normally on pre-defined criteria, that improves understanding and/or diagnosis. For example, understanding can be improved by obtaining information on material properties to assist structural assessment (e.g. yield strength of steel or compressive strength of concrete) while diagnosis can be improved by obtaining information on deterioration mechanisms to assist maintenance planning (e.g. carbonation and chloride ion ingress).

1.3 NEED FOR TESTING

1.3.1 The need for testing is normally identified during or following a General, Principal or Special Inspection or is required to support a structural assessment. In some circumstances, defects or deterioration encountered on a particular structure or component may instigate testing on similar structures/components. Examples of situations when the need for testing generally arises include:

- Investigation, and if possible quantification, of the severity and extent of a defect or damage.

- Investigation of the cause of deterioration.

- Investigation of the rate of deterioration.

- Investigation of the risk of or potential for deterioration, for example, assessing for harmful agents in a material.

- Determining the type of maintenance required and the extent of any repairs.

- Determining material characteristics, e.g. physical, mechanical and chemical.

- Providing information for a structural assessment.

1.3.2 Tests may be carried out once or repeated periodically to monitor changes in condition and performance. The Supervising Engineer, and inspection staff, should have a sound understanding of the testing techniques available and be able to identify when testing may prove beneficial. Many testing techniques require specialist equipment and training and can be expensive; as such expert advice should be sought before adopting them.

1.3.3 Specialised testing should be carried out by testing firms or laboratories fulfilling the relevant requirements of the highway authority. These will generally be those operating recognised quality assurance procedures for the relevant tests, such as those in accordance with *BS EN ISO 9000* [1] or the United Kingdom Accreditation System (UKAS). Where testing is to be undertaken by a third party, interpretation of the results should remain the ultimate responsibility of the Supervising Engineer, as this requires knowledge of factors other than just the results obtained.

1.4 LAYOUT OF PART E

1.4.1 The layout of this part of the Manual is summarised in Table E.1.

Table E.1 – Layout of Part E	
Section	**Summary of Contents**
2. The Testing Process	Provides a summary of a considered approach to assessing testing needs and developing a testing programme.
3. Summary of Testing Techniques	Summaries in tables of all the testing techniques presented in this part of the Manual giving their main characteristics. There is a table for general tests and one table for each material type, i.e. concrete, metal, masonry, timber and advanced composites.
4. General Testing Techniques	A large number of testing techniques are relevant to more than one material or structure type. This section presents summaries of these tests in order to prevent duplication in the following sections.
5. Tests on Concrete	Presents summaries of tests that are specific to concrete.
6. Tests on Metal	Presents summaries of tests that are specific to metal.
7. Tests on Masonry	Presents summaries of tests that are specific to masonry.
8. Tests on Timber	Presents summaries of tests that are specific to timber.
9. Tests on Advanced Composites	Presents summaries of tests that are specific to advanced composites.

1.4.2 The summaries provided Sections 4 to 9 describe, for each testing technique, the purpose (i.e. what is the test identifying or measuring?), the approach used (i.e. equipment involved and how it is used), the results, and any relevant considerations. The testing techniques are grouped in categories which relate to the purpose of the test, the categories are:

- ***Structural Arrangement and Hidden Defects*** – testing techniques used to establish or confirm structural/element arrangement and/or to detect hidden/internal defects or damage.

- **Distortion and Movement** – testing techniques used to measure and/or monitor distortion and movement of the structure or components.

- **Material Properties** – testing techniques used to determine the physical and mechanical material properties.

- **Deterioration Activity** – testing techniques used to detect deterioration activity.

- **Deterioration Rate** – testing techniques used to indicate or measure the rate of deterioration.

- **Deterioration Cause or Potential** – testing techniques used to assess the cause of or potential for deterioration (including tests for material chemical properties).

1.4.3   Frequently, the purpose of the test is to establish the severity and extent of a particular defect, for example, delamination survey, half-cell potential survey, and crack measurement. Therefore, a category that identifies testing techniques based on the severity and/or extent of defects or damage is not used because this is implicitly covered by the aforementioned categories.

1.4.4   In some instances there is a degree of overlap between the aforementioned categories, i.e. a testing technique applies to more than one category. In such instances the testing technique is described under the first relevant heading and subsequent occurrences of the testing technique referred to this description.

2 The Testing Process

2.1 GENERAL

2.1.1 Testing can be expensive when compared to inspections. It is therefore important to take a considered approach to planning and reviewing the testing programme (Section 2.2), identifying appropriate sample sizes and techniques (Section 2.3), recording and reporting results (Section 2.4), and evaluating results (Section 2.5). The following guidance aligns with that provided in *Management of Highway Structures: A Code of Practice* [2].

2.2 PLANNING AND REVIEWING TESTING

2.2.1 Good practice is to establish a formal process for assessing testing needs and identifying effective solutions. A suggested approach for developing a testing programme is outlined below, the four steps in the approach are:

- *Setting the objectives of testing* – clearly define the objectives of the testing.

- *Identification of testing options* – identify the alternative testing options.

- *Appraisal of testing options* – appraise and compare the identified testing options and select the most effective solution.

- *Review testing* – review and assess the suitability of the selected programme as the testing is undertaken.

2.2.2 Further advice on setting objectives, identifying options, appraising options and reviewing progress is provided in the following. The time and effort expended on the following should be appropriate to the scale and complexity of the problem at hand. For example, if a certain testing technique is accepted as providing good value for money and beneficial information then in-depth appraisal is not necessary, equally if it is a low cost/skill testing activity, that can be readily combined with a scheduled inspection for no/negligible extra cost, then in-depth appraisal is not necessary.

Setting the Objectives of Testing

2.2.3 The reasons for, and the objectives of, testing should be clearly understood, defined and documented from the outset, i.e. before devising a programme of tests. The objectives are likely to include the identification of the information required from the testing and how it will be used in the management of the structure. For example, it should be clear whether the testing is for structural strength assessment, for ongoing condition monitoring, for initial diagnosis of a problem (as a precursor to further testing), or for the development of the appropriate repair solution.

2.2.4 At this stage, appropriate information about the structure and/or defect should be compiled and reviewed in order to make an initial assessment of the need for testing. This information may include the Structure File/Maintenance Manual, drawings, inspection reports, etc. This may identify specific characteristics of the structure (e.g. structural form, material, maintenance history) that will assist/influence the development of the testing programme, or in some instances, may indicate that testing is not required.

Identification of Testing Options

2.2.5   Testing techniques are frequently identified on the basis of previous experience and engineering judgement while taking account of available guidance. Identification should also take into account advice from specialists, where appropriate. The identification exercise should compile the information required to carry out the subsequent appraisal, such as, costs, reliability, accuracy, information/results provided, etc.

Appraisal of Testing Options

2.2.6   Testing should not be considered in isolation but should be regarded as one facet of the collection of information required for the management of a highway structure. Testing should complement the drawings, inspection records, previous test data, structural form, material type and the history of the structure. It is therefore normal for testing programmes to include a range of tests. Devising a test programme is inevitably a compromise between costs and obtaining an adequate set of information. The following should be considered when appraising the testing options:

- Effective combination of testing techniques, e.g. ducts in post-tensioned concrete construction may be located by specialist probes or radar in preparation for intrusive drilling and sampling. Another example of the benefit of using different techniques in combination is where a simple test may be used to survey the whole structure or large areas of a structure, followed by more refined testing of a representative sample of locations having a particular characteristic, e.g. high chloride content.

- The risk of damage to the structure from destructive (sampling or intrusive) investigations. The removal of samples or formation of access holes for inspection and the subsequent repair should be carefully specified and supervised to avoid potentially serious damage to the structure or the creation of points of weakness that are vulnerable to deterioration in the future. For example, where information is required on strength of the materials, the following important points should be considered:

    i.   the removal of samples for strength tests may permanently weaken the structure, particularly in fatigue of steel structures;

    ii.   it is not always practicable to make sufficient tests to provide adequate data on the variability of strength, but a few tests can be useful in giving a broad indication of the quality of materials; and

    iii. the likely value of the results in relation to possible damage to the structure and whether indirect techniques of assessing strength might be more appropriate.

- The need for calibration and return visits to site. Testing techniques generally require calibration and the results may not always be definitive. Consequently a return to site after analysis of early test data may be required.

2.2.7   Experience has shown there is often a wide variation in the number and type of tests selected for the same objective. This variation results from a lack of guidance and the natural tendency to base specifications on previous contracts, which may lead to the inclusion of tests that are not strictly necessary. Care should be taken to avoid 'a shopping list' approach whereby, to obtain as much information as possible, all the techniques available for a particular problem are applied. This approach results in unwarranted expenditure that provides a surplus of data, much of which adds little to the investigation in hand. A problem solving approach, that assesses costs and benefits, should be adopted and the selection of techniques should be restricted to those that can add value to the investigation.

2.2.8   Care should always be taken when preparing a specification for testing work, and during the site work and interpretation stages, to check that all factors likely to influence the interpretation of results are understood and allowed for. Prior approval should be obtained from the Supervising Engineer when destructive testing is required.

Reviewing Testing

2.2.9   A cautious approach should be adopted for the implementation of testing. Where appropriate, it is advisable to test a small, but representative, area/part of a structure initially and use this to assess the usefulness of the technique and results. Testing programmes can only be provisional, and may require amendment as a result of initial testing and interpretation. Staged testing, permitting interpretation of results between each stage, should be considered in all cases. However, the use of staged testing could lead to significantly increased access costs and this should be taken into account in these considerations.

2.3   SAMPLING

2.3.1   Some testing techniques are related to a specific feature or defect, for example Demec gauges to monitor crack growth or thermograph to detect voids in a beam, while others can be readily and cost effectively applied to larger areas of a structure, for example delamination survey to detect separation of concrete cover or a covermeter survey to locate reinforcing bars. However, in many circumstances a sampling approach is more appropriate, or required, because:

- The resources required by the testing technique (labour, plant, material and time) prohibit comprehensive or extensive coverage.

- The testing technique is destructive thereby limiting the number of samples that can be removed, for example, material cores for laboratory testing.

- Comprehensive testing is not required because a sample will provide sufficiently accurate information on the issue being investigated.

2.3.2   The identification of an appropriate sample size normally requires a trade off between the above and consideration of the expected variability of the test criteria. For example, less concrete cores are required for compression strength analysis if the variability across an element is considered to be low; this information may be know from construction records, e.g. slump test, concrete cube test, etc. However, the sample should be statistically significant and as such it is beneficial to have knowledge of basic statistical techniques when determining a sampling plan and interpreting the sample data [13].

2.3.3　All sampling and testing should be carried out by testing firms or laboratories operating recognised quality assurance procedures, such as those in accordance with *BS EN ISO 9000* [1] or the United Kingdom Accreditation System (UKAS).

2.4　RECORDING AND REPORTING OF TEST RESULTS

2.4.1　A report, containing full details of the test, should be provided by those undertaking the tests. The report should be clearly set out and follow a logical sequence; it should include, but not be restricted to:

- The objectives of the testing.

- Name and full contact details of the organisation and any sub-contractors undertaking the testing, and where appropriate, include names of relevant personnel, their qualifications and experience.

- The date, time and prevailing weather conditions, including temperature, during the test.

- A full description of the equipment used including details of calibration, accuracy, reliability and repeatability.

- A copy of the technique statement which should include, but not be restricted to, methodology, location of measurements, location referencing system, grid size (where appropriate), number of measurements at each location and a copy of any risk assessment (where appropriate). The location referencing system should be in sufficient detail to clearly identify the test locations and to enable the tests to be repeated at the same location, if required, in the future.

- The raw and, if appropriate, filtered test data.

- An interpretation of the data.

- Any other observations that are considered relevant.

- Videos, photos, diagrams and sketches, as appropriate, that support the above points and help the reader understand details of the testing.

2.4.2　Results should be presented in a format that can be clearly understood and related to the construction form, material and other pertinent characteristics of the structure. The interpretation of test results should use the data obtained from testing and any background material available. Where appropriate a diagrammatic interpretation of the results should be provided.

2.5　EVALUATION OF TEST RESULTS

2.5.1　All the information relevant to the problem under investigation should be assembled before attempting to evaluate the results of testing. The information to hand should include details of the structure and results of inspections and any previous testing.

2.5.2   Some tests provide factual data, e.g. properties of materials taken from the structure, depth of cover to reinforcement and paint thickness. Other tests require specialist interpretation, e.g. electrode potential measurements or the results of radar surveys. The specialists making the interpretation should be provided with any additional information that may assist them, e.g. structural drawings showing the layout of the reinforcement helps the interpretation of radar surveys.

2.5.3   When specialists are employed to carry out testing, either on site or in the laboratory, they may be best placed to also provide a practical interpretation of the results. However, the ultimate responsibility for the interpretation of tests and any consequent maintenance action remains with the Supervising Engineer, as the final assessment requires knowledge of other factors, e.g. local factors and maintenance history.

2.5.4   It may sometimes appear desirable to undertake further testing before coming to a decision on what action to take. Careful consideration should be given as to whether additional testing will be of benefit. Testing does not always provide quantitative information on structural condition and the interpretation nearly always includes an element of engineering judgment.

3 Summary of Testing Techniques

3.1 GENERAL

3.1.1   A wide range of testing techniques is presented in Sections 4 to 9. Section 4 presents testing techniques that are either relevant to more than one material type or not dependent on material type, whereas Sections 5 to 9 present testing techniques that are specific to material types (i.e. concrete, metal, masonry, timber and advanced composites). All sections are sub-divided under headings that relate to the purpose of the testing technique (see paragraph 1.4.2). The purpose of this section is to present the testing techniques in a series of tables (see Table E.2-Table E.7) that can be used to quickly identify testing techniques, and also set out the key characteristics of the tests using the symbols defined below:

Aston University

- **ND** – Non-destructive testing technique

- **De** – Destructive testing technique

- **Lb** – Includes laboratory tests

- **Op** – Specialist/skilled operators required

- **Sp** – Specialist interpretation of results required

3.1.2   In Table E.2-Table E.7, 'destructive' refers purely to the anticipated effect that testing would have on a structure, e.g. a trial pit adjacent to the structure is non-destructive as the structure is not damaged.

3.2 GENERAL TESTS

3.2.1 Table E.2 summarises testing techniques that are either relevant to more than one material type or not dependent on material type, and gives their main characteristics.

Table E.2 – General Tests			
Test Category	**Testing technique**	**Paragraph(s)**	**Test Characteristics**
Structural Arrangement and Hidden Defects	Observation and Simple Investigation	4.2.1	ND
	Delamination Survey	4.2.2 - 4.2.4	ND
	Trial Pit Excavation	4.2.5 - 4.2.7	ND
	Breakout and Coring	4.2.8 - 4.2.11	De, Lb
	Endoscopic Investigation	4.2.12 - 4.2.13	ND
	Thermography	4.2.14 - 4.2.15	ND, Op, Sp
	Ground Penetrating Radar	4.2.16 - 4.2.18	ND, Op, Sp
	Radiography	4.2.19 - 4.2.22	ND, Op, Sp
	Sonic Techniques	4.2.23	ND, Op, Sp
Distortion and Movement	Observation and Simple Investigation	4.3.2 - 4.3.3	ND
	Precision Survey	4.3.4 - 4.3.5	ND, Op
	Crack Width Comparator	4.3.6 - 4.3.7	ND
	Crack Width Microscope	4.3.8	ND
	Tell-tale Crack Monitor	4.3.9 - 4.3.10	ND
	Vernier Callipers or Demec Gauge	4.3.11	ND
	Strain Gauges	4.3.12	ND, Op
	Inductive Displacement Transducers	4.3.13	ND, Op
	Optical Fibre Sensors	4.3.14 - 4.3.15	ND, Op, Sp
	Dynamic Response	4.3.16	ND, Op, Sp
	Load Tests	4.3.17	ND, Op, Sp
Special: Ground Investigation	Trial Pits, Boreholes and In-situ or Laboratory Tests	4.8.2	ND, Lb
	Groundwater Levels	4.8.4 - 4.8.5	ND
	Ground and Groundwater Constituents	4.8.6	ND, Lb
	Slip Surface Identification	4.8.7 - 4.8.9	ND, Op
	Partial Demolition	4.8.10	De
	Ground Movement Measurements	4.8.11 - 4.8.14	ND
Special: Continuous Monitoring	Continuous Monitoring	4.9.1 - 4.9.5	ND, Op

3.3 TESTS FOR CONCRETE

3.3.1 Table E.3 presents testing techniques that are specific to concrete, and gives their main characteristics.

Table E.3 – Tests for Concrete			
Test Category	**Testing technique**	**Paragraph(s)**	**Test Characteristics**
Structural Arrangement and Hidden Defects	Observation and Simple Investigation	4.2.1	ND
	Delamination Survey	4.2.2	ND
	Trial Pit Excavation	4.2.5 - 4.2.7	ND
	Breakout and Coring	4.2.8 - 4.2.11	De, Lb
	Endoscopic Investigation	4.2.12 - 4.2.13	ND
	Thermography	4.2.14 - 4.2.15	ND, Op, Sp
	Ground Penetrating Radar	4.2.16 - 4.2.18	ND, Op, Sp
	Radiography	4.2.19 - 4.2.22	ND, Op, Sp
	Sonic Techniques	4.2.23	ND, Op, Sp
	Covermeter	5.2.2 - 5.2.3	ND
	Impact Echo	5.2.4	ND, Op, Sp
Distortion and Movement	Observation and Simple Investigation	4.3.2 - 4.3.3	ND
	Precision Survey	4.3.4 - 4.3.5	ND, Op
	Crack Width Comparator	4.3.6 - 4.3.7	ND
	Crack Width Microscope	4.3.8	ND
	Tell-tale Crack Monitor	4.3.9 - 4.3.10	ND
	Vernier Callipers or Demec Gauge	4.3.11	ND
	Strain Gauges	4.3.12	ND, Op
	Inductive Displacement Transducers	4.3.13	ND, Op
	Optical Fibre Sensors	4.3.14 - 4.3.15	ND, Op, Sp
	Dynamic Response	4.3.16	ND, Op, Sp
	Load Tests	4.3.17	ND, Op, Sp
Material Properties	Rebound Hammer	5.4.1	ND
	Internal Fracture	5.4.2	De
	Penetration Resistance	5.4.3	De
	Pull-off	5.4.4 - 5.4.5	De
	Pull-out (Drilled Hole)	5.4.6 - 5.4.7	De, Op
	Break-off	5.4.8	De, Op
	Ultrasonic Pulse Velocity	5.4.9 - 5.4.10	ND, Op, Sp
	Radiometry	5.4.11 - 5.4.12	ND, Op, Sp
	Moisture Content	5.4.13	De, Lb, Op

Continued

Table E.3 – Tests for Concrete (continued)

Test Category	Testing technique	Paragraph(s)	Test Characteristics
Material Properties (continued)	Permeability and Surface Absorption	5.4.17 - 5.4.19	De, Op
	Core Compression	5.4.19	De, Lb, Op
	Density	5.4.20	De, Lb, Op
	Reinforcement Yield Strength	5.4.21	De, Lb, Op
	Petrographic Analysis	5.4.22 - 5.4.23	De, Lb, Op, Sp
Deterioration Activity	Acoustic Emission	4.5.2 - 4.5.3	ND, Op, Sp
	Resistivity	5.5.2	ND, Op
	Core Expansion (ASR)	5.5.3 - 5.5.5	ND, Op, Sp
	Thaumasite Form of Sulfate Attack	5.5.6 - 5.5.7	ND, Op, Sp
Deterioration Rate	Linear Polarisation Resistance	5.6.1	ND, Op, Sp
	Electrochemical Techniques	5.6.2	ND, Op, Sp
Deterioration Cause or Potential	Carbonation	5.7.1 - 5.7.2	De
	Half-Cell Potential	5.7.3 - 5.7.6	ND, Op, Sp
	Sulfate Resistance	5.7.7	De, Lb, Op, Sp
	Chloride Content	5.7.8	De, Lb, Op
	Diffusion	5.7.9	De, Lb, Op
	Sulfate Content	5.7.10 - 5.7.11	De, Lb, Op
	Alkali Content	5.7.12	De, Lb, Op
	Analysis of Concrete	5.7.13	De, Lb, Op
Special: Investigation of Post-Tensioned Tendons	Void Detection	5.8.4	De, Op
	Internal Examination	5.8.5	De, Op
	Steel Stresses	5.8.6 - 5.8.7	De, Op
	Concrete Stresses	5.8.8 - 5.8.11	De, Op

3.4   TESTS ON METAL

3.4.1   Table E.4 presents testing techniques that are specific to metal, and gives their main characteristics.

Table E.4 – Tests for Metal			
Test Category	**Testing technique**	**Paragraph(s)**	**Test Characteristics**
Structural Arrangement and Hidden Defects	Observation and Simple Investigation	4.2.1	ND
	Delamination Survey	4.2.4	ND
	Trial Pit Excavation	4.2.5 - 4.2.7	ND
	Breakout and Coring	4.2.8 - 4.2.11	De, Lb
	Endoscopic Investigation	4.2.12 - 4.2.13	ND
	Thermography	4.2.14 - 4.2.15	ND, Op, Sp
	Ground Penetrating Radar	4.2.16 - 4.2.17	ND, Op, Sp
	Radiography	4.2.19 - 4.2.22	ND, Op, Sp
	Sonic Techniques	4.2.23	ND, Op, Sp
	Ultrasonic	6.2.2 - 6.2.5	ND, Op
	Dye Penetrant	6.2.6 - 6.2.8	ND
	Magnetic Particle Inspection	6.2.9 - 6.2.11	ND
	Eddy Current	6.2.12 - 6.2.13	ND, Op, Sp
Distortion and Movement	Observation and Simple Investigation	4.3.2 - 4.3.3	ND
	Precision Survey	4.3.4 - 4.3.5	ND, Op
	Crack Width Comparator	4.3.6 - 4.3.7	ND
	Crack Width Microscope	4.3.8	ND
	Tell-tale Crack Monitor	4.3.9 - 4.3.10	ND
	Vernier Callipers or Demec Gauge	4.3.11	ND
	Strain Gauges	4.3.12 & 6.3.2	ND, Op
	Inductive Displacement Transducers	4.3.13	ND, Op
	Optical Fibre Sensors	4.3.14 - 4.3.15	ND, Op, Sp
	Dynamic Response	4.3.16	ND, Op, Sp
	Load Tests	4.3.17	ND, Op, Sp
Material Properties	Hardness	6.4.1 - 6.4.3	ND, Op
	Chemical Analysis	6.4.4 - 6.4.5	De, Op
	Tensile	6.4.6	De, Op
	Impact	6.4.7 - 6.4.8	De, Op

Continued

Table E.4 – **Tests for Metal** (continued)			
Test Category	**Testing technique**	**Paragraph(s)**	**Test Characteristics**
Deterioration Activity	Acoustic Emission	4.5.2 - 4.5.3	ND, Op, Sp
Deterioration Rate	Plate Thickness Measurement	6.6.1 - 6.6.2	ND or De, Op
Deterioration Cause or Potential	Exposure	6.7.2	ND or De, Lb, Op, Sp
	Manufacturing Defects	6.7.3	ND, Op, Sp
	Cyclic Loading	6.7.4	ND, Op
Special: Aluminium.	Appropriate for Aluminium	6.8.1 - 6.8.2	
Special: Paintwork & Metallic Coatings	Paint Film Thickness	6.9.4 - 6.9.5	ND or De, Op
	Adhesion	6.9.7 - 6.9.8	De, Op
	Discontinuities of the Paint Film	6.9.9	ND, Op
	Chemical	6.9.10	ND or De, Lb, Op

3.5   TESTS ON MASONRY

3.5.1   Table E.5 presents testing techniques that are specific to masonry, and gives their main characteristics.

Table E.5 – Tests for Masonry			
Test Category	**Testing technique**	**Paragraph(s)**	**Test Characteristics**
Structural Arrangement and Hidden Defects	Observation and Simple Investigation	4.2.1 & 7.2.2	ND
	Delamination Survey	4.2.2 - 4.2.4	ND
	Trial Pit Excavation	4.2.5 - 4.2.7 & 7.2.3	ND
	Breakout and Coring	4.2.8 - 4.2.11	De, Lb
	Endoscopic Investigation	4.2.12 - 4.2.13	ND
	Thermography	4.2.14 - 4.2.15	ND, Op, Sp
	Ground Penetrating Radar	4.2.16 - 4.2.18	ND, Op, Sp
	Radiography	4.2.19 - 4.2.22	ND, Op, Sp
	Sonic Techniques	4.2.23	ND, Op, Sp
Distortion and Movement	Observation and Simple Investigation	4.3.2 - 4.3.3	ND
	Precision Survey	4.3.4 - 4.3.5	ND, Op
	Crack Width Comparator	4.3.6 - 4.3.7	ND
	Crack Width Microscope	4.3.8	ND
	Tell-tale Crack Monitor	4.3.9 - 4.3.10	ND
	Vernier Callipers or Demec Gauge	4.3.11	ND
	Strain Gauges	4.3.12	ND, Op
	Inductive Displacement Transducers	4.3.13	ND, Op
	Optical Fibre Sensors	4.3.14 - 4.3.15	ND, Op, Sp
	Dynamic Response	4.3.16	ND, Op, Sp
	Load Tests	4.3.17	ND, Op, Sp
Material Properties	Compressive Strength	7.4.1 - 7.4.3	De, Lb, Op
	Pull-Out	7.4.4	De, Op
	Flat Jack	7.4.5	De, Op
	Flexural Bond Strength	7.4.6	De, Op
	Shove (In-situ Shear)	7.4.7	De, Op
	Physical & Chemical Properties of Mortar	7.4.8	De, Lb, Op
Deterioration Activity	Acoustic Emission	4.5.2 - 4.5.3	ND, Op, Sp
Deterioration Rate	Inspections	7.6.1 - 7.6.2	ND
Deterioration Cause or Potential	Weathering - Inspections	7.7.1 - 7.7.2	ND
	Change of Loading – Inspections or Monitoring	7.7.1 - 7.7.2	ND

3.6 TESTS ON TIMBER

3.6.1 Table E.6 presents testing techniques that are specific to timber, and gives their main characteristics.

Table E.6 – Tests for Timber			
Test Category	**Testing technique**	**Paragraph(s)**	**Test Characteristics**
Structural Arrangement and Hidden Defects	Observation and Simple Investigation	4.2.1	ND
	Delamination Survey	4.2.2 - 4.2.4	ND
	Trial Pit Excavation	4.2.5	ND
	Breakout and Coring	4.2.8 - 4.2.7	De, Lb
	Endoscopic Investigation	4.2.12 - 4.2.13	ND
	Thermography	4.2.14 - 4.2.15	ND, Op, Sp
	Ground Penetrating Radar	4.2.16 - 4.2.18	ND, Op, Sp
	Radiography	4.2.19 - 4.2.22	ND, Op, Sp
	Sonic Techniques	4.2.23	ND, Op, Sp
	Hammer Tapping	8.2.2	ND
Distortion and Movement	Observation and Simple Investigation	4.3.2 - 4.3.3	ND
	Precision Survey	4.3.4 - 4.3.5	ND, Op
	Crack Width Comparator	4.3.6 - 4.3.7	ND
	Crack Width Microscope	4.3.8	ND
	Tell-tale Crack Monitor	4.3.9 - 4.3.10	ND
	Vernier Callipers or Demec Gauge	4.3.11	ND
	Strain Gauges	4.3.12	ND, Op
	Inductive Displacement Transducers	4.3.13	ND, Op
	Optical Fibre Sensors	4.3.14 - 4.3.15	ND, Op, Sp
	Dynamic Response	4.3.16	ND, Op, Sp
	Load Tests	4.3.17	ND, Op, Sp
Material Properties	Microscopy	8.4.1	De, Lb, Op
	Hardness	8.4.2	ND, Op
	Tensile & Compressive Strength	8.4.3	De, Lb, Op
	Resonance, Vibration or Mechanical Impedance	8.4.4	ND, Op, Sp
	Glue & Fastener Strength	8.4.5	De, Lb, Op

Continued

Table E.6 – Tests for Timber (continued)			
Test Category	**Testing technique**	**Paragraph(s)**	**Test Characteristics**
Deterioration Activity	Radiography	4.2.19 - 4.2.22	ND, Op, Sp
	Acoustic Emission	4.5.2 - 4.5.3	ND, Op, Sp
	Probe	8.5.2 - 8.5.3	De or De, Op
	Energy Absorption	8.5.4	De, Op
	Moisture Content	8.5.5	De, Lb, Op
	Bioassay	8.5.6	De, Lb, Op, Sp
	Biochemical	8.5.7	De, Lb, Op, Sp
	Ultrasonic	8.5.8	ND, Op
Deterioration Rate	Inspections	8.6.1	ND
Deterioration Cause or Potential	Rot	8.7.1	De, Lb, Op, Sp
	Fungal Attack	8.7.1	De, Lb, Op, Sp
	Insect Attack	8.7.1	De, Lb, Op, Sp

3.7 TESTS ON ADVANCED COMPOSITES

3.7.1 Table E.7 presents testing techniques that are specific to advanced composites, and gives their main characteristics.

Table E.7 – Tests on Advanced Composites

Test Category	Testing technique	Paragraph(s)	Test Characteristics
Structural Arrangement and Hidden Defects	Observation and Simple Investigation	4.2.1	ND
	Delamination Survey (Pat Test)	4.2.2 - 4.2.4 & 9.1.2	ND
	Trial Pit Excavation	4.2.5 - 4.2.7	ND
	Breakout and Coring	4.2.8 - 4.2.11	De, Lb
	Endoscopic Investigation	4.2.12 - 4.2.13	ND
	Thermography	4.2.14 - 4.2.15	ND, Op, Sp
	Ground Penetrating Radar	4.2.16 - 4.2.18	ND, Op, Sp
	Radiography	4.2.19 - 4.2.22	ND, Op, Sp
	Sonic Techniques	4.2.23	ND, Op, Sp
Distortion and Movement	Observation and Simple Investigation	4.3.2 - 4.3.3 & 9.1.2	ND
	Precision Survey	4.3.4 - 4.3.5	ND, Op
	Crack Width Comparator	4.3.6 - 4.3.7	ND
	Crack Width Microscope	4.3.8	ND
	Tell-tale Crack Monitor	4.3.9 - 4.3.10	ND
	Vernier Callipers or Demec Gauge	4.3.11	ND
	Strain Gauges	4.3.12	ND, Op
	Inductive Displacement Transducers	4.3.13	ND, Op
	Optical Fibre Sensors	4.3.14 - 4.3.15	ND, Op, Sp
	Dynamic Response	4.3.16	ND, Op, Sp
	Load Tests	4.3.17	ND, Op, Sp
Material Properties	Seek specialist advice		
Deterioration Activity	Seek specialist advice		
Deterioration Rate	Seek specialist advice		
Deterioration Cause or Potential	Seek specialist advice		

4 General Testing Techniques

4.1 GENERAL

4.1.1 Sections 5 to 9 present testing techniques that are specific to material types, this section presents testing techniques that are either relevant to more than one material type or not dependent on material type. Where relevant, the text describes which material types the testing technique is relevant to; alternatively the tables provided in Section 3 may be used to identify which testing techniques are relevant to a specific material type.

4.2 STRUCTURAL ARRANGEMENT AND HIDDEN DEFECTS

4.2.1 Observation, measurement and simple investigation should be carried out initially before the more specific tests described below are used.

Delamination Survey

4.2.2 A delamination survey is used to identify regions of material that have detached, or partially detached, from the parent mass of material, typically in concrete and masonry structures.

Brookes Specialist Contractors Ltd

4.2.3 The region to be investigated is tapped with a light hammer (it has been found that a ½ kg geologist's hammer is a suitable tool for this purpose) at approximately 100mm intervals. Good, sound material produces a ringing noise whilst delaminated areas produce a hollow sound. The extent of delamination may be plotted on a drawing and/or on the material surface and a photographic record made. In masonry structures, hammer tapping may also detect arch ring separation in arch barrels, irregularities within individual stones or bricks and any loose stones or bricks.

4.2.4   Care must be taken not to inflict any further damage to the structure. However, the opportunity should be taken to remove any areas of loose material that pose a threat to the public. A delamination survey is relatively quick and easy to perform, and it is good practice to carry out a delamination survey as part of a Principal Inspection.

Trial Pit Excavation

4.2.5   The excavation of trial pits (for example, in surfacing and arch fill, adjacent to foundations or adjacent to or behind retaining walls) is one of the most common techniques of investigation carried out to structures, particularly masonry structures. Trial pits are often required to confirm details of the structure shown on as-built records or to establish the details where no or inadequate records exist. As such, trial pits are frequently used alongside non-destructive testing techniques (e.g. Thermography, Ground Penetrating Radar and Radiography, described in paragraphs 4.2.14-4.2.22), which may be used to identify suitable locations for trial pits, or the trial pits may be used to confirm findings from these tests. In almost all cases, some excavation is required to confirm the features indicated by such tests.

4.2.6   The excavated ground may be carrying load or providing structural support. Hence, when planning trial pits due care and consideration should be given to structural integrity, possibly by limiting the number and/or size of excavations; for example, avoiding trenches across the full width of a bridge, or providing additional support to the structure when the excavation is made. The fill over an arch imparts strength and stability to the structure and removing some of this fill may endanger the stability of the structure unless this is strictly controlled. Trial pits in front of a retaining structure should be sited and sized to cause minimum loss of structural support. The approval of the Supervising Engineer should be sought before embarking on trial pits on or adjacent to any structure. Excavation should then be carried out in such a way as not to damage the structure or any services or drains in the vicinity.

4.2.7   Reinstatement of excavated material should be carried out in accordance with good engineering practice and any relevant statutory and regulatory requirements; for example, the Specification for the *Reinstatement of Openings in Highways* [3] may be a relevant consideration in some instances. In particular, when dealing with highway structures, any waterproofing system that has been removed or damaged should be reinstated and care should be taken to ensure that a new path has not been created for water ingress.

Break Out and Coring

4.2.8   Break out and coring physically removes material and exposes inner and/or hidden parts of the structure. This technique damages the structure but enables direct inspection and measurement of features exposed by the holes and provides samples for further testing. This technique can provide valuable information on structural arrangement and assist in the detection of suspected/hidden features. This approach is normally used for masonry and concrete structures, although holes drilled in metal elements can be used to determine thickness, e.g. sheet pile walls and hollow cast iron members.

4.2.9   To avoid unnecessary damage, exploratory holes may be drilled during an initial investigation, e.g. of geometrical details. Damage can be restricted in certain circumstances by taking the necessary measurements via these holes, for example, insertion of poker type covermeters for locating internal prestressing tendons. In some cases it may be prudent to check thickness at several points, for example, around the perimeter of a hollow core cast iron column as parts may have moved during casting.

4.2.10   Break out is most commonly carried out using small hand-held or machine mounted impact breakers. Break out is usually confined to a small area, sufficient to locate and inspect the desired feature, but on occasion may be used to expose relatively large areas. Break out and coring may be used in conjunction with other site tests, such as carbonation depth (see paragraphs 5.7.1- 5.7.2) or to obtain samples for laboratory testing. The area to be broken out must be selected with care to avoid unwarranted weakening of or unnecessary damage to the structure, for example:

- **Concrete** – impact breakers may damage reinforcing bars, this may be avoided by using high pressure water jetting techniques.

- **Masonry** – where the structure is extensively cracked, the mortar soft or friable, or the stones loose, care must be taken to avoid causing damage to the structure from the vibrations of the impact breaker. Hand tools can prove more effective and cause minimal damage when small areas are to be broken out, for example, when removing one brick or stone.

4.2.11   Coring is normally used to obtain samples for laboratory testing, but may also be used for exposing hidden features. Cores are extracted using diamond tipped drills; they normally have a diameter of between 38mm and 150mm. Great care needs to be exercised since the drills will cut through all common highway structure materials, for example, reinforcement bars or tendons as well as concrete. In masonry structures cores may be extracted to determine the thickness of masonry, particularly in arch rings or abutment walls where access may be restricted for excavation from above. Examination of the core and of the hole can also indicate the quality of the masonry behind the first layer of the ring and may be used to confirm the depth of cracks.

Endoscopic Examination

4.2.12   Endoscopes enable close examination of hidden, obstructed and/or enclosed parts of a structure, which otherwise cannot be viewed by a visual inspection. Endoscopes can be inserted into small holes in a structure. These may be naturally occurring gaps or crevices, as a result of damage or deterioration, or specially drilled holes to enable insertion of the endoscope. Care should be taken when drilling access holes to avoid unnecessary damage, for example, to prestressing tendons or reinforcing bars. Endoscopes typically consist of:

Allen-Vanguard

- Either a rigid or flexible viewing tube containing one or more optical fibre systems and, possibly, a channel for mechanical devices. Flexible endoscopes are particularly useful in difficult situations as they can manoeuvre around obstructions

- A light delivery system to illuminate the area under inspection. The light source is normally outside the element/structure and is typically directed via an optical fibre system.

- A lens system transmitting the image to the viewer. Provision can also be made for attachment to a television monitor and the endoscope may be linked to cameras or video recorders to provide records of the inspection.

4.2.13   Endoscope examination is suitable for all structure types. Examples of its application include examination of:

- The interior of partially grouted post-tensioning ducts to detect voids (it is sometimes possible to detect that a significant length of duct has voids from a single access point, also see paragraph 5.8.4).

- The interior of hollow elements, for example, voided slabs or beams, steel box sections and tubular sections.

- Bearings and joints that are difficult to access and areas/surfaces behind cladding.

- Sides of holes, cavities and internal voids, e.g. ring separation in a masonry arch.

Thermography

4.2.14   Thermograph can be used to locate and indicate the extent of voids, delamination and cracking, and other structural features, for example, reinforcement location and spacing, buried structures such as spandrel walls and counterforts, arch ring separation, and wet areas. The technique uses sensitive infra-red equipment to measure surface temperature differentials, i.e. all objects emit infra-red radiation, the amount of which increases with temperature. Thermal imaging cameras are used to display the temperature, with variations shown as different colours.

4.2.15   Thermal imaging can be used from a distance to efficiently cover large surfaces, but definition of hidden features is on a comparative basis only and varies according to circumstances, for example, on a reinforced concrete bridge deck soffit a 'cold spot' would typically indicate delamination. The equipment is very sensitive and can be influenced by weather conditions, e.g. sunshine, wind and moisture variations.

Ground Penetrating Radar

4.2.16   Ground Penetration Radar (GPR) can be used to investigate internal defects and structural arrangement by identifying interfaces between materials. GPR examines the reflections of short duration electromagnetic pulses produced by interfaces between materials with differing dielectric constants [4]. Contact with the material surface is not required and large areas may be surveyed quickly.

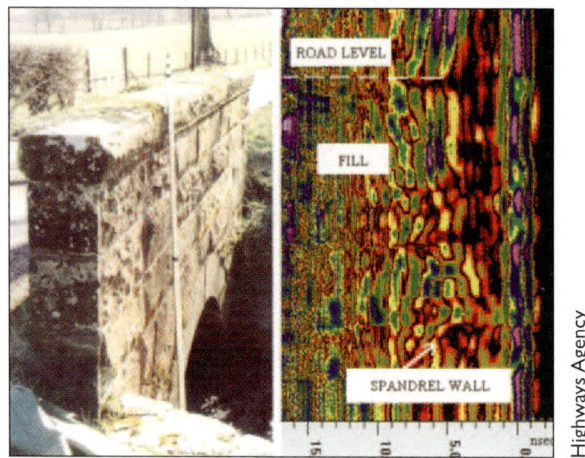

4.2.17　The results are presented in the form of a trace of reflected signals, but their interpretation is complex and may take a significant time to undertake. GPR should be undertaken by specialist organisations as highly skilled personnel with specific equipment are required. This can be an efficient technique for locating defects/hidden features as instant printouts of results can quickly identify areas for further investigation, although performance may be adversely affected when wet fill is present. Principal areas of application include:

- Investigation of internal structure, for example, of masonry arch bridges.

- Identification and location of voids, cracks and delaminations.

- Identification and location of reinforcing bars and prestressing tendons.

- Determination of the thickness of layers, walls, slabs, etc.

- Location of services.

4.2.18　Further guidance on the GPR, including its application to different material types, is provided in *BA 86* [4].

Radiography

4.2.19　Radiography is used to detect voids in locations where there is access to both sides of the material and is particularly suitable for concrete elements and metal elements. There are two sources of radiation: X-rays, which tend to be used in large static facilities, although portable X-ray generators are available; and radioactive isotopes, which are smaller and more easily portable (radiography is similar to radiometry, see paragraphs 5.4.11-5.4.12, although a more powerful radioisotope is used as the gamma ray source). Radioactive isotopes are continuously active and when not required must be contained in lead shielded carriers. The technique of application depends upon the shape of the element to be tested, and the type and location of any features/defects expected.

4.2.20　The beam of radiation is directed through the element onto a photographic film held against the opposing face; hence the need for access to both sides of the element. The defects show up as darker areas on the developed film because more radiation has passed through. High and low density materials will produce light and dark areas respectively on the film, i.e. steel shows up on the negative as lighter than concrete because its higher density impedes the radiation to a greater extent. Therefore, an internal void such as porosity in welds or castings or slag inclusions will absorb less radiation than the surrounding steel. The photographic film provides a permanent record of the defect showing the size and shape of the defect in two dimensions.

4.2.21   Radiography provides direct pictorial evidence of the interior of the element, and these records can be geometrically related to the part under investigation. It has the ability to determine internal defect size and shape and has a universal acceptance in codes by the engineering profession in general. Some of the disadvantages associated with radiography include:

- It is essential to have access to both sides of the element.

- Only the area of defect normal to the source can be observed, no indication of the defect position in the through thickness direction is given. However in welds, certain defects tend to occur in specific areas and this knowledge can help in the identification of size and position of defects.

- Detectability depends upon the volume of the defect; hence tight cracks normal to the radiation can be missed as they do not cause a significant difference in radiation energy at the film.

- The projected size of the defect observed depends upon its orientation to the radiation, which can lead to false information of the true size of a defect.

- The thickness of concrete penetrated is limited to about 500mm with a gamma ray source. High energy X-rays are more suitable for examining concrete thicknesses up to 1m.

- It is not able to detect corrosion sites with any degree of accuracy although broken wires and cables can be detected.

4.2.22   Due to the risk of radiation damage to personnel, testing must be carried out in shielded enclosures or the surroundings cleared of all personnel during the testing period. Strict safety precautions are imposed and highly specialised equipment is required. This test should be carried out by specialist organisations. Further details are contained in *BS1881: Part 205* [5].

Sonic Techniques

4.2.23   Sonic techniques, such as Sonic Transmission and Sonic Tomography, may be helpful in detecting voids or changes in the internal structure, although these are difficult to interpret. The principle of sonic techniques is that one face of the structure is struck with a hammer and the impact is recorded by an adjacent accelerometer. A second accelerometer on the opposite face of the structure records the arrival of the transmitted compression wave, and the time between transmission and reception is calculated. By repeating this procedure, variations in the transmission time over the structure can be plotted. Transmission time depends on the density of the material and the presence of voids through which the wave will not travel. Waves travel around voids, lengthening the transmission time. Interpretation of the results is often difficult and the procedure can only be used where there is access to both sides of the structure, at piers for example.

4.3   DISTORTION AND MOVEMENT

4.3.1   There are a number of different techniques available to measure the distortion and movement of structures and elements, ranging from the simple and straightforward to the complex. Several techniques are suitable for continuous or remote monitoring (see Section 4.9). Some test techniques, such as those used to measure strain and deflection, may be taken during the passage of known test vehicles to verify calculations, or for a period of normal traffic loading to determine actual field stresses.

Observation, Measurement and Simple Investigation

4.3.2   The structural integrity of the majority of structures depends upon the physical dimensions and profiles of the structure and the condition of the materials as well as the adjacent ground. These properties and any changes in them can usually be assessed by observation, measurement and simple investigation. The equipment required is little more than a testing hammer, measuring tapes, a string-line and a probe.

4.3.3   Distortion of shape, or bulging, tilting or loss of verticality can be measured with tapes and string lines. A plumb-bob can be formed from a heavy object, such as the inspector's hammer tied on the end of the string-line; alternatively a microlevel can be placed against the surface of the structure to determine the angle of the surface. Lightly tapping the surface of a structure with the testing hammer can give an indication of possible defects, whilst the probe can be used to explore cracks and crevices and to test the soundness of a component.

Precision Survey

4.3.4   The use of tapes and string lines may not be sufficiently accurate to monitor continuing distortion or movement in some cases. Improved accuracy can be provided by using Precision Survey techniques, e.g. GPS (Global Positioning System) survey using total stations. These surveys utilise the latest equipment, require experienced operators and are continually evolving, and as such are not described here. These techniques are likely to only be feasible in special circumstances and it is recommended that expert advice is sought when considering them.

4.3.5   Photogrammetry, using stereoscopic photographs, may also be used to monitor distortion and movement of large structures. Accuracy is usually not better than ± 100 or 200mm, depending on the object distance. These photographs may be digitised to aid comparisons. In all cases, it is essential to use stable datum and reference points if future comparisons are to be made.

Crack Width Comparator

4.3.6   A crack width comparator is used to measure/assess the width of a crack. The comparator is normally a transparent plastic strip or card with a series of different width lines on one side and sometimes a ruler on the other. These lines are held against the crack to assess the crack size. This piece of equipment should be carried as a matter of course by all inspectors.

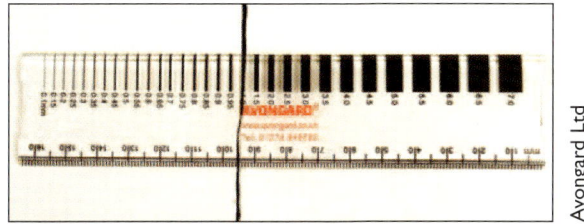

Avongard Ltd

4.3.7   The assessment of the width of fine cracks can be difficult due to the subjective interpretation of the position of the edge of a crack, which is uneven on the microscopic scale. The width of a crack at the surface can sometimes exaggerate the true width because of greater weathering. Also, damp conditions tend to accentuate crack width.

Crack Width Microscope

4.3.8   A crack width microscope is used to measure the width of a crack. It is a hand-held microscope that contains a scale; this scale is superimposed over the crack. Similar difficulties exist in interpreting crack widths as with the crack comparator technique (see above).

Tell-Tale Crack Monitor

4.3.9   Tell-tales are used to measure crack movements. Tell-tales normally consist of two plates, typically plastic, which overlap for part of their length. One plate has a grid (normally calibrated in millimetres) while the overlapping plate is transparent and marked with a cross hair or scriber. The plates are bonded to either side of the crack and the original positions recorded, preferably on a standard pro forma, and the positions then recorded at suitable intervals. The intervals between readings should reflect the knowledge of the defect and/or expected rate of change, i.e. if little is known about the defect then more frequent measurements may be preferred until a degree of confidence has been established in the rate of change.

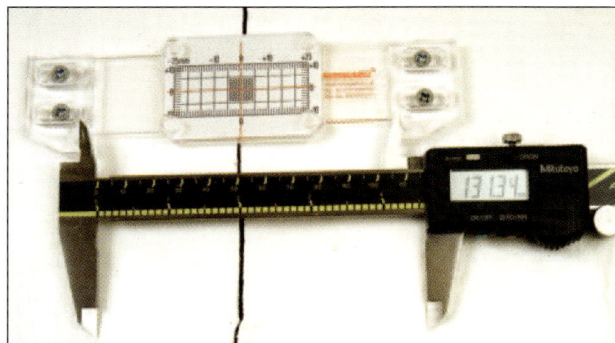

Avongard Ltd

4.3.10 A range of tell-tale crack monitors are available for different circumstances, e.g. vertical and horizontal movement, movement across cracks and out of plane movement. These tell-tales can record the movement to an accuracy of ±0.1mm. Alternatively, glass or mortar tell-tales may be used to check whether an existing crack is active (i.e. moving), i.e. the glass or mortar tell-tale, bonded to either side of the crack, will break if or when significant movement occurs. The relative movements of the two halves can be measured to determine subsequent changes. Although simple, this technique is relatively crude and should not be expected to provide accuracy better than ± 1mm. Glass tell-tales are normally used on concrete structures and mortar ones on masonry structures.

Vernier Callipers or Demec Gauges

4.3.11   Vernier callipers or Demec Gauges are used to monitor changes in crack width. Two or more studs are fixed to either side of a crack and the distance between studs on either side of the crack is measured at suitable time intervals. The intervals between readings should reflect the knowledge of the defect and/or expected rate of change, i.e. if little is know about the defect then more frequent measurements may be preferred until a degree of confidence has been established in the rate of change. Measuring points should be clearly marked and recorded to ensure that they can be replicated. This test does not measure the actual crack width but can monitor and record the change in crack width with time.

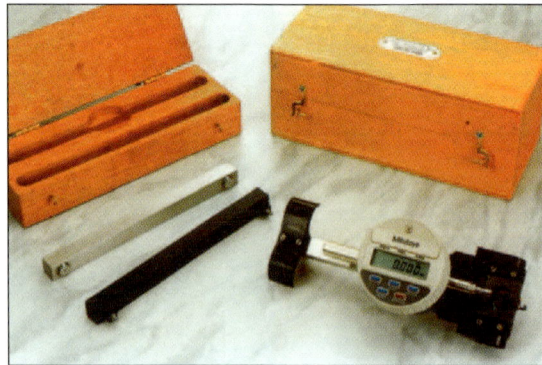

Mayes Instruments Ltd

Strain Gauges

4.3.12   Strain gauges are used to monitor the deformations (changes in the dimensions) of a structure/component caused by loading and/or temperature change. They use the principle that deformations in the structure/component are reflected in a deformation of the gauge. Strain gauges include electrical resistance strain gauges, mechanical strain gauges, and acoustic or vibrating wire gauges. Where principal stresses cannot be derived from uniaxial strains, strain gauge rosettes may need to be used. Measurement of localised strains may require the use of gauges of very short length; electrical resistance gauges are often the most suitable in such instances.

- *Electrical resistance strain gauges* – these are highly versatile and generally consist of wire or metal foils bonded to the component to be tested with special adhesives and require skilled installation; deformation of the component causes deformation of the wire/foil, which changes its electrical resistance. They can be read remotely and can be made very small. The technique is best suited for short-term measurements of strain.

- *Mechanical strain gauges* – can be beneficial for infrequent monitoring of structures over a long time scale. Deformation of the element produces a displacement over the gauge length of the instrument, which is measured using a dial gauge system. The equipment tends to be inexpensive but it is a manual technique, which depends on the skill of the operator and is less accurate than electrical strain gauges.

- *Acoustic gauges or vibrating wire gauges* – this technique is based on the principle that the resonant frequency of a taut wire varies with changes in tension. The tensioned wire is sealed in a protective tube, which contains an electromagnet to pluck the wire and record the vibration. These gauges, which have long-term stability, require skilled installation and can be read remotely.

Inductive Displacement Transducers

4.3.13　Inductive displacement transducers can be used to measure deformations in a structure/component. Inductive displacement transducers use an armature moving within two electrical coils to sense the position of the armature. They provide an accurate technique for measuring deformation, but require signal conditioning equipment. They can be read remotely but require specialist installation. The typical displacement range is +5mm, but displacements of up to +50mm may be monitored.

Optical Fibre Sensors

4.3.14　Optical Fibre Sensors can be used to measure deformations in a structure/element. Stranded optical fibre sensors work on the principle that light is lost, or attenuated, when it travels along an optical fibre and particularly when passing through areas of micro-bending. Areas of micro-bending are produced by winding three optical fibres around each other in a controlled manner for the required length of the sensor. The change in the overall length of the sensor is determined by comparison of the intensity of the light supplied with that of light emerging from the sensor as it is strained or relaxed.

4.3.15　Using the light attenuation technique, stranded optical fibre sensors have been shown to have a resolution of better than 0.02mm. This is independent of the gauge length of the sensor, which can be up to 15m. The distribution of deformation along the sensor, when fixed to the component at a number of intermediate points in its length may be obtained using an optical time domain reflectometer. This transmits light pulses (nano-seconds long) into the sensor and produces a trace of the light attenuation. Comparing a current trace with the trace prepared when the sensor was first installed allows the positions of the deformation changes to be located. This technology allows the critical sections of large structures, such as bridges, to be monitored cost effectively.

Dynamic Response

4.3.16　The response of a structure to dynamic loads, such as the movement of a known vehicle or dropping of a weight onto a bridge deck has been researched, but only with some success for simple models. Monitoring changes to dynamic response over a period of time may, however, provide an indication of structural changes. *ISO 14963* [6] provides guidelines for dynamic tests and investigations on bridges and viaducts.

Load Tests

4.3.17　Load testing of structures, in particular masonry arch bridges, has been used to compare actual performance with the theoretically predicted performance. Load testing may be used as a one-off assessment technique or it may be used to monitor changes with time to provide information on the deterioration of a structure. In all instances, experience is required to judge whether measured deflections are within acceptable limits. Guidance on load testing for assessment of strength is provided in *BA 54* [7] and the *Guidelines for the Supplementary Load Testing of Bridges* [8].

4.4 MATERIAL PROPERTIES

4.4.1 These testing techniques are specific to the material type. Details are provided in Sections 5 to 9.

4.5 DETERIORATION ACTIVITY

4.5.1 Highway structures are long life assets with, in general, slow rates of deterioration. This enables some forms of deterioration activity to be readily identified from cyclic inspection information (e.g. General and Principal Inspection pro forma), i.e. how has the severity/extent of an element changed between consecutive inspections? This is desirable to using more expensive testing techniques, but is reliant on the consistency and quality of the inspection information and the processes/systems in place that enable information to be viewed. It is therefore important that inspection staff are provided with suitable training and equipment and that information is recorded in a manner (e.g. computerised system) that enables easy comparison of consecutive inspections.

Acoustic Emission

4.5.2 Acoustic emission is used to detect crack growth as it occurs. That is, as a structure experiences an increasing load the strain energy stored within it increases and localised points may be strained beyond their elastic limit and microcracking may occur; acoustic emissions occur when these micro-cracks develop in the structure. The energy released propagates small amplitude elastic stress waves, or acoustic emissions, through the structure, which can be detected as small displacements by transducers mounted on the surface.

4.5.3 Acoustic emission has been used to monitor behaviour during load testing [9]; with the results showing good correlation with bridge displacements, and it has also been used to detect fatigue damage as an indicator of increasing levels of damage. The technique is applicable to all structure types but has been more extensively used on some structural forms and material types than others. Specialist equipment and expertise are required.

4.6 DETERIORATION RATE

4.6.1 Corrosion rate is frequently determined by assessing information compiled from successive sites visits and/or tests. The majority of testing techniques described in this manual can be used for this purpose. However, there are certain testing techniques that seek to measure the corrosion rate directly; these are normally specific to the material type. Details of such techniques are provided in Sections 5 to 9.

4.7 DETERIORATION CAUSE OR POTENTIAL

4.7.1 The techniques described in this manual are specific to the material type. Details are provided in Sections 5 to 9.

4.8 SPECIAL: GROUND INVESTIGATION

4.8.1 Global failures of abutments, piers, wing walls and retaining walls are often due to changes in the condition of the ground, changes in groundwater or inadequate initial design; the latter may be due to misinterpretations or deficiencies in the original ground investigation. In order to determine the degree to which the ground conditions have changed and to define the current loads on a structure, a ground investigation may be required.

4.8.2 The standard techniques of ground investigation consist of trial pits (see paragraphs 4.2.5-4.2.7), boreholes and in-situ or laboratory tests. The general procedures and techniques used are given in *BS 5930* [10] and details of test techniques are provided in *BS 1377* [11]. A specialist qualified engineer, with appropriate experience of ground investigations, should be consulted to identify the need and requirements for a ground investigation and to interpret the results. When specifying a ground investigation the Supervising Engineer should bear in mind the extent to which the investigation work itself might affect the stability of the structure.

4.8.3 Specific details of the wide range of ground investigation tests are outside the scope of this Manual; however, details of some tests that may be useful are described below.

Groundwater Levels

4.8.4 Groundwater pressure has a profound influence on the loading placed on and the behaviour particularly of a retaining structure. The simplest technique of determining groundwater level is by observation in a borehole or trial pit. This may involve a considerable response time for the water level to reach equilibrium, unless the ground is reasonably permeable. Observations over a period of time may be required to assess seasonal variations. The safety hazard of leaving excavations open, particularly for an extended length of time, should be considered.

4.8.5 For an accurate knowledge of groundwater levels it is preferable to use piezometers. These consist of tubing or pipe with a porous or perforated end section, placed into the ground at the appropriate level. Measurement of the water level is made by dipping or pressure measurement depending on whether a standpipe, hydraulic, electrical or pneumatic piezometer is used.

Ground and Groundwater Constituents

4.8.6 The aggressiveness of the ground and groundwater to concrete, steel, masonry and other materials forming the structure can be assessed from chemical tests on samples. Corrosive action may arise from industrial waste products that were not present at the time of construction.

Slip Surface Identification

4.8.7 Signs of surface movement or slope deformation can be monitored by inserting rows of poles or pegs into the slope. Displacements and shallow slide movements can be determined by subsequent measurement and observation of the distortion of the rows of poles or pegs. Movements at greater depths can be found by installing slip indicators as described in *HA 48* [12].

4.8.8 A slip indicator is a tube inserted into the ground that passes through a potential slip plane in order to determine the depth of movement. Movement of the soil mass distorts the tube preventing the free passage of a thin heavy rod suspended by a cord within the tube. If a rod is left at the base of the tube, it may be pulled up using the cord to the underside of the distortion. A second rod may then be lowered down the tube until it is stopped by the upper part of the distortion. The mean of the two distances gives the approximate depth of the slip plane. If the difference between the two distances is excessive, when compared to the height of slope, it may indicate that movement is occurring at more than one depth.

4.8.9   Alternatively, inclinometers provide a more sophisticated means of measuring small deformations; where an inclinometer is an instrument for measuring angles of slope (or tilt), elevation or inclination of an object with respect to the vertical (i.e. gravity). They can, however, be quite expensive and access tubes can be damaged by active slips preventing the sensor from passing down the tube.

Partial Demolition

4.8.10   The investigation of modular based structures may be carried out by partial demolition. For instance, crib walls or reinforced earth structures may be partially dismantled to enable samples of crib units or earth straps to be removed for testing, or to investigate the condition of previously buried elements. The dismantling and replacement should be designed and carried out in a safe manner that does not affect the stability of the remaining structure. Similarly, facing units may be removed in whole panels to enable examination of cavities and of the structural component behind.

Ground Movement Measurements

4.8.11   Large scale measurements can be used to monitor the interrelated response of the ground and structures as outlined in *BS 5930* [10]. Several techniques can be used in ground investigation to monitor displacements and strains associated with known or suspected ground mass movements resulting from slope or structure failures, some of these are summarised below.

4.8.12   Ground and structure movements are generally associated with stress redistribution and water pressure changes that are characteristic of the particular ground. Total stress can be monitored using total pressure cells, while normal and shear stresses can be measured by transducers. Ground movements are generally measured in terms of the displacement of points which can be positioned on the surface of the ground or within the ground mass, as discussed in paragraphs 4.8.7-4.8.9. The absolute movement of a point has to be referred to a stable datum.

4.8.13   Surface observations of ground movements can be made by a Precision Survey (also see paragraphs 4.3.4-4.3.5). An accuracy of ±5mm for distance measurements over 2000m can be achieved using electro-optical instruments. Care has to be taken to position datum points away from the effects of movements due to load and/or water changes.

4.8.14   Vertical movements can be observed by means of settlement gauges with an accuracy of ± 0.1mm. The use of telescopic tubes, inclinometers and tensioned wires anchored in drillholes can also be used for measuring strains or displacements. Lateral movements can be measured by offsets and triangulation. Rods, telescopic tubes, tensioned wires and lasers can be used.

4.9   SPECIAL: CONTINUOUS MONITORING

4.9.1   Continuous monitoring is a form of testing in which data logging technology is used to automatically monitor structures on an ongoing basis. The most common parameters monitored are temperature, strain, half-cell potential, resistivity and corrosion currents, and concrete moisture content. Data can be collected at the structure or transmitted to a remote location.

4.9.2   Monitoring systems can be designed to process data as it is being collected from the instrumentation. Hence, if the system is connected by telephone or other transmission systems, it can be designed to act as an early warning device, automatically issuing an alarm when pre-defined limits of the parameters are reached. This type of system can be used effectively as part of a risk management strategy.

4.9.3   Several key issues need to be addressed when considering the installation of continuous monitoring. The site equipment must be sufficiently robust to withstand elements of the weather and sited to minimise the risk of vandalism. The system needs to be maintained, the power source should be either battery or mains with battery back-up and the data logging capacity should be sufficient to store the required data between downloads. Data can be downloaded locally, by visiting the site, or remotely, through telephone systems. It is important to have processing systems/facilities that make full use of the potentially large volume of data provided.

4.9.4   When monitoring a structure, an accurate measure of the temperature is usually required at the same time as, for example, strain measurements. The core temperature of the element should be measured rather than surface temperature or ambient air temperature. There are several different techniques for temperature measurement; thermocouple sensors are versatile, resistance thermometers are accurate and thermistors are frequently cost-effective.

4.9.5   Monitoring systems can be used effectively with some strain gauges, e.g. electrical resistance strain gauges (see paragraph 4.3.12). Corrosion and corrosion risk can be monitored continuously by, for example, the use of the half-cell potential, resistivity and linear polarisation techniques on concrete structures (see Sections 5.5, 5.6 and 5.7).

5 Tests on Concrete

5.1 GENERAL

5.1.1 This section is relevant for all forms of concrete structures and elements, including mass concrete, reinforced concrete and prestressed concrete. One of the most common causes of deterioration in concrete structures in the United Kingdom is corrosion of the reinforcement due to contamination by chloride ions from de-icing salt. Tests relating to chloride ion ingress and the resultant corrosion are provided.

5.1.2 Coring and drilling of concrete are frequently used (refer to paragraph 4.2.8) to provide samples for a range of tests. It is important that coring or drilling be undertaken in such a manner that will minimise the adverse effects on a concrete element, i.e. the Supervising Engineer should carefully consider the location and number of samples required (guidance on sampling is provided in Section 2.3). After drilling or the removal of cores, the reinstatement should be sound, durable and of high quality.

5.1.3 The guidance produced by the Concrete Bridge Development Group, *Guide to Testing and Monitoring the Durability of Concrete Structures* [13], should be referred to for more in-depth guidance on a number of the following techniques. It also includes a table that provides guidance on the most suitable tests to confirm a provisional diagnosis of the cause of the deterioration.

5.2 STRUCTURAL ARRANGEMENT AND HIDDEN DEFECTS

5.2.1 A number of test techniques, as listed in Section 4.2 and below, are available for detecting hidden features in concrete. They may be used to determine the presence, location and dimensions of features such as voids and reinforcement, to check the location of prestressing tendons or to check the condition of reinforcement or tendons.

Covermeter

5.2.2 A covermeter is primarily used to locate steel reinforcement and to measure the thickness of the concrete cover. Where the reinforcement is not congested it may be possible to determine the bar diameter. Modern covermeters normally include a handheld unit with display panel, a probe/scanning unit and automatic data logging capability. The test is carried out by traversing the probe/scanning unit over the concrete surface. It is normal for readings to be taken on a 0.5m grid, although modern equipment enables smaller grid sizes to be readily used.

Hilti

5.2.3   Covermeter tests may be used to good effect to complement other test techniques, for example, the covermeter can be used to locate reinforcement locations/arrangements prior to a half-cell test (see paragraphs 5.7.3-5.7.6) and other tests where connection to, or avoidance of, reinforcement is necessary. *BS 1881: Part 204* [14] provides further details on the use of covermeters; manufactures tend to comply with the requirements of this British Standard.

Impact Echo

5.2.4   Impact echo techniques may be utilised for detection of hidden features, as well as for void detection. The impact-echo technique relies on propagation of a stress pulse through the concrete, which is reflected by internal discontinuities or external boundaries. The pulse is caused by a mechanical impact on the surface and the reflected waves are measured at the surface by a receiving transducer. Detailed guidance on the methodology used by impact echo and its application are provided in *BA 86* [4].

5.3   DISTORTION AND MOVEMENT

5.3.1   Details of suitable testing techniques for concrete structures are provided in Section 4.3.

5.4   MATERIAL PROPERTIES

Surface Hardness Measurement (Rebound Hammer)

5.4.1   This test assesses the compressive strength of concrete by measuring the surface hardness. It measures the surface hardness by impacting the concrete with a standard mass propelled with a standardised energy, thus causing a localised crushing. The compressive strength of the concrete can then be assessed by the use of specific correlation factors against a concrete of known strength. The results are easily influenced by many factors, such as surface carbonation hardening, inadequate rigidity, low cover and the nature of the surface finish. *BS EN 12504-2* [15] provides further details.

Internal Fracture Test

5.4.2   This test may be used to measure a combination of tensile and shear strength of concrete and is a simple and inexpensive in-situ test; however, for the assessment of compressive strength, correlation with laboratory core compression tests is required. The test involves fixing a 6mm diameter expanding wedge bolt into a drilled hole in the concrete surface to a depth of approximately 17mm. The peak force is recorded when this bolt is pulled against a standardised reaction tripod. Further details are contained in *BS 1881: Part 207* [16].

Penetration Resistance Testing

5.4.3   This test is used on site to assess the compressive strength of concrete, but correlation with laboratory core compression tests is required. A standard steel alloy probe (Windsor probe) is fired into the concrete surface using a standardised charge. The exposed length of the probe is measured to ascertain the depth of penetration, which would normally lie between 20 and 40mm. This test is quick and is particularly suitable for locations where access is difficult, although it should not be used on slender members or within 150mm of a free edge. Reinforcement must also be avoided, otherwise cracking may occur. Further details are contained in *BS 1881: Part 207* [16].

Pull-off Test

5.4.4   This test is used to assess concrete compressive and tensile strength and is performed in-situ, but correlation with a laboratory core compression test sample is required. A circular steel disk is bonded to the concrete surface using an epoxy or polyester resin adhesive. The force required to pull-off the disk with a piece of concrete attached is a measure of the tensile strength of the concrete. On site at least six pull-off tests are required, although where this test is conducted in the laboratory, only three are usually required.

5.4.5   This test is straightforward for site use and may not require a power supply. Careful preparation is required to ensure a satisfactory bond between the disk and concrete. Partial coring may be used to eliminate the surface effects. This test may also be used to test the bond strength of repairs. Further details are contained in *BS 1881: Part 207* [16].

Pull-out Test (Drilled Hole)

5.4.6   This test is used on site to assess compressive, tensile and shear strength of the concrete, but correlation with a laboratory core compression test sample is required for assessing the compressive strength, although a more general calibration may be used. An under-reamed hole is drilled in the concrete and a standardised metal insert with a removable bolt is inserted. The peak force recorded when the bolt is pulled against a standardised reaction ring is a measure of a combination of the tensile and shear strength of the concrete.

5.4.7   The test depth is approximately 25mm and it should be located at least 100mm from a free edge. Drilling requires the use of a hand-held cutter with a water supply. If the loading is stopped when the peak force is reached, there will be little damage other than the hole, but the concrete test zone may be more susceptible to frost action. If loading continues, a cone of concrete will be removed. Further details are contained in *BS 1881: Part 207* [16].

Break-off Test

5.4.8   The break-off test is used to measure the compressive strength of concrete. The break-off test measures the flexural strength of concrete from the force required to break off in-situ a cylinder of concrete, the compressive strength is calculated from this force. A 55mm diameter cylindrical core is formed in the concrete by drilling to a depth of about 70mm. An adjacent slot is formed, into which a hydraulic jack is inserted so the transverse force required to break off the core can be measured. Usage has indicated a high degree of variability in results and the test also damages the concrete locally, so repairs are needed.

Ultrasonic Pulse Velocity (UPV) Testing

5.4.9 The Ultrasonic Pulse Velocity (UPV) test can be used on site to assess concrete compressive strength, but correlation with a laboratory core compression test sample is required. Transducers are placed on the concrete surface, a pulse transmitter placed on one face and a receiver on the opposite face. A timing device measures the transit time of the ultrasonic pulse through the material. If the path length is known, then the UPV can be calculated from the path length divided by the transit time. As such, UPV does not directly measure the compressive strength of the concrete but can be used as a comparative technique.

5.4.10 Access to opposite faces of the concrete member is required and an experienced operative is needed, as most concretes have pulse velocities that lie within a narrow range. The concrete is not damaged during the test although some slight staining may result from the use of couplants. Further details are contained in *BS EN 12504-4* [17].

Radiometry

5.4.11 Radiometry can be used to measure the density of concrete and can thereby provide beneficial information on durability. Radiometry is applied in-situ and measures the neutron absorption rate of the concrete. A beam of gamma rays from a radioisotope is directed into the concrete and the neutron signal is picked up with a Geiger counter. This measures the amount of radiation reflected back to the surface, providing a simple comparative indication of the density of the surface zone (up to approximately 100mm depth). This application is largely restricted to comparative assessments unless calibrated against laboratory core density tests. Measurements may also be taken through the concrete member with the detector on the opposite face.

5.4.12 Radiometry is a costly technique and requires extensive safety precautions. It is unlikely to be cost-effective for most durability investigations. Where used it should only be undertaken by specialist organisations as highly skilled personnel with specific equipment are required; their advice should be sought at an early stage.

Moisture Content Tests

5.4.13 The moisture content in concrete has a significant influence on durability, in particular corrosion rate, carbonation rate, frost attack and the ingress of contaminants, e.g. chloride ions [13]. Moisture Content tests also provide useful information for interpreting permeability test results (see paragraphs 5.4.17-5.4.18). Site tests for determining the moisture content of concrete can be divided between those that directly measure moisture content and those that measure it indirectly through relative humidity.

5.4.14 Direct measurement of moisture content include:

- **Direct Reading Meters** – moisture content can be estimated indirectly by measuring the resistance or capacitance of the concrete (i.e. probes placed into drilled holes in the concrete surface measure changes in the electrical resistance/capacitance caused by the moisture content) or by measuring the penetration of gamma rays (similar to the Radiometry techniques described in 5.4.11-5.4.12).

- **Chemical (Carbide-Acetylene) Test** – this test measures the quantity of acetylene gas produced when water (moisture in the concrete) reacts with a chemical (calcium carbide). A sample of concrete is chipped from the structure, crushed and places in the test apparatus. The test apparatus is shaken to ensure the calcium carbide comes into contact with all the moisture in the sample. The reaction generates acetylene gas, which raises the pressure in the vessel. The apparatus includes a gauge to measure the pressure, which indicates the moisture content. This test is suitable for use on site.

- **Thermography and Radar** – comparative assessments of moisture content may be made using thermography (see paragraphs 4.2.14-4.2.15) or radar (see paragraphs 4.2.16-4.2.18).

- **Microwaves and Radio Waves** – microwaves and radio waves are absorbed by water molecules. This level of absorption varies with the amount of water present, and therefore can be used to indicate moisture content.

5.4.15   Indirect measures of moisture content normally involve inserting the relative humidity probe into a drilled, or naturally occurring, hole in the concrete and sealing the hole. Tests include [13]:

- **Mechanical hygrometer** – the probe contains a hair, paper or synthetic fibre which changes in length as the relative humidity changes.

- **Electronic sensing element** – the electrical output of these sensors changes non-linearly as relative humidity changes.

- **Dew point** - a cyclically chilled mirror is placed within a drilled hole in the concrete surface and electronically controlled light beam sensors then detect misting of the mirror. This is the most accurate technique.

- **Timber plugs** – the timber plugs, over time, will reach a state of equilibrium with the moisture content in the surrounding concrete. The moisture content of the timber plugs can be determined using oven drying or by measuring the resistance between two probes inserted into the plug.

5.4.16   Alternatively, a laboratory test can be carried out to obtain a more accurate measurement. This involves thorough oven drying of a concrete sample and measuring the change in weight of the sample. Care needs to be exercised when removing the concrete sample, storing it and transporting it back to the laboratory in order not to affect the 'as-removed' moisture content. The data may be used as part of the density measurement.

Permeability and Surface Absorption

5.4.17   The majority of durability problems affecting concrete relate to the ingress of foreign elements into the concrete through its surface. Examples include carbonation of the concrete and attack by chloride ions from road salt or by sulfates from ground water. Concretes whose surfaces are more permeable and susceptible to water absorption are at greater risk from attack. There are a number of techniques to determine the water absorption and permeability of the concrete surface, these include:

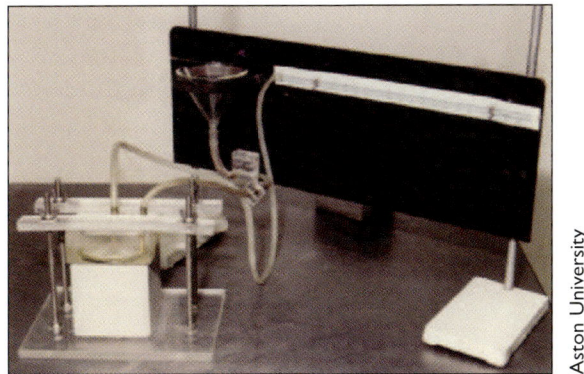

Aston University

- **Initial Surface Absorption Test** [18] – requires a constant head of water to be brought into contact with the concrete surface and the rate of absorption is recorded using a capillary tube. However, this test is not suitable for testing soffits as it is not possible to exclude air from the sealed cup placed on the concrete surface.

- **Modified Figg Air Test** [19] – air permeability testing requires the drilling of a small hole which is sealed with a liquid rubber compound and a hypodermic needle inserted. The air in the hole is evacuated until it reaches a pre-determined pressure and then the hole is sealed. The time taken for the pressure to return to a higher specified level, as determined by a manometer, may be taken as a measure of the air permeability.

- **Capillary Absorption** [13] – the rate at which a sample of concrete, in laboratory conditions, absorbs water is measured (through weight change) and plotted against the square root of time. This should be linear and the results can be used to determine porosity, sorptivity (water absorption) and pore size.

- **Autoclam Test** [20] – the apparatus is clamped to the test area on site. The apparatus puts water or air, maintained at a constant pressure, in contact with the concrete surface and measures the rate of water absorption or rate of decay of air pressure.

5.4.18 The equipment used to carry out these tests is relatively inexpensive and widely available. The tests can be carried out quickly and results are instantly available. However, these tests are particularly influenced by the existing moisture content and should be carried out under conditions as dry as practically possible, usually after a period of sustained dry weather.

Core Compression Test

5.4.19 The compressive strength of the concrete in an existing structure may be measured by direct compression testing of a core cut from the structure. For best results a core diameter of 150 mm should be used, but reasonable results can be obtained from smaller cores. The length/diameter ratio of the core after trimming should be between 1:1 and 1.2:1. Care is needed to avoid reinforcement when cutting the core, since not only would the structure be damaged but also the accuracy of the test would be compromised. Further details are contained in *BS EN 12504-1* [21] and *BS 6089* [22].

Density Measurement of Core Sample

5.4.20   Density measurement of cores can provide useful information when interpreting other test results. The density of a sample can be readily measured in the laboratory from the known mass and displaced volume of water. Care should be taken when storing and transporting the sample from site. It is important to distinguish between 'in-situ' density, saturated density and oven dried density; the 'in-situ' density is usually the most appropriate, although oven dried density is usually required for the alkali content test (see paragraph 5.7.12). Test details are contained in *BS 1881: Part 114* [23].

Reinforcement Yield Strength Test

5.4.21   This test provides information on the yield strength of the reinforcing bars (see Volume 1: Part B: Section 4: paragraph 4.3.9 for a definition of yield strength). This requires the removal of at least a 600mm length of bar from the concrete. The sample is then subjected to a tensile load in the laboratory from which the yield stress is derived. The removal of a significant piece of reinforcement requires careful consideration as this may adversely affect the structural integrity of the element. For this reason testing of reinforcement cut out from structures is not widely practised.

Petrographic Analysis

5.4.22   Petrographic analysis can be a valuable tool in the assessment of the composition and durability of concrete and is essential for the reliable diagnosis of many forms of deterioration [13]. It is commonly used to assist in the diagnosis of alkali-silica reaction, but it may also be used as an aid to the diagnosis of freeze-thaw action, sulfate attack, attack by acids, leaching, carbonation and cement paste and aggregate shrinkage.

5.4.23   The test requires the removal of a core from the concrete member. The core is then sliced into thin sections and examined under a petrological microscope using polarised light, providing information on the composition and micro-structure of the concrete. Ultra-violet light can be used to aid the identification of crack patterns. This analysis must be carried out by specialists who should be consulted, where feasible, before cores are removed from the structure, as test locations are critically important.

5.5   DETERIORATION ACTIVITY

5.5.1   Section 4.5 provides details of acoustic emission which is suitable for concrete.

Resistivity Measurement

5.5.2   Resistivity measurement [24] provides an indication of areas of likely corrosion activity, although it should not be relied upon exclusively and should be used in conjunction with other test techniques, for example it may supplement half-cell potential tests. The test is quick to perform and measures the potential drop between probes mounted on the surface of the concrete to be tested. Interpretation of the results requires experience, is based on broad categories and is largely comparative in site applications.

Core Expansion (AAR) Test

5.5.3   The core expansion test has been devised as an aid to the diagnosis of the presence of alkali-silica reaction. The core sample should be taken from the central zone of the concrete member where the AAR reaction is most severe. The core is tested by storing it in a continuously saturated (but not immersed) state, in a temperature controlled environment. The change in length due to the expansive forces of AAR is measured over a period of time, which ideally should be of some 12 months duration. A significant expansion of the core sample indicates that the concrete component may be at risk from further expansion due to the swelling of the existing AAR gel that has already formed or from the formation and swelling of new AAR products.

5.5.4   The standard temperature at which the test is undertaken is 20°C. However, an accelerated test can be undertaken at storage temperature of 38°C. A minimum of three months is required to provide useful data for the accelerated test, although it should be noted that the greater expansion is often observed when the lower temperature, longer duration, technique is used.

5.5.5   On completion of the expansion test, the sample should be subject to petrographic analysis (see paragraphs 5.4.22-5.4.23) for further diagnosis and examination of the formation of any gel. A detailed description of the recommended technique is contained in the *Diagnosis of Alkali-Silica Reaction* [25].

Thaumasite Form of Sulfate Attack

5.5.6   In order to confirm the presence of the thaumasite form of sulfate attack (TSA) or to determine the area and depth of concrete affected, concrete samples need to be subjected to laboratory analysis. Techniques which are suitable for confirmation of the presence of thaumasite include:

•   Petrographic analysis (optical microscopy) (see paragraphs 5.4.22-5.4.23);

•   X-ray diffraction [13]; and

•   Scanning electron microscopy.

5.5.7   The chemical analysis of concrete for sulfate content (see paragraphs 5.7.10-5.7.11) cannot, in isolation, differentiate between TSA and other forms of sulfate attack. A combination of chemical analysis and the above test techniques is the most powerful diagnostic tool. Further information is provided in the report of the Thaumasite Expert Group [26]. Testing should always be carried out by laboratories with appropriate experience and which have been instructed in the diagnosis of TSA.

5.6   DETERIORATION RATE

Linear Polarisation Resistance

5.6.1   The linear polarisation resistance technique [27] is a technique of assessing the rate at which corrosion is taking place. It is typically used in conjunction with the resistivity (see paragraph 5.5.2) or half-cell potential (see paragraphs 5.7.3-5.7.6) tests once areas at risk from corrosion have been identified. The technique measures the electrical resistance of reinforcement against a small impressed current (approximately ±10 to 20mA) applied about the at-rest potential obtained from the half-cell measurement. However, it is unable to distinguish between severe localised (pitting) corrosion and more extensive general corrosion.

Electrochemical Techniques

5.6.2   In addition to linear polarisation resistance, a number of other electrochemical techniques have been used for evaluating the corrosion rates of steel reinforcement in concrete structures. These include:

- AC impedance;

- AC harmonic analysis;

- Electrochemical noise;

- Electrical impedence spectroscopy; and

- Galvanostatic pulse transient analysis.

5.6.3   Gowers and Millard [28] have studied four of these techniques plus Linear polarisation resistance, and concluded that none gives a precise value of corrosion rate for steel in concrete. They advise that results from any of these techniques should be used with care, taking account of the limitations of each technique. They conclude that Linear Polarisation Resistance, AC impedance, AC harmonic analysis and Galvanostatic pulse transient analysis are all useful techniques for obtaining direct measurement of corrosion rates, but that Electrochemical noise is a less accurate technique. They suggest that AC impedance, AC harmonic analysis and Electrochemical noise are likely to be restricted to laboratory use for the immediate future. Linear polarisation resistance is routinely used (see paragraph 5.6.1) but the Galvanostatic pulse transient analysis offers a number of advantages and could be used more widely.

5.7   DETERIORATION CAUSE OR POTENTIAL

Carbonation Test

5.7.1   The carbonation test is used to measure the depth of carbon dioxide penetration into concrete. The protection given by concrete against corrosion of the reinforcement is reduced if the concrete becomes carbonated. Carbonation proceeds inwards from exposed surfaces of the concrete, generally at a very slow rate. The depth of this carbonation front may be determined by the use of a chemical indicator, usually phenolphthalein solution.

Corrosion Prevention Association

5.7.2   A sample of the concrete is broken off to provide a fresh surface onto which the indicator is sprayed. Pink areas represent uncarbonated, highly alkaline concrete, while clear areas represent the zone of carbonation. Alternatively, a small hole can be progressively drilled and dust samples placed on filter paper saturated with indicator solution until the first 'pink' indication appears; the depth of the drill hole where the last sample was removed from indicates the depth of the carbonation front. A general rule of thumb is that the depth of carbonation is proportional to the square root of the age of the structure.

Half-Cell Potential Measurement

5.7.3   This widely used technique identifies areas of reinforcement that may be at higher risk from corrosion, but it does not predict or identify areas of actual corrosion or the corrosion rate. The test equipment measures the potential difference between the steel reinforcement and a reference half-cell (usually copper/copper sulfate). The interpretation of the degree of risk of corrosion is based largely on experience. Isopotential contour maps can be produced over an appropriate part of the structure indicating areas of differing possible corrosion risk: areas with 'whirlpools' of closely spaced contours represent areas where active corrosion can be expected. As a general guideline, the corrosion risk is often taken to be less than 10% for a potential more positive than -200mV and greater than 90% for a potential more negative than -350mV. It should be noted that these criteria are specific to the type of structure. The test is highly sensitive to changes in moisture content and temperature.

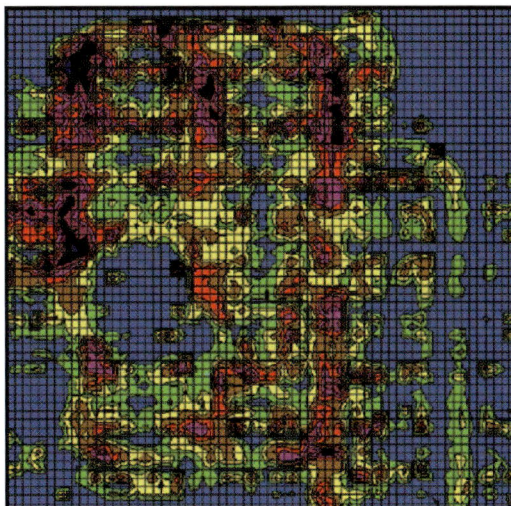

MG Associates

5.7.4   Testing should normally be carried out in a series of test areas, generally 2m x 1m, and within each test area half-cell potentials should be taken on a 500mm grid. Test areas should be positioned where chloride contamination is expected to be the most severe or where there are signs of deterioration. The following guidelines are suggested:

- ***Piers, Abutments, Columns and Crossheads*** – for elements below a deck joint test areas should be positioned close to the top of the element where leakage or staining has occurred. Where substructures are subject to traffic spray, test areas should be positioned close to ground level; it is often appropriate to test on the 'leading' face of a pier. It is normal for two to four test areas to be investigated on an abutment or pier subject to both leakage and spray

- ***Wingwalls and Retaining Walls*** – where these elements are subject to traffic spray, test areas should be positioned close to ground level. For long walls close to traffic lanes, it may be appropriate to repeat the test areas at intervals of about 10m.

- ***Parapets*** – for reinforced concrete parapets subject to traffic spray, test areas should be positioned on the traffic face. For long parapets it may be appropriate to repeat the test areas at intervals of about 10m.

5.7.5   The test is usually supported by chloride sampling where chloride induced corrosion is being investigated. Dust samples for chloride analysis should be taken from positions of high negative half-cell potentials or where the potentials change rapidly over a short distance. Depth of cover should be measured over the test areas and carbonation depth should be measured at representative locations. Where appropriate, permanent connections to the reinforcement should be made to facilitate half-cell testing.

CMT Instruments Limited

5.7.6   Currently available equipment includes 'single cell' equipment, portable 'wheel', and 'multi-cell' devices. These latter two may include computerised data recording and presentation, which can greatly assist large scale surveys. Further details on this technique have been published elsewhere [29, 30, 31 and 32].

Sulfate Resistance of Concrete

5.7.7   This test is used to determine the resistance of concrete to sulfate attack; where the sulfate may be within the concrete mix (e.g. aggregate) or from an external source (e.g. clay sub-soils and de-icing salts). Sulfate attack may lead to deterioration of the concrete (e.g. cracking and a progressive disintegration) and pitting corrosion of reinforcement [33]. The sulfate resistance of a concrete sample can be deduced from a microscopic examination of the cement grains. Dust samples from the concrete are set in a resin, which is then sectioned and polished to expose the cement grains. Chemical etching of the grains is used to reveal the mineral phases characteristic of the type of cement present. These can be identified under a microscope. This examination requires specialist laboratory techniques and the results should be regarded as indicative but not absolute.

Chloride Content

5.7.8   This test is used to determine the degree of penetration of chlorides into concrete and/or the concentration of chlorides in the concrete; high chloride content indicates a high corrosion potential or a high likelihood of corrosion activity. The penetration/content is measured by taking samples of concrete dust from differing depths by drilling a hole of about 20mm diameter into the concrete member. The powder is analysed in the laboratory by the technique described in *BS 1881: Part 124* [34]. A quick simplified technique available as a site test is described in the Building Research Establishment *Information Paper IP21/86* [35], although this is not as accurate as the laboratory test. The rate at which chlorides can diffuse through a concrete member is also an important factor in estimating durability; see paragraph 5.7.9 for an appropriate test.

Diffusion Test

5.7.9   The rate of movement of chloride through a concrete specimen can be used to determine the diffusion coefficient, which provides beneficial information in assessing the durability of the concrete and developing a suitable maintenance strategy. The test requires a 100mm diameter concrete core, which is fully water saturated under vacuum and then immersed in a concentrated chloride solution at a specified temperature. The depth of chloride ion diffusion into the core can then be obtained by incremental grinding and potentimetric titration. This technique is mainly used in research.

Sulfate Content

5.7.10   This test is used to determine the sulfate content of concrete; where the sulfate may be within the concrete mix (e.g. aggregate) or have originated from an external source (e.g. clay sub-soils and de-icing salts). Powdered concrete test samples are taken from the concrete, usually by drilling. The sample is tested in the laboratory by removing various constituents chemically and then using barium chloride solution to obtain a sulfate precipitate. Test details are contained in *BS 1881: Part 124* [34].

5.7.11   The presence of sulfates within the matrix may not mean that sulfate attack has occurred in the concrete member but indicates that it is at risk; petrographic analysis may be required for a more positive diagnosis (see paragraphs 5.4.22-5.4.23). Where internal sulfate attack is suspected from delayed ettringite formation, this can only be diagnosed by detailed petrographic analysis. For the diagnosis of the thaumasite form of sulfate attack, see paragraphs 5.5.6-5.5.7.

Alkali Content

5.7.12   The content of alkali in concrete has relevance to the initiation of alkali-aggregate reactions. Dust samples obtained by site drillings of the concrete or from cores should be tested in the laboratory. The alkali content is determined by flame emission or atomic absorption spectrophotometric techniques. The oven dry density of the concrete is required in order that the alkali content may be expressed in terms of kg/m^3 of concrete.

Analysis of Concrete

5.7.13   A range of tests are available for testing the constituents of concrete including the cement content, aggregate content, aggregate grading, original water content and type of cement. These tests are described in *BS 1881: Part 124* [34].

5.8   SPECIAL: INVESTIGATION OF POST-TENSIONED TENDONS

5.8.1   This section presents additional testing techniques suitable for post-tensioned concrete bridges with tendons grouted in ducts. The techniques adopted should commence with a simple visual examination and routine surface and material tests. Progression to more involved techniques, described below, may be justified if there is evidence of tendon corrosion and a risk of sudden failure. The investigation should be undertaken in accordance with the guidance in *BA 50* [36], and further useful guidance on assessing the condition of ducts in post-tensioned concrete is provided in *BA 86* [4].

5.8.2   An indication of the general corrosion of the reinforcement in the concrete around the tendons may be taken as indicative of the potential for corrosion occurring in the prestressing steel. Therefore the techniques for determining corrosion risk in reinforced concrete, as outlined elsewhere in Section 5, provide a valuable precursor to the use of techniques specific to post-tensioned tendons. In particular, chloride content of the concrete should be determined (see paragraph 5.7.8).

Highways Agency

5.8.3   Non-destructive testing may be used to assist in the detection of voids in the tendon ducts (see Section 5.2). However, if no voids are found this does not preclude the possibility of corrosion occurring. An assessment of the need for further investigation should be undertaken in accordance with *BA 50* [36]. If voids are detected and the conditions within the concrete are conducive to corrosion of the steel, then internal examination of the tendons should be undertaken. The following outlines the principal techniques available for detection of voids, internal examination of ducts and determination of steel and concrete stresses.

Void Detection

5.8.4   The detection of voids in post-tensioning ducts is important in isolating potential areas where corrosion of the tendon may occur, although, experience to date has indicated that voids are more common than resulting corrosion [4]. The techniques of detection can be non-destructive and a guide to the use of such techniques is included in *BS 1881: Part 201* [37]. Determining the position of any voids, prior to an internal examination to ascertain the condition of the tendon, should restrict the degree of damage caused to the structure. However, the only certain technique of determining the tendon condition is by exposing it for visual inspection. Techniques available for void detection include the following:

- **Endoscopic Examination** – refer to paragraphs 4.2.12-4.2.13 for details of endoscopic examination. The endoscope may be inserted into a small diameter access hole drilled through the concrete. Where possible a hole should then be made in the duct using a hand-held chisel. The number and position of the holes drilled will depend upon the cable profile and the ease of access for drilling. Holes should be drilled as close to the anchorages as possible, at mid-span and at either side of high points where voids are most likely to have formed. The major limitation of this technique is the difficulty of access for the drilling of holes into the duct. Drilling should be closely supervised to prevent damage to the tendon. An endoscope survey carried out using small diameter inspection holes drilled at carefully chosen locations offers a reliable technique of checking for the presence of voids and the condition of tendons. However, this can only be inspected locally to the drilled hole.

- **Air pressure testing** – the techniques of air pressure and vacuum testing enable the volume and continuity of voids and leakage into a duct to be determined. Access has to be gained to the top of the duct by drilling 25mm diameter holes through the concrete. A small hole should then be made in the duct using a chisel. The number and position of the holes drilled will depend upon the cable profile and the ease of access for drilling. Holes should be drilled as close to the anchorages as possible, at mid-span and at either side of any high points where voids are most likely to have formed. The continuity of any voids found is determined by evacuating each hole in turn and measuring any pressure change at the remaining holes. The volume of the voids is estimated by using a water gauge connected to the evacuated holes and measuring the height of water drawn up a perspex tube. Leakage out of the duct is measured by pressurising it and measuring the input flow rate required to maintain a set pressure. Errors can arise in the estimation of the volume of the voids if there is leakage from the duct. In addition, the pressure within a partially grouted duct after it has been evacuated is not uniform. The major limitation of this technique is the difficulty in making holes into the duct, particularly at the end anchorages where the amount of bursting steel makes drilling difficult. In all cases, drilling holes should be carried out with utmost care to ensure that the tendon is not damaged. No information can be gained on the condition of the tendon using this technique of testing alone, therefore an endoscope survey or an open visual examination should also be undertaken.

- **Ground Penetrating Radar (GPR)** – (see paragraphs 4.2.16-4.2.18) is not, at present, able to determine voids or tendon condition in structures with metallic ducts. The ducts mask any reflection from within.

- **Impact-echo** – see paragraph 5.2.4. No information can be gained on the tendon condition using this technique.

- **Reflectometry** – this technique utilises the relationship between the electrical parameters and defects in tendons. A high frequency signal is input at one end of a tendon and received at the same end. Changes in impedance are used to indicate the type of deterioration in terms of a reduction in tendon section and voids in the surrounding grout. The major limitation of the technique is the need to have access to the end of the tendon. Where there are a number of faults in the tendon, the first fault encountered may mask the others. The specialised equipment and difficulties in interpreting the results makes it essential that experienced operators perform this type of work. Any exposure of the end anchorages of the tendons must be carried out with utmost care.

Internal Examination

5.8.5   Once voids and potential corrosion of post-tensioning tendons have been identified, the most direct way of establishing the degree of damage is by an internal examination of the duct. As outlined below, there are a number of ways by which access can be gained to post-tensioning ducts for an internal examination. The degree of damage caused to the structure will depend upon the technique of exposure chosen. Any exposure of the tendon should be carried out with utmost care to ensure that the tendon is not damaged.

- **Break Out** – also see paragraphs 4.2.8 to 4.2.10 for details of hand-held and small machine mounted impact breakers. The damage caused to the concrete section may be irregular and strict control must be exercised over the exposure of the ducts. Exposure of the tendon should be carried out using a hand-held chisel to ensure no damage is caused. These techniques should be limited to the exposure of easily accessible tendons in order to limit the level of damage caused. In addition, micro-cracking of the exposed concrete surface layers is likely and subsequent repair procedures will need to take this into account.

- **Coring** – also see paragraph 4.2.11. Coring offers an effective way of exposing post-tensioning ducts, however, the use of this technique relies upon the operative knowing when the core drill has made contact with the duct, in order to stop drilling. An alternative is the use of a drill with an automatic cut-out which should stop once the duct is reached. A hand-held chisel should be used to expose the tendon. Every effort should be made to ensure that the cooling water does not penetrate the duct.

- **High pressure water jetting** – water jets can be used to expose lengths of post-tensioning ducts after rotary drilling of pilot holes has established the precise location. However, every effort should be made to ensure that the water cannot penetrate the duct; if it is suspected that ducts are not watertight this technique of exposure should not be used. Cutting of the concrete using this technique is likely to produce irregular shaped holes. Water jetting with a grit additive offers better control for concrete cutting, but it should not be used since it can also cut through the prestressing steel.

- **Grit Blasted Holes** – dry grit blasting can be used to form access holes in the concrete in a similar manner to high pressure water jetting. The use of a vacuum pump to carry away the debris negates the problems caused by dust and fly back of material. Caution has to be exercised since the duct and tendon may be damaged, however the lack of water makes this a more attractive option in some situations.

Steel Stresses

5.8.6   Determination of steel stresses at critical sections may be used to give an indication of the local and general levels of residual prestress. The current levels of stress in individual prestressing tendons can be determined directly by hole drilling in conjunction with electrical resistance strain gauges. The technique relies on the principle of stress-relief and is an adaptation of the centre-hole technique (see *BA 50* [36]). The diameter of the hole drilled may be 1-3mm, depending upon the form of the prestressing tendon. The centre-hole technique gives a very accurate picture of the particular wires drilled, but the results must then be extrapolated to the whole tendon. The stress measured is the total stress; it is necessary to subtract locked-in manufacturing stresses. Steel stress measurements on grouted tendons represent conditions for about 1m on either side of the test location and a representative number should be taken if reliable average values are required.

5.8.7   The centre-hole technique, although partially destructive, may be used on a bridge in service. Alternatively, the residual prestress may be determined by a complete release of stress in selected wires or strands. Several wires in a tendon may be cut by closely controlled use of a hacksaw if access is possible. The main benefits of this technique are cheapness and direct measurements of the residual pre-strain in individual wires or tendons. It is necessary to think very carefully before cutting any wires on a bridge in service.

Concrete Stresses

5.8.8   Determination of concrete stresses at critical sections may be used to give an indication of the local and general levels of residual prestress. Specialist techniques have been developed to provide in-situ measurements of concrete stresses using instrumented coring and slot-cutting techniques (see *BA 50* [36]); these are summarised in the following:

- **Coring** – cores of 75-150mm diameter are drilled into the concrete and the stress release determined using carefully selected strain gauge patterns. A likely lower-bound value for the concrete modulus may be determined from compression tests on the concrete cores. Additionally a hydraulic jacking system may be inserted into the hole produced by the coring process, to provide an in-situ determination of the elastic modulus. This form of test provides an in-plane measurement of the modulus and represents an upper-bound composite value created by the surrounding prestressing tendons, reinforcement and concrete. The main advantage of the instrumented coring technique for the assessment of existing levels of concrete stress is that principal concrete stresses can be determined in both magnitude and direction. Although total concrete stresses are obtained, the likely magnitude and presence of self-equilibrating internal concrete stresses due to secondary effects can also be determined during coring.

- **Slot-cutting** – the slot-cutting technique for concrete stress determination may be carried out in several ways. Narrow slots, 300-500mm in length, can be produced using a diamond saw mounted on a travelling rig and cooled by water. Strain measurements taken across the slot on either side of the slot can be converted into stresses using laboratory calibrations and the local elastic modulus determined by hydraulic jacking or compression tests from a complementary core test. An alternative technique of slot cutting employs a mounted air-cooled diamond saw in conjunction with very thin semi-circular flat jacks which are used to restore the stress state uni-axially. The main advantage of this form of slot-cutting technique using pressure compensated jacks is that a value of elastic modulus is not required to be known.

5.8.9 The concrete core and slot-cutting techniques give a useful guide to the overall concrete total stress levels at a given point. Where parallel beams are connected by in-situ concrete or mortar joints, the slot-cutting techniques cause minimal damage to the structure and can provide a measurement of the effect transverse prestress across the longitudinal joint. Both coring and slot cutting have a stress raising effect and require careful making good to prevent ingress of water and road salts.

5.8.10 Cutting of the reinforcement in a structure should be avoided. Apart from local damage to the structure, severance of highly stressed reinforcement may cause debonding and local micro-cracking at the concrete surface. Any strain gauge measurements on the surface are therefore likely to be invalidated.

5.8.11 The various techniques have different levels of sensitivity and the selected techniques should be appropriate to the anticipated levels of concrete stress and the accuracy required in the determination. It is very important to realise that concrete stress measurements do not provide a means for identifying local strand failures in a section. Therefore, the proper use of concrete stress measurements is to provide an indication of the global levels of residual prestress in a deck.

6 Tests on Metal

6.1 GENERAL

6.1.1 This testing techniques described in this section are relevant for a range of metal types, including steel, cast iron, wrought iron, and aluminium (additional techniques specific to aluminium are presented in Section 6.8). Many metal elements on highway structures are protected by paint or other metallic coatings in order to improve durability and/or appearance. Section 6.9 presents a range of tests suitable for paint and metallic coatings.

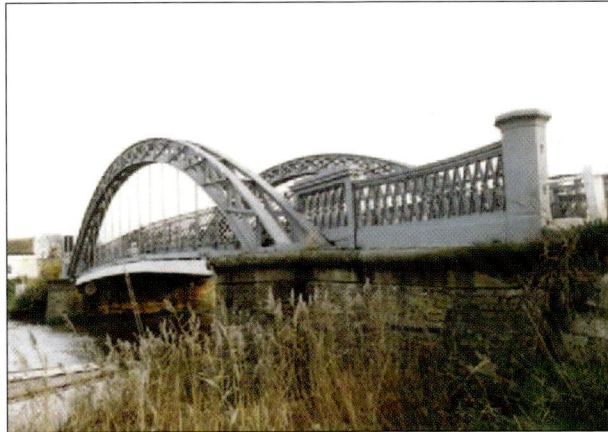

Cambridgeshire CC

6.1.2 Before embarking on destructive sampling and testing, particularly for cast iron or wrought iron members, the need for the testing and the usefulness of the results should be established clearly (refer to Section 2 for guidance on developing a testing programme). If the structure is listed, consent may be needed before taking any samples. Cast iron and wrought iron tend to be variable in consistency and strength, thus a test specimen may not be representative of critical parts of the structure.

6.1.3 Two important issues relating to metal structures are initial identification of the metal type and obtaining samples, these are dealt with in the following.

Identification

6.1.4 A common initial concern in testing metals is identification of the metal type; this is important since they have different properties. Initial identification can often be made by visual inspection and by testing the magnetic property (since aluminium is non-ferrous, i.e. magnets do not stick to aluminium). Details of visual criteria for identifying cast iron, wrought iron and steel are provided in Volume 1: Part F: Appendix F.

Obtaining Samples

6.1.5 Information about obtaining samples from steel can be found in *BS 970: Part 1* [38] and *BS EN 10025 (for general information)* [39 and 40]; *BS EN 10002: Part 1 (for tensile testing)* [41]; *BS EN 10045: Parts 1 and 2 (for Charpy impact testing)* [42 and 43]; *BS EN ISO 6506: Parts 1, 2 and 3* [44]; *BS EN ISO 6507: Parts 1, 2 and 3* [45] and *BS EN ISO 6508: Parts 1, 2 and 3 (for Brinell, Vickers and Rockwell hardness testing respectively)* [46]. Information about obtaining samples from welds can be found in *BS EN 875* (for impact testing) [47] and *BS EN 910* (for bending tests) [48].

6.1.6 Techniques for removing samples include sampling drilling swarf, scrapings, hacksaw and manual effort, interlocking drilled small holes and flame cutting (which requires at least an additional 10mm around each face to allow for the removal of the heat affected zone). It is of prime importance that the taking of such samples does not weaken the structure by the reduction of the cross section or the introduction of sharp corners that may cause stress concentration, or embrittlement following heating.

6.2 STRUCTURAL ARRANGEMENT AND HIDDEN DEFECTS

6.2.1 In addition to the testing techniques described in Section 4.2 the following testing techniques are also suitable for metal structures.

Ultrasonic Testing

6.2.2 Ultrasonic testing is used to identify the thickness of metal and to detect defects in the metal; the former can be used to measure loss of thickness, for example, in a sheet pile wall. It makes use of ultra high frequency sound waves in the range of 2-5MHz. The principle of the operation is that electrical energy is converted into sound energy by a transducer. This sound wave passes into the component, usually via a coupling agent such as water or oil, the wave is reflected by the back face and returns to the transducer. It is then re-converted into electrical energy and both pulses, the initial generation and the return, are displayed on an oscilloscope. As sound travels at a constant speed the use of a time base on the oscilloscope allows the thickness of the steel in the area under test to be determined. Any defect present between the two surfaces of the steel will interrupt the sound beam and cause a reflected pulse to occur at a location intermediate between the impulse and return signals.

6.2.3 A large variety of transducers have been developed which allow the majority of components and defect locations to be examined. There are two main types of transducer (or probes as they are more commonly known): compression probes that provide a vertical sound beam and angled probes that introduce the beam at a variety of fixed angles.

6.2.4 The advantages of ultrasonic testing are its portability, sensitivity and ability to detect planar defects at depth such as lamellar tearing, hydrogen cracking, solidification cracking and lack of fusion defects. The system is less reliable however in identifying surface and minor defects, for example fine liquidation cracks, small slag entrapments and where porosity is less than 5mm diameter, when misleading results are often produced.

6.2.5 Ultrasonics is a sensitive and versatile inspection tool but interpretation of the results is dependent upon the skill of the operator to recognise and report indications of defects. Test details are described in *BS 7448: Part1* [49], *BS EN 10228-3* [50] and *BS EN 12680-1* [51].

Dye Penetrant Testing

6.2.6 Dye penetrant testing is probably the most commonly used site technique for defect detection on steel structures. It provides information on the surface condition of the steel such as the presence of cracks and surface impregnations, particularly those associated with welding.

6.2.7 The technique involves the spray application of a dye, which is typically red, onto a thoroughly cleaned surface, usually to a grit blasted standard. The dye is allowed to soak into any surface at the defects and cracks. The surface is then wiped clean and sprayed with a chalk based developing fluid. This absorbs the dye remaining in the defects, which shows up as red lines on a chalk background. Modifications to the system can include the use of penetrants of different viscosity to detect different sized cracks, wet rather than dry developers and penetrants which fluoresce under ultra-violet light, enabling the detection of smaller defects.

6.2.8 The technique is inexpensive and easily applied and interpreted, allowing small defects to be detected that are not visible to the naked eye. The principle disadvantage however is that only surface defects can be detected, thereby limiting the usefulness of the technique for defect depth determination: it merely demonstrates that a defect is present. Further details of dye penetrant testing are provided in *BS EN 571-1* [52].

Magnetic Particle Inspection

6.2.9 This is essentially a similar process as dye penetrant testing. However it can detect defects that are sub-surface, i.e. 1-2mm deep. When a steel component is magnetised, any defects present produce local disruptions of the lines of force generated at the metal surface. The maximum disruption occurs when the lines run perpendicular to the defect. Therefore the component must be magnetised from two mutually perpendicular directions for maximum effect.

6.2.10 On site, the metal is magnetised locally by the use of two current carrying rods, usually of copper or aluminium, which are placed on the surface. The surface may also be magnetised using a permanent magnet or electro magnet. The rods produce a circular magnetic field about each contact point when the current flows between them. The surface is then sprayed with a solution containing iron powder. If the magnetic field is undisturbed by any surface or subsurface irregularities, the iron powder is aligned with the field in a uniform film. If a defect is present, the field is disturbed and a concentration of magnetic lines of force will occur forming a concentration of iron powder. This concentration will show the defect during visual inspection. A surface crack is normally indicated by a line of fine particles that follow the outline of the crack, whilst a subsurface irregularity is shown by a fuzzy build-up of the fine particles on the surface near the defect. Surface defects are detected transverse to the field. Therefore, by moving the rods across the surface to be examined, defects at any orientation can be detected. Test details are described in *BS EN 1369* [53], *BS EN ISO 9934-1* [54] and *PD 6513* [55].

6.2.11 The major advantages to this technique are its relative portability, the minimum skills required to operate it and its ability to detect very fine cracks. Although it is considered to be a simple technique for smooth flat-plate surfaces with flush welds it is less simple with more complex structural arrangements. Also, the test zone is left in a magnetised condition that can affect subsequent welding operations.

Eddy Current Testing

6.2.12 Eddy current testing can be used to detect defects as small as approximately 0.05mm to 0.1mm deep (assuming very good surface finish). The test is similar to the magnetic particle inspection technique, except the defect is detected by a perturbation in an electrical field rather than the magnetic field. The technique uses a coil carrying alternating current that produces an eddy current in a conductor nearby and in turn creates impedance in the exciting coil. The impedance produced is dependent upon the nature of the conductor and the exciting coil, the magnitude and the frequency of the current and the presence or absence of the discontinuities in the conductor. The equipment is calibrated and the coil is scanned over the surface of the area to be examined, any defects present produce a characteristic change in impedance which is read from a meter. Further details are provided in *BS EN 12084* [56].

6.2.13 The technique has limited application essentially because only simple geometries can be examined. Complex geometries change the impedance readings in themselves and thus mask the effect of defects. It does have the advantage that defects can be detected at depth and defect size can also be estimated from the response of the area examined. However, the technique is generally used on non-ferritic applications, and for special applications, such as paint dry film thickness measurements.

6.3 DISTORTION AND MOVEMENT

6.3.1 The distortion and movement techniques described in Section 4.3 are generally applicable to metal structures and elements. Strain gauges, as described below, can prove beneficial in the assessment of fatigue.

Strain Gauges – fatigue of metal structures

6.3.2 Strain measurements are frequently required for the assessment of residual fatigue life of steel members. Measurements should normally be made in the vicinity of fatigue susceptible details but outside areas of stress concentrations caused by geometrical discontinuities such as a change in section, an aperture or a welded attachment. Details of strain gauges are provided in paragraph 4.3.12.

6.4 MATERIAL PROPERTIES

Hardness Testing

6.4.1 This test provides an approximate value of the ultimate tensile strength of the metal (steel, cast iron and wrought iron). Hardness testing can be carried out in-situ using a portable Brinell hardness tester. The diameter of the imprint made, when a hardened steel ball is pressed against a smooth surface with a known force, is measured. The figure obtained is then correlated very approximately with the ultimate (not yield) strength and as such is a useful measure of consistency or variability. The result however should be regarded as a guide only and the hardness survey on its own should not be relied upon as a direct measure of strength.

6.4.2 The ranges of tensile strengths of the various grades of structural steel can overlap and it may not be possible therefore to identify positively the grade of steel from its hardness number. They can however provide reassurance on consistency and often highlight material anomalies, but they should be calibrated against specimens of known strengths.

6.4.3 Standard laboratory hardness tests include Brinell, Vickers and Rockwell testing and can be used to verify the results obtained from the portable in-situ testing equipment. Test details for each of the testing techniques are described in *BS EN ISO 6506 Parts 1, 2 and 3* [44], *BS EN ISO 6507: Parts 1, 2 and 3* [45] and *BS EN ISO 6508: Parts 1, 2 and 3* [46] respectively. The test technique for cast iron is described in *BS EN 1561* [57].

Chemical Analysis

6.4.4 Chemical analysis by taking drillings (drilling swarf) or scrapings can be used to check the weldability of metals or to provide further information on the type and associated physical properties of metals. For structural steels, it is usually sufficient to test for carbon, silicon, manganese, sulphur and phosphorus. Care must be taken in collecting such samples to avoid contamination and to be sure that only the materials from the member under consideration are collected for analysis.

Metallography

6.4.5 Metallographic examinations can provide useful information on the internal structure of a material, e.g. porosity and microstructure. A metallographic microscope, or other suitable equipment, that provides observations and measurements on a screen, at a suitable level of magnification (×100 to ×200), is required. This normally requires a sample with one flat surface of approximately 10mm × 10mm.

Tensile Testing

6.4.6 The information provided by tensile testing includes yield strength, ultimate tensile strength, modulus of elasticity and elongation at failure; this information is particularly important for the strength assessment of highway structures. Test details are described in *BS EN 10002: Part 1* [41]. Ideally the test requires the removal of samples approximately 200mm x 100mm from structural members but in practice, and at some extra cost, samples of 100mm x 50mm or smaller can be used. The cross-section of the samples may be circular, rectangular, square or other form. In addition, a wedge penetration test and a cylinder splitting test have been developed for assessing the tensile strength of cast iron specimens [58].

Impact Tests

6.4.7 Brittleness and notch ductility can be measured by Charpy (beam) or Izod (cantilever) testing which measure the energy required to fracture a standard notched specimen with a blow from a pendulum. The former tends to be the most versatile since the results are obtained over a range of temperatures. Test details for the Charpy test are described in *BS EN 10045: Parts 1 and 2* [42 and 43]. Test details for the Izod test are described in *BS 131: Part 1* [59].

6.4.8 The sample size required for the Charpy impact test is 55mm x 10mm x 10mm for either the U or V notch test. Sample sizes for the Izod test are 70mm for one notch, 98mm for two notches and 126mm for three notches, with the cross section being 10mm x 10mm.

6.5 DETERIORATION ACTIVITY

6.5.1 The acoustic emission technique, described in Section 4.5, is suitable for metal structures and elements.

6.6 DETERIORATION RATE

6.6.1 Tests that directly measure the rate of metal deterioration (corrosion) have not been identified for inclusion in this Manual. The normal approach for measuring the corrosion rate of metal structures includes:

- Measurement of plate thickness (e.g. flange/web) and/or corrosion/pit depth, and comparison with the original design or manufactured thickness. This can give an indication of the rate of thickness loss if the original construction date is know. However, this rate can be misleading if the corrosion initiated after the date of construction; if this is the case then a suspected date of corrosion initiation can be used. Drilling of elements, as described in paragraph 4.2.8, can be used to measure the thickness of elements where only one side is accessible. Ultrasonic testing described in paragraphs 6.2.2-6.2.5 can also be used.

- A series of measurements of plate thickness (e.g. flange/web) and/or corrosion/pit depth over a period of time, thereby enabling the corrosion rate to be directly calculated (without the need to refer to the construction date or a known/suspected corrosion initiation date). The frequency of the measurements will depend on the suspected rate of corrosion, in some case they may need to be spread over a period of years.

6.6.2 Formal long term programmes of thickness measurements on a selected number of structures or elements, say as part of the Principal Inspection, can provide highly beneficial information for Whole Life Cost analysis and determination of service life.

6.7 DETERIORATION CAUSE OR POTENTIAL

6.7.1 The main causes of deterioration of metal structures and elements are exposure, manufacturing defects (including the constituent materials of the metal) and cyclic loading.

Exposure

6.7.2 Corrosion of metal is normally caused by exposure, either to the naturally occurring environment or additional man-made pollutants. As such, the cause of corrosion is normally failure of the protective coatings. Details of testing techniques that assess the integrity of protective coatings are provided in Section 6.9

Manufacturing Defects

6.7.3 Some of the testing techniques described in Section 6.2 can be used to identify manufacturing defects such as voids or inclusions, although sampling is required to ascertain the makeup of the inclusion. The testing techniques described in Section 6.4 can be used to assess the make up of metal and to identify any constituents parts that may be causing, or have the potential to cause, deterioration.

Cyclic Loading

6.7.4 The cyclic loading of metal elements causes deterioration through fatigue, i.e. metal structures normally have a certain fatigue life defined by a number of loading cycles. Strain gauges can be used to assess the remaining fatigue life, see paragraph 6.3.2.

6.8 SPECIAL: ALUMINIUM

6.8.1 The techniques used for testing steel and iron structures are generally applicable to aluminium alloys, except those relying on magnetic properties, such as magnetic particle testing. There are a wide range of aluminium alloys on the market, with specifications overlapping, and only some of these are used in the construction industry. Removal of samples for chemical analysis or metallographic examination (see paragraphs 6.4.4-6.4.5) is therefore unlikely to be conclusive; if details from drawings or fabrication records are not available then mechanical load testing of large specimens is required in addition to chemical and metallographic analysis.

6.8.2 Stress corrosion may be identified by cracks that propagate with little associated macroscopic plastic deformation. It should not be assumed that the crack visible at a surface is representative of the behaviour below the surface. Metallographic examination of samples may be required to confirm the source and type of cracking. In some cases laboratory testing of samples may be undertaken to investigate susceptibility to further stress corrosion cracking. However, reproducing environmental and stress conditions is difficult and results are unlikely to be of much practical use.

6.9 SPECIAL: PAINTWORK AND METALLIC COATINGS ASSESSMENT

6.9.1 The condition of the paintwork and documentation of the extent of corrosion is important to ensure that the protective coatings, where applied, provide the required performance. Inspections should make an overall judgement as to the condition of the paint based on the state of the majority of the surfaces, not on localised areas of corrosion. Detailed paint inspections should only be undertaken by experienced specialist personnel.

6.9.2 It is beneficial to obtain the historic records of the structure wherever possible; these will provide details of the paint system(s), the date(s) of application and the nature of surface preparation used prior to application. Where the records are not available, techniques listed below can be used to enable the approximate number of paint layers present to be identified and assist in identifying paint characteristics to determine the paint system present and the presence and types of surface contaminants.

6.9.3 Some of these techniques may be suitable where the systems are zinc coated or aluminium coated steel. It is advisable however to check with specialist inspectors and experienced personnel which testing techniques and equipment are applicable to these types of systems. For example the electronic gauges used for the thickness measurement of organic coatings may be used for metallic systems with the use of the appropriate probe attachment.

Paint Film Thickness Measurements

6.9.4 A range of instruments are available for the measurement of dry paint film thickness and, if used properly and correctly calibrated, they provide reasonably accurate measurements. The difficulty is in deciding the number of measurements, which should be taken to indicate the true average thickness. Good practice suggests that measurements should initially be taken at the rate of one per square metre. If the results of testing 10 square metres indicate that the coating is reasonably uniform and no measurement is below the specified minimum, then the rate of measurement can be continued. If wide variations exist in the first group of readings the number of measurements should be increased. If the uniformity of the coating persists over a large number of the 10 square metre areas, then the number of measurements may be reduced. Where the specification calls for measurements to be taken over specific areas this must be carried out, alternatively, the judgement of an experienced inspector should be relied upon.

6.9.5   Further examination of the paint film may be carried out using a paint inspection gauge. The gauge will cut a V in the paint film through to the steel. The cut is inclined on one side to enable individual coats of paint to be identified and if there is sufficient contrast between the coats, the thickness of each coat can be determined. The gauge can also be used to examine any irregularities such as under or interface corrosion creep.

6.9.6   Paint flakes may be removed to confirm with the aid of a stereo microscope the total or individual paint film thickness.

Adhesion Tests

6.9.7   The adhesion of paint coatings both to the substrate and interstitially should be checked when chemically cured paints are used. Spot checks can be made by the cross hatch adhesion test or the cross cut adhesion test, both of which attempt to lift the film from an incision made into the paint film. Test details for the cross-cuts test are described in *BS EN ISO 2409* [60].

6.9.8   On site, the technique often used is that which measures the perpendicular force required to remove a 'dolly' adhered to the surface by using a quick curing resin. This pull off test can be performed using the Elcometer or HATE apparatus, test details are described in *BS EN ISO 4624* [61]. Generally however, these tests only provide an indication of the potential problems and further investigations are often required to substantiate the results.

Discontinuities of the Paint Film

6.9.9   Discontinuities such as pin holes or pores that may be present even in multi-coat systems can be detected using a 'holiday' detector, which may be either low voltage or high voltage DC. The former is generally suitable for films less than 500 microns whilst the latter is suitable for thick non-conductive films.

Chemical Tests

6.9.10   Chemical tests can include distilled water swabbings of the surface to determine the type and concentration of surface contaminants such as chloride, ammonia and sulfate. Paint flakes may need to be removed and then may be used for chemical analysis to confirm the presence of lead.

7 Tests on Masonry

7.1 GENERAL

7.1.1 The testing techniques described in this section are relevant for all types of masonry construction, e.g. brick or block, and material, e.g. manufactured bricks, sandstone, granite, etc. Masonry structures have rarely been constructed in recent years so the majority of those on the highway network are of a considerable age, typically greater than 50 years old. As such, it is not uncommon for the records, including as-built drawings, to be missing. Therefore a frequent use of testing is to confirm the structural arrangement of masonry structures, for example, masonry arch bridges. Further guidance on testing and investigation of masonry arch bridges is provided in *Masonry Arch Bridges: Condition Appraisal and Remedial Treatment* [62].

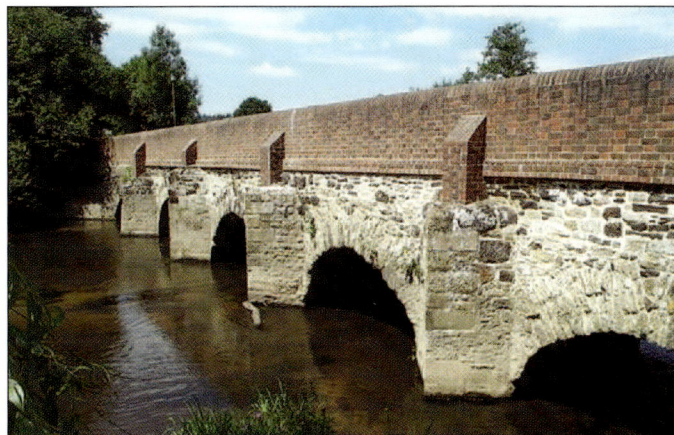

Surrey CC

7.2 STRUCTURAL ARRANGEMENT AND HIDDEN DEFECTS

7.2.1 The testing techniques described in Section 4.2 are generally relevant to masonry structures. In addition, the following considerations are relevant.

Observation, Measurement and Simple Investigation

7.2.2 General details are provided in paragraphs 4.3.2 to 4.3.3. For masonry structures, a probe, less than 10mm wide, is required for testing the condition of mortar joints by scraping and probing. Longer and/or thinner probes may be required to explore the depth of hollow areas and cracks. Inspectors should be aware that crevices in masonry bridges may be roosting sites for bats or nest sites for birds. It is an offence to damage or destroy such sites. Guidance on the appropriate procedure for inspection of structures that may contain bat roosts or birds' nests is provided in Volume 1: Part C: Section 3.11.

Trial Pit Excavation

7.2.3 General details, including guidance on the number and/or size of excavations are provided in paragraphs 4.2.5-4.2.7. In masonry bridges, excavation of surfacing and fill is typically carried out to determine the thickness of the arch ring and spandrel walls; to confirm the presence and details of waterproofing, internal spandrel walls and services; and to determine the type and condition of the fill and any damage caused to the arch during the installation of services.

7.3 DISTORTION AND MOVEMENT

7.3.1   The testing techniques described in Section 4.3 are generally relevant to masonry structures.

7.4 MATERIAL PROPERTIES

Compressive Strength

7.4.1   Although most masonry is not as strong as structural concrete, the proportions of masonry structures mean that the compressive strength of the material is not often a critical factor. Only occasionally, for example when assessing the strength of a highly stressed arch, is it necessary to determine the compressive strength of masonry. This is fortunate, since there is no simple way of assessing the basic compressive strength of masonry in-situ.

7.4.2   The strength of masonry depends on the mortar as well as the brick or stone; if compressive strength data is required then testing of samples from the structure is the only reliable technique. It should be remembered that materials and hence strength may vary widely over a structure. To estimate the strength of masonry it is necessary to remove a whole section of masonry and for it to be tested in accordance with *BS 5628: Part 3* [63].

7.4.3   Stone blocks may have cores taken from them which can then be tested in accordance with the guidelines for testing concrete cores given in *BS EN 12504-1* [21] and *BS 6089* [64]. Specimens should be tested in the direction in which they are normally stressed in the structure. Whole bricks can be tested by the techniques described in *BS 3921* [65].

Pull-Out Test

7.4.4   This test provides an indication of the compressive strength of mortar. The test, which is described in the Building Research Establishment *Digest 421* [66], is straightforward and measures the force required to shear out a small cylinder of the mortar. A 30mm deep pilot hole is drilled in the mortar into which a helically-threaded stainless steel wall-tie is driven. This is then jacked out and the load recorded. This load can then be related to the compressive strength by comparison with calibration curves. Weaker mortars may not need the drilled pilot hole. The test is suitable for testing materials of strengths up to about 7N/mm².

Flat Jack Test

7.4.5   This test is used to measure the in-situ compressive stress in masonry and is described in the Building Research Establishment *Digest 409* [67]. A small slot is cut in the face of the structure normal to the expected stress direction. Strain gauges are positioned around the slot prior to cutting and monitored when the slot is cut. A flat jack is then inserted and pressurised until the gauges return to their original values prior to the stress relief caused by cutting the slot. The pressure then represents the stress in the structure at the position of the cut. This test has a long history and has been proven to be reliable.

Flexural Bond Strength Test

7.4.6   A bond wrench is a simple tool for the in-situ testing of the bond of masonry units to mortar. The test is described in the Building Research Establishment *Digest 360* [68], which also describes the development of a bond wrench known as BRENCH. A basic bond wrench is a long lever that is clamped to a brick or stone at one end. The other end has a load applied until the brick is prised free from the mortar joint immediately below it. The load at which this occurs is a measure of the bond strength of the masonry and can be converted to a stress using a calibration chart. This enables an assessment of the lateral bending strength of an existing panel or parapet wall to be made.

Shove Test (In-situ Shear)

7.4.7   The in-situ shear strength of the mortar bed joint may be tested in accordance with ASTM *Standard C1197* [69]. The technique involves removing a masonry unit to provide access for a hydraulic jack and removal of the joint on the opposite side of the unit. The jack then pushes against the isolated unit in order to break the bed joints. The test gives a measure of the shear resistance of a masonry wall or parapet and can be used comparatively as a quality indicator. The relationship between sliding shear strength and vertical stress would also need to be investigated to indicate behaviour at different levels in the structure.

Physical and Chemical Tests of Mortar

7.4.8   Physical and chemical tests can provide useful information about the physical properties and durability (e.g. constituent components) of mortars used in masonry structures. Mortars may be sampled and tested in the laboratory to determine physical and chemical properties; details of suitable procedures are provided in *BS 4551* [70]. It should be realised that a number of different batches of mortar may have been used during construction. The mortar used for final pointing, or for subsequent repairs is unlikely to be the same as the mortar used to construct the core of the structure.

7.5   DETERIORATION ACTIVITY

7.5.1   The acoustic emission technique, described in Section 4.5, is suitable for masonry structures and elements.

7.6   DETERIORATION RATE

7.6.1   Test that directly measure the rate of masonry deterioration have not been identified for inclusion in this Manual. Deterioration is normally slow and significant changes in surface condition are normally picked up through standard General and Principal Inspections.

7.6.2   However, due consideration should be given to masonry arch structures that carry railways. If there is a change in use of the railway line, e.g. changes from freight to suburban, there is the potential for the structure to rapidly deteriorate and potentially fail [71]. In such instances, a combination of inspections and monitoring may prove effective.

7.7 DETERIORATION CAUSE OR POTENTIAL

7.7.1 The main causes of deterioration in masonry structures are weathering and change in loading conditions although the presence of de-icing salts can increase the effect of weathering due to freeze-thaw action.

- *Weathering* – weathering of some masonry blocks can be due to the use of inappropriate mortar in the joints. Deterioration of masonry structures due to weathering is normally a slow process. Changes in condition caused by weathering can usually be effectively identified through General and Principal Inspection information. As such, testing techniques that seek to directly measure the deterioration potential of weathering are not required as the effects are normally evident to the Inspector/Supervising Engineer or obvious from a change in condition of the structure.

- *Loading* – changes in loading (as discussed in paragraph 7.6.2), can cause deterioration of masonry structures, and in some instances this can be a rapid deterioration. If masonry structures are to be subject to a significant change in loading (weight and/or frequency), the Supervising Engineering should give due consideration to the approach required to monitor the safety and functionality (performance) of the structure, e.g. daily/weekly/monthly inspections or remote monitoring.

7.7.2 Masonry structures, in particular the mortar, can be susceptible to rapid deterioration if exposed to water runoff, water seepage through a bridge deck and chemicals. Inspection staff should pay particular regard to the paths taken by runoff/seepage, especially if these change overtime, and record them on the inspection pro-forma. In the case of chemical spillages, laboratory tests may be required to investigate the impact of the chemical on the mortar and the masonry.

8 Tests on Timber

8.1 GENERAL

8.1.1 Timber structures are not common on the United Kingdom highway network, although there are a significant number on the Public Right of Way (PRoW) Network. Recently, there has been a renaissance in the construction of timber footbridges as focal points, especially in conservation areas. The following provides a summary of the testing techniques suitable for timber structures; further guidance is provided by the Timber Research and Development Association (TRADA) [72].

8.2 Structural Arrangements and Hidden Defects

8.2.1 The testing techniques described in Section 4.2 are generally relevant to timber structures. In addition, the following technique is relevant.

Hammer Tapping

8.2.2 The soundness of timber members can be assessed by tapping with a light hammer. Timber which is decayed will produce a dull sound as opposed to a ringing noise. Hammer tapping can also be used to detect loose connections.

8.3 DISTORTION AND MOVEMENT

8.3.1 The testing techniques described in Section 4.3 are generally relevant to timber structures.

8.4 MATERIAL PROPERTIES

Microscopy

8.4.1 If an experienced inspector, or individual who specialises in timber, is unable to identify the species, microscopy may be used. Microscopy requires the recovery of samples but allows the wood matrix/grains to be analysed in detail, enabling the species to be readily identified. Microscopy can also be used to investigate fungal types present in the timber. The presence of fungal strands in a timber is not necessarily a cause for concern; instead it is the severity, spread and current activity level of fungal decay which are of importance.

Hardness

8.4.2　Indicative surface condition and material properties of timber can be derived from a hardness test. A hardness (Janka or Brinnell) test typically involves measuring the indentation caused by a spherical object pressed into the surface of the timber under a known load.

Tensile and Compressive Strength Test

8.4.3　Where this information is not available, tensile and compressive strength tests can provide important information for the assessment of timber structures. Tests for determining tensile and compressive strength, and other mechanical properties, of timber components are described in *BS EN 1193* [73]. These tests normally require a test specimen to be taken from the component.

Resonance, Vibration or Mechanical Impedance

8.4.4　These techniques can be used to determine material properties, such as the modulus of elasticity. When an object is caused to vibrate, e.g. by an impact, the frequency at which the highest amplitude of vibration occurs is normally the object's first resonant frequency. The natural resonant frequency depends on numerous factors, including modulus of elasticity, density, shape and fixing conditions. Thus, if other factors are known, a resonance test may be carried out to determine the modulus of elasticity [72]. This technique can be used for simple components, such as beams, and can prove beneficial in grading timbers.

Glue and Fastener Tests

8.4.5　There are a variety of tests that can be carried out on glues (e.g. in laminates and connections) and fasteners used in timber. Delamination and shear tests of glue lines are described in *BS EN 391* [74] and *BS EN 392* [75] respectively. Details of a pull-through resistance test for timber fasteners are provided in *BS EN 1383* [76].

8.5　DETERIORATION ACTIVITY

8.5.1　In addition to the following techniques, radiography (paragraphs 4.2.19-4.2.22) and acoustic emission (paragraphs 4.5.2-4.5.3) can also be used to identify deterioration in timber components.

Probe Tests

8.5.2　In its simplest form, a probe test consists of inserting a probe, such as a knife blade or small screwdriver, into the timber and assessing the resistance. Areas suffering from decay or insect attack offer less resistance than sound areas. Using an appropriate probe, this technique can be valuable in detecting decay behind an apparently sound surface. Clearly, the technique only gives comparative results.

8.5.3　There are many variations on the probe test. Typical ones include the 'Pilodyn' (see paragraph 8.5.4), which is similar to the Windsor Probe used for assessing concrete strength (see paragraph 5.4.3). The penetration distance can be correlated to the degree of decay and strength loss. Another variation is the incremental borer, which enables timber samples to be taken at measured depths. The disadvantages common to all probe tests are that they only pick up defects at the test location. There may be other defects that are not detected and the overall capacity of the timber member is not assessed.

Energy Absorption

8.5.4   Energy absorption tests are used to detect decay of the timber. Enegry absorption tests include:

- ***Pilodyn Test*** – this is a surface-quality measuring device. A blunt pin is fired, with known energy, into the timber surface and depth of penetration measured. The depth of penetration provides an indication of fibre strength and the extent of externally visible decay. This test cannot detect internal decay when the surface is sound and is sensitive to a number of factors, including density, moisture content and angle to the grain.

- ***Decay Detecting Drill and Densitomat*** – these test techniques are based on resistance drilling. The Decay Detecting Drill uses a 200mm long, 1mm diameter drill bit, measuring the rate of penetration under nominally constant speed and pressure, while the Densitomat measures the torque required to maintain a constant rate of penetration of the drill bit [72]. Both techniques allow information to be electronically stored. Both techniques detect density changes, but other effects, such as changes in moisture content, angle to the grain, and splits can give spurious results. It is therefore important that results are interpreted with care and specialist advice is sought. A similar, but less scientific approach is to use a hand drill whereby the inspector has to 'feel' for the loss of resistance.

Moisture Content

8.5.5   Moisture content indicates the risk, or presence, of decay (e.g. rot or insect attack) and also influences the rate at which any existing decay progresses. Electrical test techniques are commonly used to measure the moisture content of the timber, typically in the range of 8 to 25% [72]. Resistance meters can be used to measure the electrical resistance at the surface of the timber, capacitance meters measure the dielectric constant of the timber, while radio frequency power loss meters measure the electrical impedance. All these parameters can be correlated to the moisture content of the timber. Internal resistance meters are more expensive but measure the electrical resistance of the body of the timber. This provides an indication of the presence of internal decay and can be applied relatively easily over the whole member.

Biological Test - Bioassay

8.5.6   Bioassay techniques are used to determine the presence of fungal decay and the effectiveness of preservative treatments [72]. This test requires a sample of material to be incubated in a suitable medium, for a defined period, to monitor the development of any fungal growth. Expert interpretation of the results is required, as they relate only to the sample removed.

Biochemical - Indicator Dyes and Selective Stains

8.5.7   Indicator dyes, which change colour based on the pH (acidity or alkalinity) of the object, can be used to detect the presence of fungi. Careful interpretation of the results is required because other factors, such as natural variations in the timber and the presence of metallic fixings, may also produce responses.

Ultrasonic Tests

8.5.8 Sonic test techniques measure the velocity of a sound wave through the timber element; a lower velocity indicates the presence of decay (see paragraphs 5.4.9-5.4.10 and 6.2.2-6.2.5 for more details on the application of ultrasonic tests). The measurements are quick and relatively simple to make, but interpretation of the results requires judgement and can be time consuming. They are best used on a comparative basis, looking for significant changes within a timber, or by comparing readings from similar species of timber.

8.6 DETERIORATION RATE

8.6.1 Observing and monitoring change in condition overtime, aided by the techniques described in Section 8.5, is the most effective technique for determining the rate of deterioration of timber components.

8.7 DETERIORATION CAUSE OR POTENTIAL

8.7.1 Common forms of timber deterioration are rot, fungal attack and insect attack. While it may be possible to readily diagnose some forms of attack on site, it may be necessary to send specimens (of the timber, fungus and/or insects) to specialist laboratories for assessment. This can be expensive, but can provide beneficial information for diagnosing the problem.

9 Tests on Advanced Composites

9.1.1 Advanced composites are a relatively new structural material and testing techniques are continually evolving and being developed. Therefore details of tests (apart from those given below) are not included here. If it is necessary to undertake other tests on advanced composites, it is recommended that specialist advice is sought. Further guidance on testing and investigation of structures that have been strengthened by the use of advance composites is provided in *Strengthening Concrete Structures with Fibre Composite Materials: Acceptance, Inspection And Monitoring* [77] and in *Repair and Maintenance of FRP Structures* [78].

9.1.2 The testing techniques described in Section 4.2 and 4.3 are generally relevant for advanced composite structures bearing in mind the following considerations.

- *Visual Examination* – visual examination of advanced composite elements should be carried out in sufficient detail to identify defects such as surface erosion, delamination, impact damage or loose components.

- *Pat Test* – the pat test is used to detect delaminations within the body of the composite material. The sound produced by a gentle tap on the material differs in delaminated areas compared with sound areas. This is analogous to the hammer test for delamination, see paragraphs 4.2.2-4.2.4.

10 References for Part E

1. *BS EN ISO 9000-1 Quality Management and Quality Assurance Standards - Guidance for Selection and Use*, British Standards Institution.

2. *Management of Highway Structures: A Code of Practice*, TSO, 2005.

3. *Specification for the Reinstatement of Openings in Highways*, 2nd Edition, Department for Transport, 2002.

4. *BA 86 Advice Notes on the Non-Destructive Testing of Highway Structures*, DMRB 3.1.7, TSO.

5. *BS 1881: Part 205 Testing Concrete – Recommendations for Radiography of Concrete*, British Standards Institution.

6. *ISO 14963 Mechanical Vibration and Shock – Guidelines for Dynamic Tests and Investigations on Bridges and Viaducts*, ISO.

7. *BA 54 Load Testing for Bridge Assessment*, DMRB 3.4.8, TSO.

8. *Guidelines for the Supplementary Load Testing of Bridges*, National Steering Committee for the Load Testing of Bridges, Institution of Civil Engineers, T Telford, London, 1998.

9. *Acoustic Emission Observations on a Stone Masonry Bridge Loaded to Failure*, Hendry AW & Royles R, Proceedings of the 2nd International Conference on Structural Faults and Repair, Engineering Technics Press, 1985.

10. *BS 5930 Code of Practice for Site Investigations*, British Standards Institution.

11. *BS 1377 Techniques of Test for Soils for Civil Engineering Purposes*, British Standards Institution,.

12. *HA 48 Maintenance of Highway Earthworks and Drainage*, DMRB 4.1.3, TSO.

13. *Technical Guide 2: Guide to Testing and Monitoring the Durability of Concrete*, Concrete Bridge Development Group, Concrete Society, Slough, 2002.

14. *BS 1881: Part 204 Testing Concrete – Recommendations on the Use of Electromagnetic Covermeters*, British Standards Institution.

15. *BS EN 12504-2 Testing Concrete in Structures – Non-destructive testing: Determination of rebound number*, British Standards Institution.

16. *BS 1881: Part 207 Testing Concrete – Recommendations for the Assessment of Concrete Strength by Near-to-Surface Tests*, British Standards Institution.

17. *BS EN 12504-4 Testing Concrete in Structures – Determination of Ultrasonic Pulse Velocity*, British Standards Institution.

18. *BS 1881: Part 208 Testing Concrete – Recommendations for the Determination of the Initial Surface Absorption of Concrete*, British Standards Institution.

19. *Improvements to the Figg Technique of Determining the Air Permeability of Concrete*, Cather R, Figg JW, Marsden AF & O'Brien TP, Magazine of Concrete Research 36, 129, 1984.

20. *In-Situ Testing of Near Surface Concrete: The Foundation ror Service Life Prediction*, Long AE, Basheer PA & Rankin GI, Insight, Journal of the British Institute of Non-Destructive Testing, 39, July 1997, pp. 432-487.

21. *BS EN 12504-1 Testing of Concrete in Structures – Taking, Examining and Testing in Compression*, British Standards Institution.

22. *BS 6089 Guide to Assessment of Concrete Strength in Existing Structures*, British Standards Institution.

23. *BS 1881: Part 114 Testing Concrete – Techniques for determination of Density of Hardened Concrete*, British Standards Institution.

24. *Reinforced Concrete Resistivity Measurement Techniques*, Millard SG, Proc. Inst. Civ. Engrs, 91, March 1991.

25. *The Diagnosis of Alkali-Silica Reaction*, 2nd Edition, British Cement Association, Slough, 1992.

26. *The Thaumasite Form of Sulfate Attack: Risks, Diagnosis, Remedial Works and Guidance on New Construction*, Thaumasite Expert Group, Department of the Environment, Transport and the Regions, London, 1999.

27. *Permanent Corrosion Monitoring, Construction Repair: Concrete Repairs 5*, Broomfield JP, 1996.

28. *Electrochemical Techniques for Corrosion Assessment of Reinforced Concrete Structures*, Gowers KR & Millard SG, Proc. Instn Civ. Engrs Structs & Bldgs, 134, May 1999, pp 129-137.

29. *Technical Report 26: Repair of Concrete Damaged by Reinforcement Corrosion*, Concrete Society, Slough, 1984.

30. *Techniques for Evaluating Reinforced Concrete Bridge Decks*, Van Daveer J, ACI Journ., 1975, p. 697.

31. *Application Guide No. 9: The Half-cell Potential Technique of Locating Corroding Reinforcement in Concrete Structures*, Vassie PR, Transport Research Laboratory, Crowthorne, 1991.

32. *ASTM C876-80 Standard Test Technique for Half Cell Potentials of Reinforcing Steel in Concrete*, American Society for Testing and Materials, Philadelphia.

33. *Digest 363: Sulfate and Acid Resistance of Concrete in the Ground*, New Edition, Building Research Establishment, Watford, 1996.

34. *BS 1881: Part 124 Testing Concrete – Techniques for Analysis of Hardened Concrete*, British Standards Institution.

35. *IP 21/86: Determination of the Chloride and Cement Contents of Hardened Concrete*, Building Research Establishment, Watford, 1986.

36. *BA 50 Post-Tensioned Concrete Bridges: Planning, Organisation and Techniques for Carrying Out Special Inspections*, DMRB 3.1.3, TSO.

37. *BS 1881: Part 201 Testing Concrete – Guide to the Use of Non-destructive Techniques of Test for Hardened Concrete*, British Standards Institution.

38. *BS 970: Part 1 Specification for Wrought Steels for Mechanical and Allied Engineering Purposes – General Inspection and Testing Procedures and Specific Requirements for Carbon, Carbon Manganese, Alloy and Stainless Steels*, British Standards Institution.

39. *BS EN 10025: Part 1 Hot Rolled Products of Structural Steels – General Technical Delivery Conditions*, British Standards Institution.

40. *BS EN 10025: Part 2 Hot Rolled Products of Structural Steels – Technical Delivery Conditions for Non-Alloy Structural Steels*, British Standards Institution.

41. *BS EN 10002: Part 1 Tensile Testing of Metallic Materials – Method of Test at Ambient Temperature*, British Standards Institution.

42. *BS EN 10045: Part 1 Charpy Impact Test on Metallic Materials*, British Standards Institution.

43. *BS EN 10045: Part 2 Charpy Impact Test on Metallic Materials*, British Standards Institution.

44. *BS EN ISO 6506: Parts 1, 2 and 3 Metallic Materials – Brinell Hardness*, British Standards Institution.

45. *BS EN ISO 6507: Parts 1, 2 and 3 Metallic Materials – Vickers Hardness Test*, British Standards Institution.

46. *BS EN ISO 6508: Parts 1, 2 and 3 Metallic Materials – Hardness Test,* British Standards Institution.

47. *BS EN 875 Destructive Tests on Welds in Metallic Materials – Impact Tests, Test Specimen Location, Notch Orientation and Examination*, British Standards Institution.

48. *BS EN 910 Destructive Tests on Welds in Metallic Materials – Bend Tests*, British Standards Institution.

49. *BS 7448: Part 1 Method for Determination of KIc Critical CTOD and Critical J Values of Metallic Materials*, British Standards Institution.

50. *BS EN 10228-3 Non-destructive Testing of Steel Forgings – Ultrasonic Testing of Ferritic or Martensitic Steel Forgings*, British Standards Institution.

51. *BS EN 12680-1 Founding: Ultrasonic Examination – Steel Castings for General Purpose*, British Standards Institution.

52. *BS EN 571-1 Non-destructive Testing: Penetrant Testing – General Principles*, British Standards Institution.

53. *BS EN 1369 Founding: Magnetic Particle Inspection*, British Standards Institution.

54. *BS EN ISO 9934-1 Non-Destructive Testing: Magnetic Particle Testing – General Principles*, British Standards Institution.

55. *PD 6513 Magnetic Particle Flaw Detection – A Guide to the Principles and Practice of Applying Magnetic Particle Flaw Detection in Accordance with BS 6072*, British Standards Institution.

56. *BS EN 12084 Non-Destructive Testing: Eddy Current Testing – General Principles and Guidelines*, British Standards Institution.

57. *BS EN 1561 Founding – Grey Cast Iron*, British Standards Institution.

58. *Appraisal of Existing Structures*, 2nd Edition, Institution of Structural Engineers, London, 1996.

59. *BS 131: Part 1 The Izod Impact Test of Metals*, British Standards Institution.

60. *BS EN ISO 2409 Paints and Varnishes – Cross-cut Test*, British Standards Institution.

61. *BS EN ISO 4624 Paints and Varnishes – Pull-off Tests for Adhesion*, British Standards Institution.

62. *Masonry Arch Bridges: Condition Appraisal and Remedial Treatment*, CIRIA C656, 2006.

63. *BS 5628 Part 3 Code of Practice for Use of Masonry – Materials and Components, Design and Workmanship*, British Standards Institution.

64. *BS 6089 Guide to Assessment of Concrete Strength in Existing Structures*, British Standards Institution.

65. *BS 3921 Specification for Clay Bricks*, British Standards Institution.

66. *Digest 421: Measuring the Compressive Strength of Masonry Materials: The Screw Pull-Out Test*, Building Research Establishment, Watford, 1997b.

67. *Digest 409: Masonry and Concrete Structures: Measuring In-situ Stress and Elasticity Using Flat Jacks*, Building Research Establishment, Watford, 1997a.

68. *Digest 360: Testing Bond Strength of Masonry*, Building Research Establishment, Watford, 1991.

69. *ASTM C1197-92 In-situ Measurement of Masonry Deformability Properties Using the Flat Jack Technique*, American Society for Testing and Materials, Philadelphia.

70. *BS 4551 Mortar: Methods of Test for Mortars: Chemical Analysis and Physical Testing*, British Standards Institution.

71. *Arch Bridge Assessment – How Tests and Analysis Compare*, Harvey B, Proceedings of the 1st International Conference on Advances in Bridge Engineering: Bridges - Past, Present and Future; Brunel University, West London, June 2006.

72. *Wood Information Sheet 4 – 23*, TRADA, August 2004.

73. *BS EN 1193 Timber Structures: Structural Timber and Glued Laminated Timber – Determination of Shear Strength and Mechanical Properties Perpendicular to the Grain*, British Standards Institution.

74. *BS EN 391 Glued Laminated Timber: Delamination Test of Glue Lines*, British Standards Institution.

75. *BS EN 392 Glued Laminated Timber: Shear Test of Glue Lines*, British Standards Institution.

76. *BS EN 1383 Timber Structures: Test Methods – Pull-through Resistance of Timber Fasteners*, British Standards Institution.

77. *Technical Report No 57: Strengthening Concrete Structures With Fibre Composite Materials – Acceptance, Inspection And Monitoring*, Concrete Society, 2003.

78. *Repair and Maintenance of FRP Structures*, Halliwell S & Suttie E, BRE Good Repair Guide GRG 34, 2003.

Part F
Appendices

This Part contains a selection of information which collectively support the guidance provided in Parts A to E of the Manual.

Appendix A
Addresses of Environmental and Other Organisations

STATUTORY NATURE CONSERVATION ORGANISATIONS (SNCO):

English Nature:

Northminster House
Peterborough
PE1 1UA
Tel: 01733 455000

Countryside Council for Wales:

Plas Penrhos, Ffordd Penrhos
Bangor, Gwynedd
LL57 2LQ
Tel: 01248 385500

Scottish Natural Heritage:

12 Hope Terrace
Edinburgh
EH9 2AS
Tel: 0131 554 9797

Department of the Environment for Northern Ireland:

Environment and Heritage Service
Commonwealth House
35 Castle Street
Belfast, BT1 1GU
Tel: 028 9025 1477

OTHER STATUTORY ORGANISATIONS:

Environment Agency:

Rio House, Waterside Drive
Aztec West, Almondsbury
Bristol, BS12 4UD
Tel: 01454 624400

EMERGENCY HOTLINE:
0800 80 70 60
This is the head office and will be able to provide contact details for appropriate regional offices.

Scottish Environment Protection Agency:

Erskine Court
The Castle Business Park
Stirling
SK9 4TR
Tel: 01786 457 700

Ministry of Agriculture, Fisheries and Food:

Environmental Protection Division
Nobel House, 17 Smith Square
London, SW1P 3JR
Tel: 020 7270 8000

Welsh Office Agriculture Department:

Cathays Park
Cardiff
CF1 3NQ
Tel: 029 2082 5111

Scottish Executive Rural:

Affairs Department
Pentland House,
47 Robbs Loan,
Edinburgh, EH14 1TY
Tel: 0131 556 8400

Department of Agriculture for Northern Ireland:

Dundonald House
Upper Newtownards Road
Belfast, BT4 3SB
Tel: 028 9065 0111

WILDLIFE PROTECTION ORGANISATIONS:

British Trust For Ornithology (BTO):

The Nunnery, Nunnery Place
Thetford
Norfolk
IP24 2PU
Tel: 01842 750050

Bat Conservation Trust:

15 Cloisters House
8 Battersea Park Road
London, SW8 4BG
Tel: 020 7627 2629

BAT HELPLINE: 020 7627 8822

Royal Society for Nature Conservation - Wildlife Trusts Partnership:

The Green, Witham Park
Waterside South,
Lincoln, LN5 7JR
Tel: 01522 544400

They will be able to provide contact details for the appropriate county or regional wildlife trust except in Northern Ireland.

Ulster Wildlife Trust:

Crossgar Nature Centre
3 New Line
Crossgar
Tel: 028 4483 0282

Royal Society For The Protection Of Birds (RSPB):

UK Headquarters
The Lodge, Sandy,
Bedfordshire
SG12 2DL
Tel: 01767 680551

Scottish Headquarters
17 Regent Terrace
Edinburgh
EH7 5BN
Tel: 0131 557 3136

Wales Office
Bryn Aderyn, The Bank
Newtown , Powys
SY16 2AB
Tel: 01686 626678

Northern Ireland Office
Belvoir Park Forest
Belfast
BT8 4QT
Tel: 028 9049 1547

COUNTY BIOLOGICAL RECORDS OFFICES:

The address of the appropriate biological records office can be obtained from the relevant local authority except in Northern Ireland where CEDaR and the Vertebrate Recorder should be contacted.

CEDaR:
Ulster Museum
Botanic Gardens
Belfast   BT9
Tel: 028 9038 3144

The Vertebrate Recorder:
Ulster Museum
Botanic Gardens
Belfast   BT9
Tel: 028 9038 3144

Appendix B
Example of Personal Safety Check List

B.1 **OVERVIEW**

B.1.1 Various situations are covered below but in all situations work should not commence or continue unless it is safe to do so.

B.2 **VISITS TO SITES OFF THE PUBLIC HIGHWAY**

Before Leaving Office

- Be familiar with relevant Company Health and Safety Notices, in particular No _____;

- Plan work and prepare a risk assessment or read and understand the risk assessment prepared by others for the work;

- Arrange with occupier/owner for permission to enter properties/land;

- Consider the appropriate number of staff to carry out the work;

- If it is proposed to work alone and in isolation, have additional risks been assessed and suitable procedures put in place? See Company Health and Safety Notice No_____;

- Consider need for communication links back to the office;

- Complete Job Record Form.

Generally

- Wear PPE appropriate to the situation;

- Work during hours of daylight;

- Do not work outside during adverse weather.

Hazards

- Be aware of hazards you may encounter on site, such as: contaminated land, hazardous rubbish, potholes/caves, physical attack by animals, soft sand or mud, redundant or active military areas.

On Location

- Park in a safe place;

- Walk - do not run;

- Look where you are going;

- Take care when climbing up or down;

- Avoid walking along the edge of drops, etc;

- Use recognised access ways, footpaths, paths, tracks, gates, stiles, etc.;

- Avoid disturbing animals;

- Do not cross mud flats or dried river beds unless you are sure that they are safe;

- If weather changes, consider revision or abandonment of work;

- Do NOT enter a confined space unless authorised and deemed competent to do so;

- Take heed of any warning signs.

B.3   VISITS TO HIGHWAYS OPEN TO TRAFFIC

Before Leaving Office

- Be familiar with relevant Company Health and Safety Notices, in particular No _____;

- Plan work and prepare a risk assessment or read and understand the risk assessment prepared by others for the work;

- Refer to Chapter 8 of the Traffic Signs Manual and any Notes for Guidance for particular traffic controls;

- Consider the appropriate number of staff to carry out the work, especially if planning to work alone;

- If it is proposed to work alone and in isolation, have additional risks been assessed and suitable procedures put in place? See Company Health and Safety Notice No_____;

- Complete Job Record Form.

Generally

- Work during hours of daylight;

- Do not work when visibility is poor;

- Do not work during adverse weather.

Stopping

- Think before parking; do not obstruct sight lines or accesses, including those for emergency vehicles;

- Be aware of potential traffic hazards;

- Use flashing beacon on car as appropriate.

Working

- Be alert at all times;

- Wear appropriate high visibility clothing at all times;

- Wear PPE appropriate to the situation;

- Walk - do not run;

- When possible, walk on the right hand side of a single carriageway road, facing oncoming traffic;

- Try not to walk or stand close to moving traffic;

- Do not cross carriageways on busy fast roads;

- Take care when climbing up or down embankments or cuttings;

- Do not carry anything in your hands when climbing a ladder;

- Do NOT enter a confined space unless authorised and deemed competent to do so.

On a Motorway

In addition to the above items:

- Give Police advance notice of work to be carried out;

- Use flashing beacon when stopping;

- Do NOT cross the carriageway;

- Do NOT cross slip roads near the motorway.

Appendix C
Permit to Work in Confined Spaces

C.1.1   An example of a suitable Permit to Work in Confined Spaces is given below:

PERMIT NO _____

PERMIT TO WORK IN CONFINED SPACES - SAFE WORKING PROCEDURES

ATTACHMENT TO JOB RECORD FORM Ref No. ....

COPY TO BE HANDED TO PERSON IN CHARGE OF SITE AND DISPLAYED ON SITE

1.   WORKPLACES TO BE VISITED (additional information)

LOCATION OF CONFINED SPACE ....

Work to be carried out. ....

....

....

Access Equipment. ....

....

Sketch of confined space and location of personnel.

2.   IDENTIFICATION OF HAZARDS

Complete a Risk Assessment based on the following hazards.

A. Access or egress

B. Hazardous Atmosphere:

1. Oxygen deficiency

2. Toxic gases

3. Dust

C. Flammable or explosive gases or liquids:

1. Fire

2. Explosion

D. Other Hazards:

1. Noise

2. Heat

3. Cold

4. Chemical

5. Rust

6. Radiation

7. Electrical

8. Flooding

9. Machinery

10. Physical obstruction

11. Remote controls

12. Welding and flames

13. Psychological

14. Bacterial infection

15. Climatic factors

16. Physical injury

17. Lack of light

18. Other, please specify

....

....

---

3.   SAFE WORKING PROCEDURES

a.   Means of rendering safe sources of ignition.

....

b.   Is breathing apparatus required (Yes/No) and why?

....

c.   Warning notification and signs.

1.   Who else needs to be notified and are any other approvals required?

....

2.   What warning signs need to be installed?

....

d.   What hazardous material needs to be removed and how?

....

e.   Proposed method of ventilating area.

....

f.   Monitoring of gases.

Proposed procedure and durations between monitoring of gases.

Oxygen   Yes/No   ....

Flammable gases   Yes/No   ....

Toxic gases   Yes/No   ....

g.   Other proposed safe working procedure.

....

....

h.   Emergency rescue procedure.

1.   Personnel. For each person on site state name, location and duties.

Entry Member   ....

Entry Member   ....

Standby   ....

Other   ....

2.   For the hazards identified, state the proposed rescue procedures.

....

....

---

4.   PERSONAL PROTECTIVE EQUIPMENT (PPE)

1.   Protective clothing / gloves

2.   Hard hat

3.   Safety boots

4.   Goggles

5.   Escape BA

6.   Compressed air breathing apparatus

7.   Resuscitation equipment

8.   Safety harness/line

9.   Winch plus frame

10.   Hearing protection

11.   Lighting (explosion proof)

12.   Certified electrical apparatus including insulated hand tools

13.   Air quality monitor

14.   Blower and ducts

15.   Two way radios

16.   Portable telephone

17.   Standby fire bridge

18.   Fire fighting equipment

19.   First aid kit

SECTION A   AUTHORISING OFFICER

I certify that I have examined the proposed methods of working in this confined space and have satisfied myself that the above particulars are correct and approve this work permit for the period as shown at the start of this work permit.

Authorising Officer: ....

Date: ....

Designation: ....

SECTION B   REQUEST FOR EXTENSION

The work has not been completed and permission to continue is requested.

Line Manager: ....

Date: ....

SECTION C   ACCEPTANCE OF EXTENSION

I have re-examined the work site/plant detailed above and confirm that the certificate may be extended to be valid from .... hours on .... to .... hours on ....

Authorising Officer: ....

Date: ....

SECTION D   COMPLETION OF WORK

I have inspected/tested the work covered by this permit and certify that the work has been/not been completed. Reason for work has not been completed should be stated and recommendation made for further work.

Inspector: ....

Date: ....

On completion this Permit to Work must be returned to the Authorising Officer.

SECTION E   ACCEPTANCE OF COMPLETION

Authorising Officer: ....

Date: ....

Appendix D
Specially Protected Species

D.1   GENERAL

D.1.1   The lists contained in this Appendix may be subject to review and consequent change.

D.2   ANIMAL SPECIES SPECIALLY PROTECTED UNDER THE WILDLIFE AND COUNTRYSIDE ACT, 1981 (ENGLAND, SCOTLAND AND WALES ONLY).

D.2.1   These species are protected under Schedule 5 of the *Wildlife and Countryside Act* [1]. Badgers and their setts now enjoy comparable protection under the *Protection of Badgers Act* [2]. Deer and seals are not so fully protected, but have their own legislation.

Scientific Names	English Names	Notes
Mammals		
Cetacea	All dolphins, porpoises, whales	
Felis sylvestris	Wildcat	
Lutra lutra	Otter	
Martes martes	Pine marten	
Muscardinus avellanarius	Dormouse	
Odobenus rosmarus	Walrus	
Sciurus vulgaris	Red squirrel	
Vespertilionidae and Rhinolophidae	All bats	
Reptiles		
Anguis fragilis	Slow worm	Killing, injuring and sale only
Cheloniidae and Dermochelyidae	All turtles	
Coronella austriaca	Smooth snake	
Lacerta agilis	Sand lizard	Killing, injuring and sale only
L. vivpara	Viviparous lizard	Killing, injuring and sale only
Natrix natrix	Grass snake	Killing, injuring and sale only
Vipera berus	Adder	Killing, injuring and sale only
Amphibians		
Bufo bufo	Common toad	Sale only
B. calamita	Natterjack toad	
Rana temporaria	Common frog	Sale only
Triturus cristatus	Warty (great crested) newt	
T. helveticus	Palmate newt	Sale only
T. vulgaris	Smooth newt	Sale only

Scientific Names	English Names	Notes
Fish		
Acipenser sturio	Sturgeon	
Alosa alosa	Allis shad	Killing, injuring, and taking only
Coregonus albula	Vendace	
C. lavaretus	Whitefish	
Lota lota	Burbot	
Butterflies		
Apatura iris	Purple emperor	Sale only
Argynnis adippe	High brown fritillary	
Aricia artaxerxes	Northern brown angus	Sale only
Boloria euphrosyne	Pearl-bordered fritillary	Sale only
Carterocephalus palaemon	Checkered skipper	Sale only
Coenonympha tullia	Large heath	Sale only
Cupido minimus	Small blue	Sale only
Eurodryas aurinia	Marsh fritillary	Sale only
Erebia epiphron	Mountain ringlet	Sale only
Hamearis lucina	Duke of Burgundy fritillary	Sale only
Hesperia comma	Silver-spotted skipper	Sale only
Leptidea sinapis	Wood white	Sale only
Lycaena dispar	Large copper	Sale only
Lysandra bellargus	Adonis blue	Sale only
L. coridon	Chalkhill blue	Sale only
Maculinea arion	Large blue	
Mellicta (Melitaea) athalia	Heath fritillary	
Melitaea cinxia	Glanville fritillary	Sale only
Nymphalis polychloros	Large tortoiseshell	Sale only
Papilio machaon	Swallowtail	
Plebejus argus	Silver-studded blue	Sale only
Strymonidia pruni	Black hairstreak	Sale only
Strymonidia w-album	White-letter hairstreak	Sale only
Thelca betulae	Brown hairstreak	Sale only
Thymelicus acteon	Lulworth skipper	Sale only

Scientific Names	English Names	Notes
Moths		
Acosmetia caliginosa	Reddish buff	
Hadena irregularis	Vipers' bugloss	
Pareulype berberata	Barberry carpet	
Siona lineata	Black-veined	
Thalera fimbrialis	Sussex emerald	
Thetidia smaragdaria	Essex emerald	
Zygaena viciae	New Forest burnet	
Beetles		
Chrysolina cerealis	Rainbow leaf beetle	
Curimopsis nigrita	Mire pill beetle	Damage/obstruction of place of shelter/ protection
Graphoderus zonatus	Water beetle	
Hydrochara caraboides	Lesser silver water beetle	
Hypebaeus flavipes	Beetle	
Limoniscus violaceus	Violet click beetle	
Paracymus aeneus	Water beetle	
Bugs		
Cicadetta montana	New Forest cicada	
Grasshoppers and crickets		
Decticus verrucivorus	Wart-biter	
Gryllotalpa gryllotalpa	Mole cricket	
Gryllus campestris	Field cricket	
Dragonflies		
Aeshna isosceles	Norfolk aeshna	
Spiders		
Dolomedes plantarius	Fen raft spider	
Eresus niger	Ladybird spider	
Crustaceans		
Austropotamobius pallipes	Atlantic (white-clawed) crayfish	Taking and sale only
Chirocephalus diaphanus	Fairy shrimp	
Gammarus insensibilis	Lagoon shrimp	
Triops cancriformis	Apus	

Scientific Names	English Names	Notes
Sea-mats (bryozoa)		
Victorella pavida	Trembling sea-mat	
Molluscs		
Caecum armoricum	DeFolin's lagoon snail	
Catinella arenaria	Sandbowl snail	
Margaritifera margaritifera	Pearl Mussel	Killing and injuring only
Myxas glutinosa	Glutinous snail	
Paludinella littorina	Lagoon snail	
Tenellia adspersa	Lagoon sea slug	
Thyasira gouldi	Northern hatchet-shell	
Worms (Annelida)		
Alkmaria romijni	Tentacled lagoon-worm	
Armandia cirrhosa	Lagoon sandworm	
Hirudo medicinalis	Medicinal leech	
Sea anemones and allies (Cnidaria)		
Edwardsia ivelli	Ivell's sea anemone	
Eunicella verrucosa	Pink sea-fan	Killing, injuring, taking possession and sale only
Nematostella vectensis	Startlet sea anemone	

D.2.2   Protection for wild animals on Schedule 5 (Under Section 9):

- Part 1 – intentional killing, injuring, taking.

- Part 2 – possession or control (live or dead animal, part or derivative).

- Part 4(a) – damage to, destruction of, obstruction of access to any structure or place used for shelter or protection.

- Part 4(b) – disturbance of animal occupying such a structure or place.

- Part 5(a) – selling, offering for sale, possession or transport for purpose of sale (live or dead animal, part or derivative).

- Part 5(b) – advertisement for buying or selling such things.

D.3 PROTECTED PLANT SPECIES

D.3.1 These species are protected under Schedule 8 of the *Wildlife and Countryside Act* [1]. It is an offence for anyone intentionally to pick up root or destroy any of the wild plants listed in Schedule 8, or even to collect their flowers and seeds.

Scientific Names	English Names
Vascular plants	
Ajuga chamaepitys	Ground pine
Alisma gramineum	Ribbon-leaved water-plantain
Allium sphaerocephalon	Round-headed mallow
Althaea hirsuta	Rough marsh mallow
Alyssum alyssoides	Small alison
Apium repens	Creeping marshwort
Arabis alpina	Alpine rock-cress
A. stricta	Bristol rock-cress
Arenaria norvegica	Norwegian sandwort
Artemisia campestris	Field wormwood
Bupleurum baldense	Small hare's ear
B. falcatum	Sickle-leaved hare's ear
Calamintha sylvatica	Wood calamint
Carex depauperata	Starved wood-sedge
Centaurium tenuiflorum	Slender centaury
Cephalanthera rubra	Red helleborine
Chenopodium vulvaria	Stinking goosefoot
Cicerbita alpina	Alpine sow-thistle
Corrigiola litoralis	Strapwort
Cotoneaster integerrimus	Wild cotoneaster
Crassula aquatica	Pigmyweed
Crepis foetida	Stinking hawk's-beard
Cynoglossum germanicum	Green hound's-tongue
Cyperus fuscus	Brown galingale
Cypripedium calceolus	Lady's slipper orchid
Cystopteris dickieana	Dickie's bladder fern
Dactylorhiza lapponica	Lapland marsh orchid
Damasonium alisma	Starfruit
Diapensia lapponica	Diadpensia
Dianthus gratianopolitanus	Cheddar pink
Epipactis youngiana	Young's helleborine
Epipogium aphyllum	Ghost orchid
Equisetum ramosissimum	Branched horsetail

Scientific Names	English Names
Vascular plants (continued)	
Erigeron borealis	Alpine fleabane
Eriophorum gracile	Slender cottongrass
Eryngium campestre	Field eryngo
Filago lutescens	Red-tipped cudweed
F. pyramidata	Broad-leaved cudweed
Fumaria martinii	Martin's ramping-fumitory
Gagea bohemica	Early star of Bethlehem
Gentiana nivalis	Alpine gentian
G. verna	Spring gentian
Gentianella anglica	Early gentian
G. ciliata	Fringed gentian
G. uliginosa	Dune gentian
Gladiolus illyricus	Wild gladiolus
Gnaphalium luteoalbum	Jersey cudweed
Halilmione pedunculata	Stalked orache
Hieracium attenuatifolium	Weak-leaved hawkweed
H. northroense	Northroe hawkweed
H. zetlandicum	Shetland hawkweed
Himantoglossum hircinum	Lizard orchid
Homogyne alpina	Purple colt's-foot
Lactuca saligna	Least lettuce
Limosella australis	Welsh mudwort
Liparis loeselii	Fen orchid
Lloydia serotina	Snowdon lily
Luronium natans	Floating water-plantain
Lychnis alpina	Alpine catchfly
Lythrum hyssopifolia	Grass-poly
Melampyrum arvense	Field cow-wheat
Metha pulegium	Pennyroyal
Minuartia stricta	Teesdale sandwort
Najas flexilis	Slender naiad
N. marina	Holly-leaved naiad
Ononis reclinata	Small restharrow
Ophioglossum lusitanicum	Least adder's-tongue
Ophrys fuciflora	Late spider orchid
O. sphegodes	Early spider orchid
Orchis miliataris	Military orchid

Scientific Names	English Names
Vascular plants (continued)	
O. simia	Monkey orchid
Orobanche caryophyllacea	Bedstraw broomrape
O. loricata	Oxtongue broomrape
O. reticulata	Thistle broomrape
Petroraghia nanteuilii	Childing pink
Phyteuma spicatum	Spiked rampian
Polygonatum verticillatum	Whorled Soloman's seal
Polygonum maritimum	Sea knotgrass
Potentilla rupestris	Rock cinquefoil
Pulicaria vulgaris	Small fleabane
Phyllodoce caerulea	Blue heath
Pyrus cordata	Plymouth pear
Ranunculus ophioglossifolius	Adder's-tongue spearwort
Rhinanthus serotinus	Greater yellow-rattle
Rhynchosinapis wrightii	Lundy cabbage
Romulea columnae	Sand crocus
Rumex rupestris	Shore dock
Salvia pratensis	Meadow clary
Saxifraga bernua	Drooping saxifrage
S. cespitosa	Tufted saxifrage
S. hirculus	Yellow marsh-saxifrage
Scirpus triquetrus	Triangular club-rush
Scleranthus perennis	Perennial knawel
Scorzonera humilis	Viper's-grass
Selinum carvifolia	Cambridge milk-parsley
Senecio paludosus	Fen ragwort
Stachys alpina	Limestone woundwort
S. germanica	Downy woundwort
Teucrium botrys	Cut-leaved germander
T. scordium	Water germander
Thlaspi perfoliatum	Perfoliate penny-cress
Trichomanes speciosum	Killarney fern
Veronica spicata	Spiked speedwell
V. triphyllos	Fingered speedwell
Viola persicifolia	Fen violet
Woodsia alpina	Alpine woodsia
W. ilvensis	Oblong woodsia

Scientific Names	English Names

Mosses

Acaulon triquetrum	Triangular pygmy-moss
Barbula cordata	Cordate beard-moss
B. glauca	Glaucous beard-moss
Bartramia stricta	Rigid apple-moss
Bryum mamillatum	Dune thread-moss
B. schleicheri	Schleicher's thread-moss
Buxbaumia viridis	Green shield-moss
Cryphaea lamyana	Multi-fruited river-moss
Cyclodictyon laetevirens	Bright green cave-moss
Ditrichum cornubicum	Cornish path-moss
Drepanocladus vernicosus	Slender green feather-moss
Grimmia unicolor	Blunt-leaved grimmia
Hypnum vaucheri	Vaucher's feather-moss
Micromitrium tenerum	Millimetre moss
Mielichhoferia mielichhoferi	Alpine copper-moss
Orthotrichum obtusifolium	Blunt-leaved bristle-moss
Plagiothechium piliferum	Hair silk-moss
Rhynchostegium rotundifolium	Round-leaved feather-moss
Saelania glaucescens	Blue dew-moss
Scorpidium turgescens	Large yellow feather-moss
Sphagnum balticum	Baltic bog-moss
Thamnobryum angustifolium	Derbyshire feather-moss
Zygodon forsteri	Knothole moss
Zygodon gracilis	Nowell's limestone moss

Liverworts

Adelanthus lindenbergianus	Lindenberg's leafy liverwort
Geocalyx graveolens	Turpswort
Gymnomitrion apiculatum	Pointed frostwort
Jamesoniella undulifolia	Marsh earwort
Leiocolea rutheana	Norfolk flapwort
Marsupella profunda	Western rustwort
Petalophyllum ralfsii	Petalwort
Riccia bifurca	Lizard crystalwort
Southbya nigrella	Blackwort

Scientific Names	English Names
Lichens	
Bryoria furcellata	Forked hair-lichen
Buellia asterella	Starry breck-lichen
Caloplaca luteoalba	Orange-fruited elm-lichen
Caloplaca nivalis	Snow caloplaca
Catapyrenium psoromoides	Tree catapyrenium
Catillaria stricta	Upright mountain cladonia
Collema dichotomum	River jelly lichen
Gyalecta ulmi	Elm gyalecta
Heterodermia leucomelos	Ciliate strap-lichen
Heterodermia propagulifera	Coralloid rosette-lichen
Lecanactis hemisphaerica	Churchyard lecanactis
Lecanora achariana	Tarn lecanora
Lecidea inops	Copper lecidea
Nephroma arcticum	Arctic kidney-lichen
Pannaria ignobilis	Caledonian pannaria
Parmelia miniarum	New Forest parmelia
Parmentaria chilensis	Oil-stain parmentaria
Peltigera lepidophora	Ear-lobed dog-lichen
Pertusaria bryontha	Alpine moss pertusaria
Physcia tribacioides	Southern grey physcia
Pseudocyphellaria lacerata	Ragged pseudocyphellaria
Psora rubiformis	Rusty alpine psora
Solenopsora liparina	Serpentine solenopsora
Squamarina lentigera	Scaly breck-lichen
Teloschistes flavicans	Golden hair-lichen
Stoneworts	
Chara canescens	Bearded stonewort
Lamprothamniun papulosum	Foxtail stonewort

D.3.2   Protection for wild plants (Under Section 13):

- Part 1(a) – intentional picking, uprooting or destruction of plants on Schedule 8.

- Part 1(b) – unauthorised intentional uprooting of any wild plant not included on Schedule 8.

- Part 2(a) – selling, offering for sale, possession or transport for the purpose of sale any plant (live or dead, part or derivative) on Schedule 8.

• Part 2(b) – advertisement for buying or selling such things.

D.4 RARE SPECIES OR VULNERABLE BIRDS LISTED IN ANNEX I OF *EC DIRECTIVE 79/409/EEC ON THE CONSERVATION OF WILD BIRDS* [6] THAT REGULARLY OCCUR IN THE UK

Arctic tern	Honey buzzard	Sandwich tern
Avocet	Kingfisher	Scottish crossbill
Barnacle goose	Leach's petrel	Short-eared owl
Bewick's swan	Little tern	Slavonian grebe
Bittern	Marsh harrier	Snowy owl
Black tern	Montagu's harrier	Spotted crake
Black-throated diver	Mediterranean gull	Stone curlew
Capercaillie	Merlin	Storm petrel
Chough	Nightjar	White-fronted goose
Common tern	Osprey	(Greenland race)
Corncrake	Peregrine	White-tailed eagle
Dartford	Warbler Red-backed shrike	Whooper swan
Dotterel	Red kite	Woodlark
Golden eagle	Red-necked phalarope	Wood sandpiper
Golden plover	Red-throated diver	Wren (Fair Isle race only)
Great Northern diver	Roseate tern	
Hen harrier	Ruff	

Regularly occurring migratory species are afforded the same protection under Article 4.2 of the Directive.

D.5 WILD BIRDS LISTED ON SCHEDULE 1 OF THE *WILDLIFE & COUNTRYSIDE ACT 1981* [1]

Avocet	Golden oriole	Red-backed shrike
Barn owl	Goshawk	Red kite
Bee-eater	Great Northern diver	Red-necked phalarope
Bearded tit	Green sandpiper	Red-throated diver
Bewick's swan	Greenshank	Redwing
Bittern	Gyr falcon	Roseate tern
Black-necked grebe	Harriers (all species)	Ruff
Black redstart	Hen harrier	Savi's warbler
Black-tailed godwit	Hobby	Scarlet rosefinch
Black tern	Honey buzzard	Scaup
Black-throated diver	Hoopoe	Scottish crossbill
Black-winged stilt	Kentish plover	Serin
Bluethroat	Kingfisher	Shorelark
Brambling	Lapland bunting	Short-toed treecreeper
Cetti's warbler	Leach's petrel	Slavonian grebe

Chough	Little bittern	Snow bunting
Cirl bunting	Little gull	Snowy owl
Common quail	Little ringed plover	Spoonbill
Common scoter	Long-tailed duck	Spotted crake
Corncrake	Little tern	Stone curlew
Crested tit	Marsh harrier	Temminck's stint
Crossbills (all species)	Marsh warbler	Whimbrel
Dartford Warbler	Mediterranean gull	White-tailed eagle
Divers (all species)	Merlin	Whopper swan
Dotterel	Montagu's harrier	Woodlark
Fieldfare	Osprey	Wood sandpiper
Firecrest	Peregrine	Wryneck
Garganey	Purple heron	Velvet scoter
Golden Eagle	Purple sandpiper	

D.6 PROTECTED ANIMAL AND PLANT SPECIES IN NORTHERN IRELAND

D.6.1 Protected species under the *Wildlife (NI) Order 1985* [3]. These lists are current subject to review and may consequently change.

Birds

D.6.2 With certain exceptions all wild birds, their eggs and nests are protected at all times. The following birds may be shot by authorised persons during the whole of the year:

Great Black-backed Gull	Feral Pigeon	Hooded/Carrion Crow
Herring Gull	Wood Pigeon	Jackdaw
Lesser Black-backed	Gull House Sparrow	Magpie
Starling	Rook	

D.6.3 The following birds may be shot by authorised persons during the season of 1 September until 31 January inclusive:

Curlew	Gadwall	Scaup
Golden Plover	Goldeneye	Shoveler
Canada Goose	Mallard	Teal
Grey-lag Goose	Pintail	Tufted Duck
Pink-footed Goose	Pochard	Wigeon

Animals

D.6.4 The following animals are protected at all times:

Badger	Red Squirrel
Bats (all species)	Brimstone Butterfly
Cetaceans (Dolphins, Porpoises and Whales)	Dingy Skipper Butterfly

Common Newt

Common Otter

Common Seal

Grey Seal

Common (or viviparous) Lizards

Pine Marten

Holly Blue Butterfly

Large Heath Butterfly

Marsh Fritillary Butterfly

Purple Hairsteak Butterfly

Small Blue Butterfly

Plants

D.6.5　The plant species listed below are specially protected in Northern Ireland.

Avens, Mountain

Barley, Wood

Betony

Broomrape, Ivy

Buckthorn, Alder

Bugle, Pyramidal

Campion, Moss

Cat's ear, Smooth

Centaury, Seaside

Cloudberry

Clubmoss, Marsh

Cowslip

Cow-wheat, Wood

Cranesbill, Wood

Cress, Shepherd's

Crowfoot, Water

Fern, Holly

Fern, Kilarney

Fern, Oak

Fleabane, Blue

Globe-flower

Grass, Blue-eye

Grass, Holy

Heath, Cornish

Helleborine, Green-flowered

Helleborine, Marsh

Moschatel, or Town Hall Clock

Mudwort

Orchid, Bee

Orchid, Bird's Nest

Orchid, Bog

Orchid, Green-winged

Orchid, Irish Lady's Tresses

Orchid, Narrow-leaved Marsh

Orchid, Small white

Oyster-plant

Pea, Marsh

Pennyroyal

Pillwort

Rosemary, Bog

Saw-wort, Mountain

Saxifrage, Purple

Saxifrage, Yellow Marsh

Saxifrage, Yellow Mountain

Sea-lavender, Rock

Sedge, Broad-leaved Mud

Sedge, Few-flowered

Small-reed, Northern

Spike-rush

Thistle, Melancholy

Violet, Fen

Violet, Water

Waterwort, Eight-stamened

Wintergreen, Serrated

Yellow Bird's-nest

Appendix E
Signs of Bats in Bridges

E.1 GRADING BRIDGES FOR BATS

E.1.1   A bridge grading system has been developed [4] in which bridges are rated according to the presence or absence of crevices suitable for bats to use as a day roost:

0   No crevices with potential for day roosting;

2   Possible suitable crevices for day roosting (indicates uncertainty about suitability of crevices);

4   Crevices suitable for day roosting;

5   Evidence of bats using the site for day roosting.

E.1.2   It is recommended that this system be adopted: bridges should be surveyed by a licensed bat warden and graded. Surveys will normally take the form of a daytime visual inspection of the bridge from ground level using binoculars.

E.2 SIGNS OF BATS IN BRIDGES

E.2.1   Bats are sometimes visible when crevices are inspected with a torch, but roosts are more often found using the following signs:

* Bats audible - if disturbed by torchlight or noise, bats may make a high pitched chattering noise.

* Staining - where sites are heavily used by bats, the stonework or concrete around the roost entrance may become stained a dark brown colour with oil from the bats' fur.

* Scratches on the stonework or concrete and surfaces worn smooth by the passage of bodies are also used as evidence of bats, but roosting or nesting birds can make similar marks.

* Droppings - bat droppings in crevices, stuck to walls below suitable crevices and on the ground below suitable crevices are useful evidence of bat roosts. They are similar in appearance to mouse droppings but can be distinguished by their crumbly texture and often by their position on vertical walls where small rodents would not be able to climb.

* Bat-fly pupae - these flies are parasitic on bats, especially the Daubenton's bat. The dark brown pin head sized pupae are found attached to the stonework or concrete where the bats roost and are fairly distinctive if seen.

Appendix F
Identification Features for Cast Iron, Wrought Iron and Steel

Identification of Cast Iron, Wrought Iron and Steel [5]			
Visual characteristic	**Grey cast iron**	**Wrought iron**	**Steel**
Surface texture (uncorroded)	'Gritting' or pitted from mould Possible 'blowholes' Possible straight lines or 'steps' from junction of half moulds (e.g. diametrically opposite along axis of hollow circular column)	Smooth	Smooth Possible millscale
Surface texture (corroded)	Uniformly rough; 'powdery' rather than 'flaky'	Possible 'delamination', with rust layers flaking off like puff pastry (flat sections) or triangular wedges (rods)	Possible 'delamination' as for wrought iron, usually on flat sections only
Fracture surface	Clean break, no tearing Crystalline, bright (new) or grey (old)	Tearing and ductile necking Fibrous (crystalline if fatigue failure) Sometimes shiny in places	Slightly fibrous or striated (crystalline if fatigue failure) Ductile necking
Element cross-sectional profile	Typically 'chunky' with relatively thick sections, often ornate or complex profile (fluted or plain hollow circular or cruciform columns, $\mid$, I, $\perp$ or polygonal beam sections)	'Crisp' profile, typically •, $\mid$, I, L, T or Z section or compound riveted section; joists and channels usually thicker than steel members	Thin 'crisp' profile, typically •, $\mid$, I, U, L or T section solid or hollow circular or rectangular columns, or compound riveted or welded section

Continued

Identification of Cast Iron, Wrought Iron and Steel [5] (continued)			
Visual characteristic	**Grey cast iron**	**Wrought iron**	**Steel**
Element elevational profile	Usually varies along length; beams often have 'fish-belly' or 'hump-backed' web profile and integral web stiffeners; columns often have ornate Classical heads with spigots and extended ledges or 'tables' supporting beams, baseplates and intermediate stiffeners; other elements often ornate and complex	Constant along length unless compound beam(s) or plate girder when web profile may vary (plate girders only) and flange plates increase in number and size towards midspan	As wrought iron; recent plate girders may have web and flange plates of various thicknesses and depths butt-welded together
	Beams and columns may have intermittent openings in web	Openings in web usually stiffened (if original) by L or T framing on all sides	As wrought-iron, also castellated and cellular beams in recent construction
Corners of element	External corners sharp, typically 90°; re-entrant corners rounded	Outer flange corners sharp, often less than 90°; 'toe' and 'root' corners rounded	As wrought iron, except for recent I-sections (sharp 90° external corners)
Flange section	Rectangular, or polygonal in beams with typically larger tension flange and small or absent compression flange	Usually tapered flanges on I-sections, thickest at web; equal flange sizes	As wrought iron, except for recent I-sections (which have parallel flanges)
	Flange width or thickness may vary along element (largest at midspan)	Constant flange section along element	Constant flange section along element
Section size	Large beams (over say 11m) often cast in sections, bolted together at flanged junctions	I-sections up to 20 inches (508mm) deep, occasionally slightly more; deeper sections invariably built-up riveted plate girders; columns built-up from I-sections, angles, and plates; small tees and channels	I-sections up to 3 feet (914mm) deep; solid circular columns up to 1 foot (305mm) diameter; hollow tubular columns up to around 18 inches (457mm) side length

Continued

Identification of Cast Iron, Wrought Iron and Steel [5] (continued)			
Visual characteristic	**Grey cast iron**	**Wrought iron**	**Steel**
Connection methods	Typically bolts (often square-headed); beams often tied together at column heads by wrought iron 'shrink rings' fitted around cast-on beam lugs; welding rare (invariably repair at later date)	Rivets for all built-up sections; bolts (often square-headed); flats, bars and rods sometimes hammer-welded together in older structures; cotters and wedges for tie-rods. Welding rare except for forge welding or as repair	Rivets (up to 1950s); bolts in clearance holes (earlier square, later hexagonal heads); welding (since 1925, wide use after 1955); close tolerance bolts (since 1918); high strength friction grip bolts (since 1950s)
Identification on element	Maker's name and location often cast onto element (e.g. on web of beam, in plaque at foot of column); occasionally load capacity also indicated	Rarely, at intervals on rolled sections, cast iron plaque sometimes attached to major elements (bridges, roof structures)	Often, at set intervals on rolled steel sections, every piece of steel bears the maker's name or trade mark. In addition to this, high tensile steel, also bears the letters 'H.T.' (to distinguish it from mild steel).

Appendix G
CSS Inspection Process

G.1 GENERAL

G.1.1 The material contained in this Appendix has been reproduced from the information originally printed in the CSS Guidance Documents for bridge inspection (*Bridge Condition Indicators* Volume 2 [7] and Volume 2 Addendum [8]). Guidance is provided for reporting the condition of structural elements observed during General and Principal Inspections. Detailed guidance is provided on the use of the inspection pro-forma for bridges, retaining walls and sign/signal gantries, classification of elements, defect type, severity and extent of damage.

G.2 BRIDGE INSPECTION PRO-FORMA

G.2.1 The layout of the two-page bridge inspection pro-forma is shown in Figure G.1 and replicated in the subsequent pages. The inspection pro-forma is divided into the following areas:

- General Bridge Data (paragraph G.3) – This area of the pro-forma is for recording general information about a bridge such as bridge name, road name, O.S. grid reference, number of spans, span length, Bridge Type Code, etc.

- Bridge Elements (paragraph G.4) – This area of the pro-forma lists all the bridge elements for which a condition score needs to be recorded.

- Element Condition Reporting (paragraph G.5) – This information is recorded on the pro-forma for each bridge element, with separate columns for 'Severity', 'Extent' and 'Defect Type'.

- The 'Work Required', 'Work Priority' and 'Cost of Work' may also be recorded against each element.

- Inspection Dates.

- Multiple Defects Reporting (paragraphs G.5.6-G.5.9) – This area of the pro-forma allows the severity/extent of up to three defects on one element to be recorded.

- Comments (paragraphs G.6.1-G.6.2) – Space is provided on the pro-forma for the Inspector and Engineer to record their comments.

- Work Required and Signing Off (paragraphs G.6.3 and G.6.4).

G.2.2 The pro-forma presented herein identifies data fields that enable the creation of a comprehensive bridge database; however, the pro-forma is not a standard form and may be altered to the needs of individual authorities. The data fields that are mandatory and must not be altered are the Bridge Type Code, the element list and the element condition. The other data fields may be altered to suit each individual authority's needs but it is recommended they form the minimum data collection requirements.

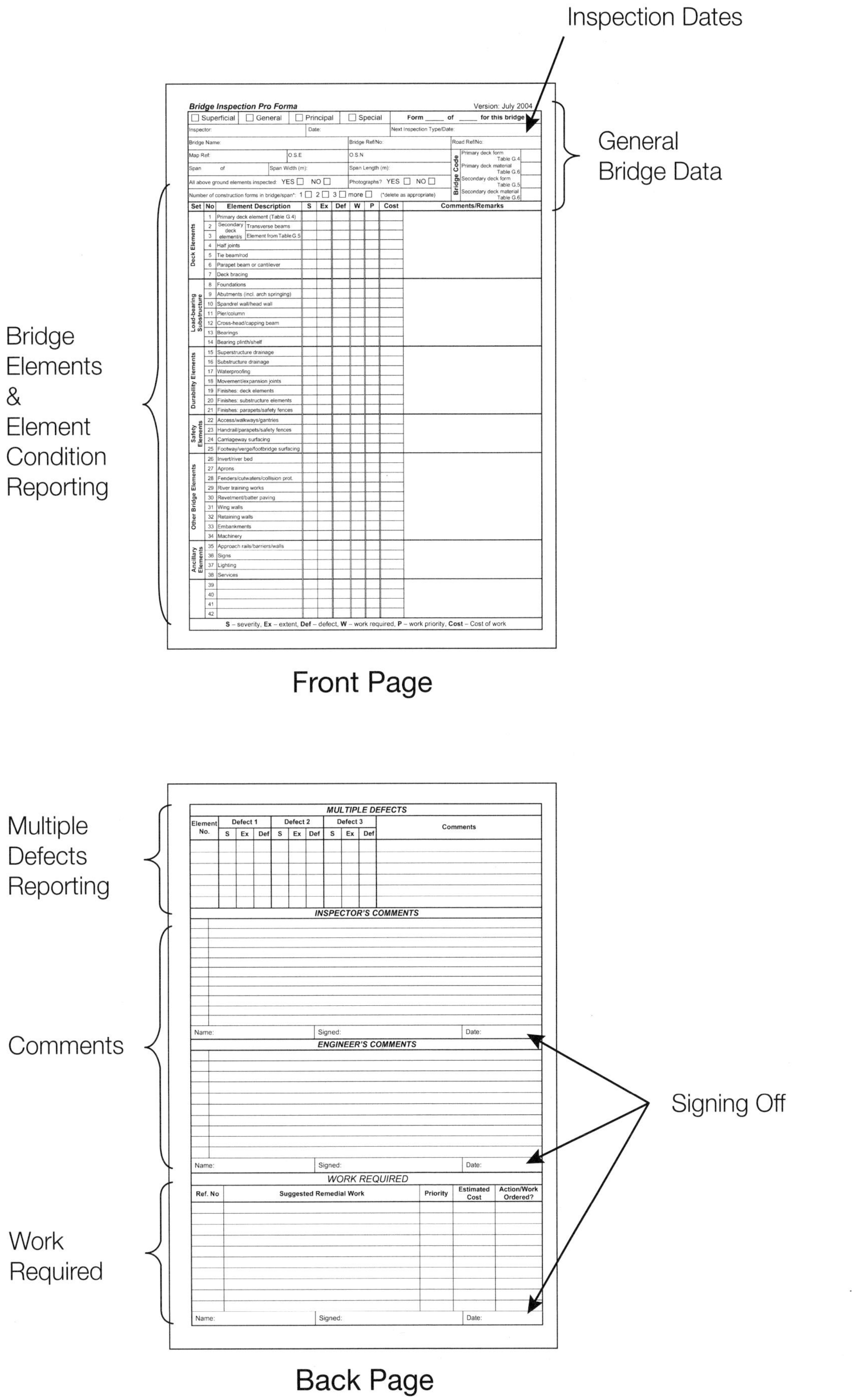

Inspection Dates

General Bridge Data

Bridge Elements & Element Condition Reporting

Front Page

Multiple Defects Reporting

Comments

Signing Off

Work Required

Back Page

Figure G.1 – Bridge inspection pro-forma layout

Bridge Inspection Pro Forma

Version: July 2004

☐ Superficial	☐ General	☐ Principal	☐ Special	**Form _____ of _____ for this bridge**

Inspector:		Date:		Next Inspection Type/Date:

Bridge Name:	Bridge Ref/No:	Road Ref/No:

Map Ref:	O.S.E	O.S.N	**Bridge Code**	Primary deck form Table G.4
Span of	Span Width (m):	Span Length (m):		Primary deck material Table G.6
				Secondary deck form Table G.5
All above ground elements inspected: YES ☐ NO ☐	Photographs? YES ☐ NO ☐			Secondary deck material Table G.6

Number of construction forms in bridge/span*: 1 ☐   2 ☐   3 ☐ more ☐   (*delete as appropriate)

Set	No	Element Description	S	Ex	Def	W	P	Cost	Comments/Remarks
Deck Elements	1	Primary deck element (Table G.4)							
	2	Secondary deck element/s — Transverse beams							
	3	Element from Table G.5							
	4	Half joints							
	5	Tie beam/rod							
	6	Parapet beam or cantilever							
	7	Deck bracing							
Load-bearing Substructure	8	Foundations							
	9	Abutments (incl. arch springing)							
	10	Spandrel wall/head wall							
	11	Pier/column							
	12	Cross-head/capping beam							
	13	Bearings							
	14	Bearing plinth/shelf							
Durability Elements	15	Superstructure drainage							
	16	Substructure drainage							
	17	Waterproofing							
	18	Movement/expansion joints							
	19	Finishes: deck elements							
	20	Finishes: substructure elements							
	21	Finishes: parapets/safety fences							
Safety Elements	22	Access/walkways/gantries							
	23	Handrail/parapets/safety fences							
	24	Carriageway surfacing							
	25	Footway/verge/footbridge surfacing							
Other Bridge Elements	26	Invert/river bed							
	27	Aprons							
	28	Fenders/cutwaters/collision prot.							
	29	River training works							
	30	Revetment/batter paving							
	31	Wing walls							
	32	Retaining walls							
	33	Embankments							
	34	Machinery							
Ancillary Elements	35	Approach rails/barriers/walls							
	36	Signs							
	37	Lighting							
	38	Services							
	39								
	40								
	41								
	42								

S – severity, **Ex** – extent, **Def** – defect, **W** – work required, **P** – work priority, **Cost** – Cost of work

MULTIPLE DEFECTS										
Element No.	**Defect 1**			**Defect 2**			**Defect 3**			**Comments**
	S	**Ex**	**Def**	**S**	**Ex**	**Def**	**S**	**Ex**	**Def**	

INSPECTOR'S COMMENTS		
Name:	Signed:	Date:

ENGINEER'S COMMENTS		
Name:	Signed:	Date:

WORK REQUIRED				
Ref. No	**Suggested Remedial Work**	**Priority**	**Estimated Cost**	**Action/Work Ordered?**
Name:	Signed:		Date:	

G.3 GENERAL BRIDGE DATA

G.3.1 The data required in this area of the pro-forma are described in Table G.1.

Table G.1 – Definition of General Bridge Data Fields	
Field	**Description of Data Required**
Form *x* of *n* for this bridge	Used to keep account of the number of inspection pro-forma used for a bridge i.e. separate pro-forma may be completed for different spans and/or different construction types within a span. x refers to this pro-forma and n to the total number of pro-forma used for this bridge.
Bridge Name	The name used for the bridge in the authority's records.
Road Name	The name used for the road in the authority's records.
Bridge Ref/No	Bridge reference used in the authority's records.
Road Ref/No	Road reference used in the authority's records.
Map Ref	Reference of map that O.S. readings are taken from.
O.S. E	Ordnance Survey grid reference, Easting.
O.S. N	Ordnance Survey grid reference, Northing.
Span *x* of *n*	Only needs to be filled in when individual spans are reported on separate pro-forma. When spans are reported separately n represents the total number of spans for the bridge and x represents which span the form relates to e.g. Span 2 of 4 refers to the second span of a four span bridge.
Span Length (m)	Used to report span length when one pro-forma is used per span of a multi span bridge, otherwise may be ignored. Some authorities may wish to collect bridge span data for all their structures if this does not exist in their records.
All above ground elements inspected	Used to determine if the inspection covered all above ground bridge elements. The inspectors should tick the 'NO' box if they are unable to survey all above ground elements due to difficulty in access, obstruction by vegetation, etc. An appropriate comment must be made on the pro-forma when an element cannot be inspected and NI (Not Inspected) recorded in the Severity or Extent column.
Photographs?	Questioning if photographs were taken during the inspection. The inspector's comments must describe which elements/bridge views were photographed.
Number of construction...	Many bridges have different construction types within, or between spans. See section on Multiple Construction Types (see paragraphs G.3.4 to G.3.8)
Bridge Type Code	Describes the structural form of the bridge, see section on Bridge Type Code (see paragraphs G.3.2 and G.3.3)

Bridge Type Code

G.3.2 There are a wide variety of bridge types in the UK, the major differences typically being between deck forms. The bridge type here is defined using a 4-key code combining the primary and secondary deck elements and their material as illustrated below (see paragraphs G.4.5-G.4.14 for element type and material lists).

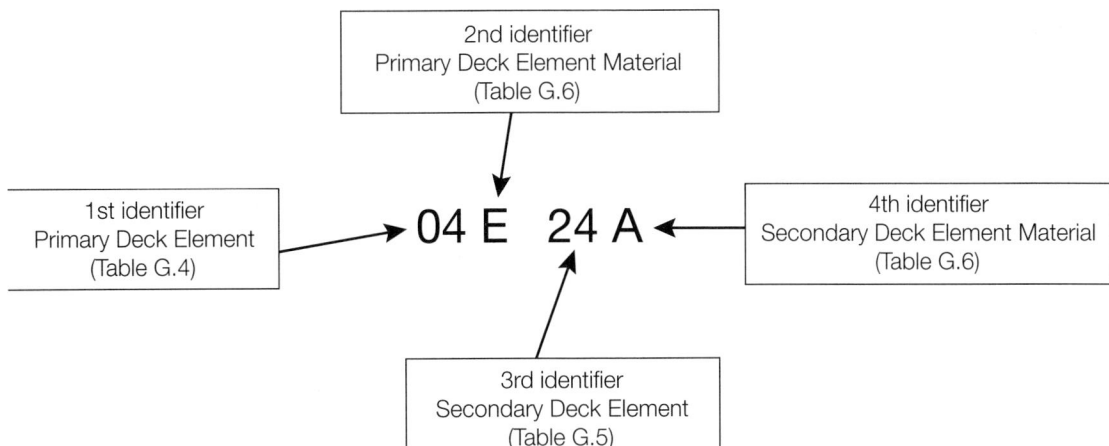

```
                        ┌─────────────────────────┐
                        │      2nd identifier       │
                        │ Primary Deck Element Material │
                        │       (Table G.6)         │
                        └─────────────────────────┘
                                   │
                                   ▼
┌──────────────────┐                            ┌──────────────────────────┐
│   1st identifier  │                            │      4th identifier       │
│ Primary Deck Element │──►  04 E   24 A  ◄──    │ Secondary Deck Element Material │
│   (Table G.4)     │                            │       (Table G.6)         │
└──────────────────┘            ▲               └──────────────────────────┘
                                │
                        ┌─────────────────────────┐
                        │      3rd identifier       │
                        │  Secondary Deck Element   │
                        │       (Table G.5)         │
                        └─────────────────────────┘
```

Examples

G.3.3   Examples of Bridge Type Code are shown below:

- **04E 24A** – a bridge composed of a reinforced concrete deck slab supported by longitudinal steel beams.

- **01K 20P** – solid spandrel brick arch.

- **10E 32E** – full through steel truss with a flat steel plate deck and transverse beams.

Multiple Construction Types

G.3.4   Some bridges can have more than one construction type, normally due to road widening, but also due to different construction types used on different spans of a multi-span bridge or within a span. When a bridge has more than one construction type a separate inspection form should preferably be used for each type if merited by the total number of elements related to it, see paragraphs G.3.7-G.3.8.

G.3.5   The inspector must tick the relevant box in the 'Number of Construction Forms' field to indicate how many are on the bridge. The inspector must clearly state on the pro-forma (e.g. in Bridge Name field and Comments field) which construction type, and part of the bridge/span, the pro-forma relates to e.g. original bridge, road widening, footpath widening, etc.

G.3.6   When more than one construction type exists it is the responsibility of the inspector to decide which elements should be recorded on each pro-forma. The following recommendations are made:

- The first pro-forma for a bridge/span should be for the original construction type and include all substructure, durability, safety, etc. elements relevant to it.

- Each additional pro-forma should report on one other construction type. The inspector should also attempt to distinguish which other bridge elements belong to the modification/widening e.g. abutments, drainage, etc. and report these on the same pro-forma.

G.3.7   When the construction type of a bridge changes from one span to the next, separate pro-forma are preferable and merited because it is relatively easy to distinguish which elements correspond to which construction type. When there is more than one construction type within a span, it is generally more difficult to distinguish which elements correspond to each type.

G.3.8   When there is more than one construction type in a single span, a separate pro-forma is preferable if five or more elements can be distinguished for each type. Otherwise the inspector should record a combined element condition on one pro-forma for an element present on more than one construction type in the span. Additional guidance on recording a combined primary or secondary deck element condition, when they are present in more than one construction type, is provided in paragraphs G.4.6 and G.4.12 respectively.

G.4   BRIDGE ELEMENTS

General

G.4.1   The bridge inspection pro-forma contains 38 predefined bridge elements categorised into: Deck Elements, Load-bearing Sub-Structure Elements, Durability Elements, Safety Elements, Other Bridge Elements and Ancillary Elements.

G.4.2   The form of the primary and secondary deck elements are defined using codes to minimise the number of elements listed on the inspection pro-forma. These codes, along with the primary and secondary deck element material type codes, are used to define the Bridge Type Code (see paragraphs G.3.2-G.3.3 on Bridge Type Code).

G.4.3   The primary deck elements are denoted using the codes defined in Table G.4, while the secondary deck elements are denoted using the codes defined in Table G.5. Material type codes are defined in Table G.6.

G.4.4   The element list shown on the bridge inspection pro-forma does not cover all the terms or element types currently used by authorities. Table G.2 and Table G.3 provide a list of 'equivalent elements' that relate to other element types/terms than those used on the pro-forma.

No.	Element description	Equivalent elements
	Table G.2 – Equivalent Elements	
1	Primary deck element	Main Beams
		Truss members
		Culvert
		Arch
		Arch Ring
		Vousoirs/Arch face
		Arch Barrel/Soffit
		Encased Beams
		Subway
		Box beam interiors
		Armco/Concrete pipe
		Portal/Tunnel portals
		Prestressing
		Sleeper bridge
		Tunnel Linings
2	Transverse Beams	
3	Secondary deck element	Concrete deck slab
		Timber deck
		Steel deck plates
		Jack Arch
		Troughing
		Stone slab (or Primary member)
		Troughing Infill
		Buckle plates
4	Half joints	
5	Tie beam/rod	
6	Parapet beam or cantilever	Edge Beams
7	Deck bracing	Diaphragms
8	Foundations	Piles
9	Abutments (incl. Arch springing)	Arch Springing
		Abutment slope
		Bank seat
		Counterfort/Buttresses
10	Spandrel wall/Head wall	Stringcourse
		Coping

Continued

Table G.2 – Equivalent Elements (continued)		
No.	**Element description**	**Equivalent elements**
11	Pier/Column	
12	Cross-head/Capping beam	
13	Bearings	
14	Bearing plinth/shelf	
15	Superstructure drainage	
16	Substructure drainage	Subway drainage Retaining wall drainage
17	Water proofing	
18	Movement/Expansion joints	Sealants
19	Painting: deck elements	Sealants Decorative appearance
20	Painting: substructure elements	Sealants Decorative appearance
21	Painting: parapets/safety fences	Sealants Decorative appearance
22	Access/Walkways/Gantries	Steps
23	Handrail/Parapets/Safety fences	Balustrade Barrier
24	Carriageway surfacing	Ramp surface Approaches
25	Footway/Verge/Footbridge surfacing	
26	Invert/River bed	Channel bedstones
27	Aprons	
28	Fenders/Cutwaters/Collision protection	Flood Barrier
29	River training works	
30	Revetment/Batter paving	
31	Wing walls	Newel
32	Retaining walls	Counterfort/Buttresses Gabions Wall
33	Embankments	Approach embankments Side slopes
34	Machinery	

Continued

Table G.2 – Equivalent Elements (continued)

No.	Element description	Equivalent elements
35	Approach rails/barriers/walls	Posts Remote approach walls
36	Signs	
37	Lighting	Subway lighting Primary lighting Secondary lighting
38	Services	Manholes Pipes Mast

Table G.3 – Other Element Relationships

Other elements	Covered by
Pointing/Arch mortar	Severity description No. 3 (see Table G.10)
Condition of masonry/brickwork	Severity description No. 3 (see Table G.10)
Condition of masonry/brickwork	Severity description No. 3 (see Table G.10)
Masonry/Brickwork	Severity description No. 3 (see Table G.10)
Vegetation	Severity description No. 5 (see Table G.10)
Decorative appearance	Severity description No. 4 (see Table G.10)
Cleanliness	Various severity descriptions
Dry Stone Wall & other walls	Corresponds to 9, 10, 11, 23, 31, 32 or 35 on the pro-forma, depending on function and location
Scour	Severity description No. 6 & 7 (see Table G.10)
Finishings	Various severity descriptions
Corrugated metal	Material codes
Leakage	Severity descriptions No. 8, 10 and 14 (see Table G.10)
Rivets and bolts	Severity descriptions No. 1 (see Table G.10)
Welds	Severity descriptions No. 1 (see Table G.10)
Arch cracks and deformation	Severity descriptions No. 3 (see Table G.10)
Fillets and haunching	Reported with element they are part of

Primary Deck Elements (or Span Primary Structural Form)

G.4.5   The Primary Deck Element is No. 1 on the bridge inspection pro-forma and is denoted using the codes defined in Table G.4. This identifies the form of the structural elements spanning in the longitudinal direction. Volume 1: Part B: Section 2: Paragraph 2.4 contains schematic illustrations of the majority of the bridge types and primary elements listed in Table G.4.

G.4.6   Some bridges contain more than one of the primary deck element types shown in Table G.4 on an individual span. Paragraph G.3.8 recommends a separate pro-forma for a construction type if five or more elements can be distinguished for it. When there are less than five elements for a construction type, or if the authority does not wish to report construction types separately, the condition score of the different primary deck element types should be recorded separately on the same pro-forma (i.e. utilising the blank rows, 39 to 42, on the pro-forma).

G.4.7   The condition of the dominant (by area, length or number, which ever is most appropriate) primary deck element should recorded in row No. 1 of the pro-forma. The blank rows, 39 to 42, on the pro-forma should be used to report the condition of the other primary deck elements. The inspector's comments must clearly state if an element is a primary deck element. The inspector should also record the approximate deck area (or proportion of deck area) served by each different primary deck element type (the proportion may be based on length or number as appropriate).

G.4.8   Multiple construction types may be used where there is more than one primary structural form within a bridge/span, see paragraphs G.3.4-G.3.8 for further guidance. However, it is recommended that a separate pro-forma is used for each construction type.

Table G.4 – Primary Deck Element Codes

Span Structural Form (Primary Deck Element)		Code
Arch	solid spandrel	01
	open/braced spandrel	02
	tied (including hangers)	03
Beam/Girder	at/below deck surface	04
	box beams (exterior & interior)	05
	half through	06
	filler beam	07
Truss	at/below deck surface (underslung)	08
	half through	09
	full through	10
Slab	solid	11
	voided	12
Culvert/Pipe/Subway	circular/oval	13
	box	14
	portal/U-shape	15
Troughing		16
Cable stayed/Suspension		17
Tunnel		18
Other		19
Multiple construction types		MC

Secondary Deck Element

G.4.9   Secondary Deck Elements are recorded in row No.'s 2 and 3 on the bridge inspection pro-forma. These are denoted using codes defined in Table G.5, which identifies the form of the structural elements spanning transversely between primary elements. On some bridges secondary deck elements may not be present, e.g. arch bridges, a code of '20' or '30' signifies 'no secondary deck element', the code used depends on whether or not transverse beams are present. No secondary deck element code is required for retaining walls.

G.4.10   Transverse beams are a very common type of secondary deck element and have been assigned their own row on the bridge inspection pro-forma, i.e. row No. 2. If transverse beams are not present codes '20' to '26' are used in the Bridge Type Code, when transverse beams are present codes '30' to '36' are used in the Bridge Type Code.

G.4.11   When transverse beams are present the elements given in Table G.5 are sometimes called 'tertiary' deck elements; if transverse beams are not present they are called 'secondary' deck elements. For simplicity, and consistency, they are called 'secondary' deck elements throughout this document whether transverse beams are present or not.

Table G.5 – Secondary Deck Element Codes		
Secondary Deck Element	**Code**	
	No Transverse Beams	**Transverse Beams**
No secondary deck element	20	30
Buckle Plates	21	31
Flat Plate	22	32
Jack Arch	23	33
Slab	24	34
Troughing	25	35
Other	26	36

G.4.12   Some bridges contain more than one of the secondary deck element types shown in Table G.5 on an individual span. Paragraph G.3.8 recommends a separate pro-forma for a construction type if five or more elements can be distinguished for it. When there are less than five elements to a construction type, or if the authority does not wish to report them separately, the condition score of the different secondary deck element types should be recorded separately on the same pro-forma (i.e. utilising the blank rows, 39 to 42, on the pro-forma).

G.4.13   The condition of the dominant (by area, length or number, which ever is most appropriate) secondary deck element should be recorded in row No. 3 of the pro-forma. The blank rows, 39 to 42, on the pro-forma should be used to report the condition of the other secondary deck elements. The inspector's comments must clearly state if an element is a secondary deck element. The inspector should also record the approximate deck area (or proportion of deck area) served by each different secondary deck element type (the proportion may be based on length or number as appropriate).

Material Type

G.4.14 The material type code of the primary and secondary deck elements is also used in defining the Bridge Type Code. The Material Type codes are given in Table G.6.

Table G.6 – Material Type Code		
Material		**Code***
Concrete	reinforced	A
	plain/mass	B
	post-tensioned	C
	pre-tensioned	D
Metal	steel	E
	cast iron	F
	wrought iron	G
	aluminium	H
	corrugated steel	I
	corrugated aluminium	J
Masonry	brick	K
	stone	L
FRP/GRP/Composite		M
Timber		N
No secondary element, so no material		P
Other		Q

*Letter O not used, avoids confusion with zero 'element type' codes

Multiple Elements

G.4.15 If one element description on the pro-forma covers several equivalent elements (see Table G.2 and Table G.3) then the condition reporting should take the condition of all of these into account.

G.4.16 The following situations are covered by one element description and one condition score on the pro-forma:

- Multiple elements of one type e.g. longitudinal beams, transverse beams, piers/columns, etc.

- Elements repeated over several spans if the whole bridge is reported on one bridge inspection pro-forma e.g. primary deck elements, abutments, invert/river bed, etc.

- Element descriptions on the bridge inspection pro-forma that cover several element types, e.g. the primary deck element description on the pro-forma covers arch barrel and voussoirs for a masonry arch bridge.

- 'Elements' that were previously treated as separate items by some authorities, e.g. pointing is now included in masonry severity description, vegetation is covered by severity descriptions, welds are covered by metalwork severity descriptions, etc. Severity descriptions are covered in paragraph G.5.

Half-joints

G.4.17 Half-joints, although not distinct elements, receive a separate entry on the inspection bridge inspection pro-forma due to their structural criticality and inherent maintenance problems. However, given that half-joints are an integral part of the primary deck element there is the possibility that defects may be double counted during the inspection. The condition of half-joints should be reported as:

- Defects on the primary element, in the immediate vicinity of the half-joint, likely to have been caused by the presence of the half-joint, e.g. defects in a region D (beam or slab depth) either side of the joint, see Figure G.2; and

- Defects to the half-joint e.g. dowel/bearing plate, filler, etc.

G.4.18 Defects used to assess the condition of the half-joint should not be included in the condition assessment of the primary deck element. A typical section through a half joint is shown in Figure G.2.

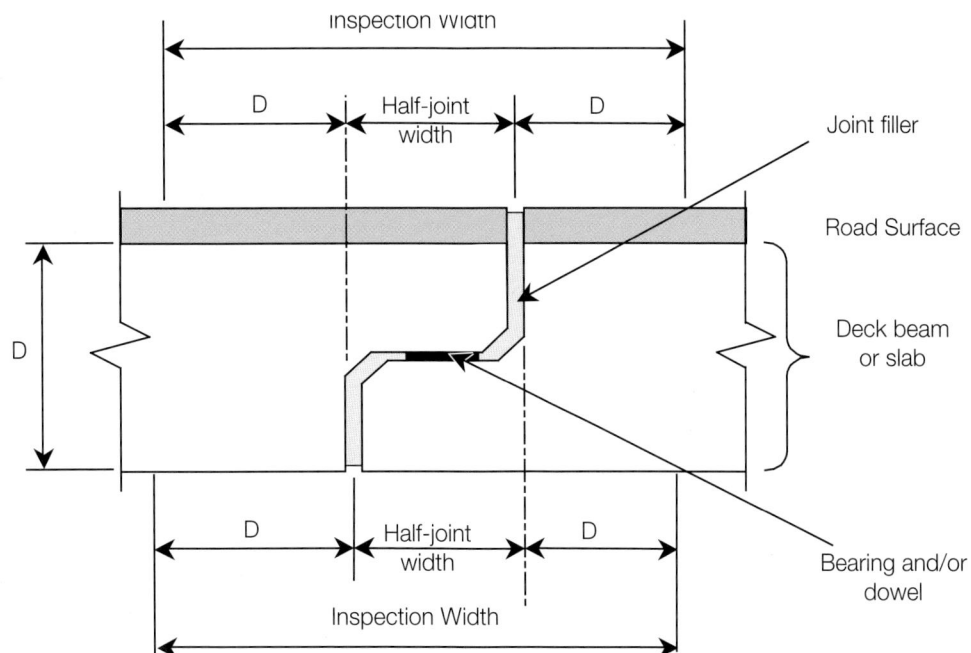

Figure G.2 – Half-joint cross-section

Blank Rows

G.4.19 Four blank rows, 39 to 42, are provided on the bridge inspection pro-forma. These may be used for any elements that are not covered by the pro-forma if the inspector regards it as important to report the condition of these elements, e.g. third party elements, fire equipment, telecommunications, smoke detectors, one-off element types, decorative elements, etc. However, it is recommended that every effort is made to report the complete bridge condition using the element descriptions already provided on the pro-forma.

G.5 ELEMENT CONDITION REPORTING

Extent and Severity Codes

G.5.1 The condition of a bridge element should be recorded in terms of the Severity of damage/defect and the spatial Extent of the damage/defect. The following definitions may be adopted to describe the Extent and Severity parameters:

- Extent: The area, length or number (as appropriate) of the bridge element affected by the defect/damage.

- Severity: The degree to which the defect/damage affects the function of the element or other elements on the bridge.

G.5.2 Both extent and severity are parameters that are used to inform decisions about maintenance planning and management. The use of separate codes for each parameter eliminates any obscurity in the distinction between, for example, a single but severe defect and extensive but superficial deterioration. Codes that may be used to describe the Extent and Severity levels are shown in Table G.7 and Table G.8 respectively.

G.5.3 Permissible combinations of Severity and Extent are shown in Table G.9. This shows that some severity/extent combinations are not permissible, such as 2A, 3A, 4A and 5A. These combinations are not permitted because it is not possible to have a Severity condition greater than 1 with an Extent description of 'no significant defect'.

Table G.7 – Extent Codes	
Code	**Description**
A	No significant defect
B	Slight, not more than 5% of surface area/length/number
C	Moderate, 5% - 20% of surface area/length/number
D	Wide: 20% - 50% of surface area/length/number
E	Extensive, more than 50% of surface area/length/number

Table G.8 – Generic Severity Descriptions	
Code	**Description**
1	As new condition or defect has no significant effect on the element (visually or functionally)
2	Early signs of deterioration, minor defect/damage, no reduction in functionality of element
3	Moderate defect/damage, some loss of functionality could be expected
4	Severe defect/damage, significant loss of functionality and/or element is close to failure/collapse
5	Extensive, more than 50% of surface area/length/number

Table G.9 – Permissible Combinations of Severity and Extent					
Extent	**Severity**				
	1	**2**	**3**	**4**	**5**
A	1A				
B		2B	3B	4B	5B
C		2C	3C	4C	5C
D		2D	3D	4D	5D
E		2E	3E	4E	5E

G.5.4   More detailed guidance on severity descriptions for different construction material and defect types is contained in Table G.10. These descriptions do not cover all element or defect types but provide general guidance on the identification of severity states. Many of the severity states in Table G.10, contain a number of descriptions for each item, e.g. metalwork has four possibilities in severity state 3. The element condition only needs to satisfy one of these possibilities to be categorised as severity state 3. Volume 2: Part B provides photographic examples of some of the defects described in Table G.10.

G.5.5   Table G.8 gives the generic severity descriptions and must be used as the primary source for defining severity. Table G.8 should be used to assess those materials, elements and defect types not covered by Table G.10. It is considered that if Table G.10 is used in conjunction with Table G.8 a more consistent approach to inspection reporting will be achieved by authorities.

Table G.10 – Severity Descriptions							
No	**Item**	**Severity**					
			1	**2**	**3**	**4**	**5**
1	Metalwork	.1	No signs of rusting or damage	Minor surface rusting	Moderate pitting	Deep pits and perforations (localised severe corrosion)	Disintegrated by corrosion mechanisms
		.2	No loss of section thickness	Minor section loss (penetration less than 5% of section)	Moderate section loss causing some reduction in functionality (penetration 5 to 20% of section thickness)	Major section loss causing significant reduction in functionality (penetration more than 20% of section)	Collapsed or collapsing
		.3	No signs of rusting or damage to bolts, nuts and rivets	Non structural bolts loose, minor corrosion of nuts and washers	Non structural bolts missing, moderate corrosion of rivet heads, nuts and washers	Structural bolts missing, rivets loose or missing, crack through bolt	Failure of element due to missing/failed bolts/rivets
		.4	No corrosion or damage of weld runs	Slight corrosion of weld run	Crack at toe of weld, moderate reduction in size of weld due to corrosion	Longitudin-ally cracked weld, major reduction in size of weld due to corrosion	Weld connection failure (longitudinal crack)

Continued

Table G.10 – Severity Descriptions (continued)							
No	**Item**	**Severity**					
			1	**2**	**3**	**4**	**5**
2	Reinforced Concrete, Prestressed Concrete & Filler Joist	.1	No spalls	Minor localised spalls exposing shear links	Major localised spalls exposing shear links and main bars with general corrosion	Joined up deep spalls exposing shear links and main bars with general and pitting corrosion	Collapsed
		.2	Hairline cracks, difficult to detect visually	Cracks and crazing in areas of low flexural behaviour (cracks less than 0.3mm)	Cracks and crazing in areas of high flexure. Cracks approx. 1mm and easily visible	Wide/deep cracks (more than 2mm). Shear cracks	Element unable to function due to structural cracks
		.3	No signs of damage to prestressing	Substandard grouting of ducts (may not be visible)	Cracks along line of prestressing duct	Exposed prestressing cables	Failed prestressing cables
		.4	No signs of delamination	Early signs of delamination e.g. cracks with rust staining	Delamination in areas of low flexural and/or shear action	Delamination in areas of high flexural and/or shear action	Failure due to delaminated bars
		.5	No signs of thaumasite or freeze-thaw attack	Slight cracking caused by thaumasite or freeze-thaw	Moderate thaumasite or freeze-thaw attack	Major thaumasite or freeze-thaw attack	Failure due to thaumasite or freeze-thaw attack

Continued

Table G.10 – Severity Descriptions (continued)							
No	**Item**	**Severity**					
			1	**2**	**3**	**4**	**5**
3	Masonry, Brickwork & Mass Concrete	.1	No evidence of deformation	Minor deformation	Moderate deformation	Major deformation	Collapsed
		.2	Pointing sound	Minor depth of pointing deteriorated	Moderate to significant depth of pointing lost, but does not appear to be rapidly disintegrating or crumbling, bricks not easily loosened	Pointing in very poor condition, severely weathered, crumbling to touch and/or significant depth loss, bricks easily loosened	Collapsed
		.3	No arch ring cracking or separation	Arch ring cracks difficult to see	Arch ring separation (gap less than 25mm)	Arch ring separation (gap greater than 25mm)	Disintegrated
		.4	No arch barrel cracks	No diagonal cracks, longitudinal cracks less than 3mm wide, lateral cracks	Diagonal cracks, longitudinal cracks greater than 3mm wide	Diagonal cracks, longitudinal cracks breaking barrel into 1m sections or less	Arch barrel failure
		.5	No cracks	Minor hairline cracks and shallow spalls	Moderate cracks (easily visible, crazing) and deep localised spalls	Major cracks and spalling	Failure due to structural cracks
		.6	No bricks/ masonry blocks missing, minor surface weathering	Few bricks/stones missing (no adjacent ones missing), major surface weathering	Moderate loss of bricks/stones	Severe loss of bricks/stones	Failure due to missing bricks/stones
		.7	No bulging, leaning or displacement	Minor bulging, leaning or displacement	Moderate bulging, leaning or displacement	Severe bulging, leaning or displacement	Collapsed or non functional

Continued

Table G.10 – Severity Descriptions (continued)							
No	**Item**	**Severity**					
			1	**2**	**3**	**4**	**5**
4	Paintwork & Protective Coatings	.1	Finishing coat sound, slight weathering	Normal weathering of finishing coat	Spots, chips and cracks of finishing coat, undercoat exposed but sound	Failure of finishing coat and spots, chips and cracks to undercoat/ substrate	All coats failed
5	Vegetation	.1	Slight to no vegetation	Minor vegetation causing no structural damage (surface mosses, small grass and weeds)	Vegetation growth on or near bridge causing minor structural damage and/or deformation e.g. roots and branches of nearby trees, small tree/plants growing on structure	Vegetation growth on or near bridge causing major structural damage and/or deformation e.g. roots and branches of nearby trees, large tree growing on structure	Failure caused by vegetation growth or a tree collapsing on the structure
		.2	Slight to no vegetation	Low depth/density of vegetation cover, easily removed, e.g. moss	Significant depth/density of vegetation, obscuring inspection, e.g. ivy	Inspection impossible due to vegetation growth but structural damage due to vegetation unlikely	Inspection of critical structural elements not possible due to density of vegetation and root systems likely to be causing structural damage

Continued

Table G.10 – Severity Descriptions (continued)							
No	Item		Severity				
			1	2	3	4	5
6	Foundations	.1	No visible settlement of structure	No visible settlement, but cracks that may be due to it	Minor settlement of structure	Major settlement of structure	Collapsed due to settlement
		.2	No visible differential movement of structure	No visible movement, but cracks that may be due to it	Minor differential movement of structure	Major differential movement of structure	Collapsed due to differential movement
		.3	No visible sliding of structure	No visible sliding, but cracks that may be due to it	Minor sliding of structure	Major sliding of structure	Collapsed due to sliding
		.4	No visible rotation of structure	No visible rotation, but cracks that may be due to it	Minor rotation of structure	Major rotation of structure	Collapsed due to rotation
		.5	No scour	Minor scour	Moderate scour	Major scour	Dangerous scour or failure
		.6	Substructure appears unaffected by foundation faults (assume no foundation faults)	Foundation faults causing minor cracks in substructure	Foundation faults causing moderate cracks in substructure	Foundation faults causing major cracks and deformation in substructure	Failure due to foundation faults
7	Invert, apron & river bed (also see 2 and 3)	.1	No scour	Minor scour	Moderate scour	Major scour	Dangerous scour or failure
		.2	No vegetation growth or silting	Vegetation growth, trapped debris and silting causing slight disruption to flow	Vegetation growth, trapped debris and silting, significant disruption to flow causing faster flow in areas of the river	Vegetation growth, trapped debris and silting, severe disruption to flow causing much faster flow in areas of the river	Failure caused by vegetation growth, trapped debris and silting

Continued

No	Item		Severity				
			1	2	3	4	5
8	Drainage	.1	In sound condition and fully functional	Mostly functional (less than 25% of cross section blocked)	Part functional (25 to 50% of cross section blocked)	Mostly non-functional (more than 50% of cross section blocked)	Totally blocked / non-functional / broken
		.2	Causing no staining	Causing minor staining	Cleaning of staining required	Urgent cleaning required	Urgent & frequent cleaning
		.3	No structural damage	Causing minor structural damage	Causing moderate structural damage	Causing major structural damage	Causing severe damage to adjacent elements
		.4	No blockage of weep holes, outlets	Minor blockage of weep holes, outlets	Moderate blockage of weep holes, outlets	Major blockage of weep holes, outlets	Non functioning weep holes
9	Surfacing	.1	Little to no wear and weathering	Minor wear/ weathering	Moderate wear/ weathering	Major wear/ weathering	Dangerous
		.2	No crazing, tracking or fretting	Minor crazing, tracking and/or fretting	Moderate crazing, tracking and/or fretting	Major cracks, tracking and/or fretting	Complete break up
		.3	Dense	Poor texture	Open texture	Very open texture	Dangerous
		.4	Sound	Cracks in top layer	Top layer breached	Deep cracks and potholes	Top layer completely missing
		.5	Not slippery	Starting to become slippery	Definitely becoming slippery	Slippery	Dangerous
	Flagged surface	.6	No defects	Trips < 5mm	Cracked flags Trips >5mm and < 10mm	Trips >10mm and <20mm	Trips > 20mm

Continued

No	Item			Severity				
				1	**2**	**3**	**4**	**5**
10	Asphaltic plug	.1	Sound	Minor debonding between plug and road	Moderate debonding between plug and road	Major debonding between plug and road	Dangerous	
		.2	Sound	Slight loss of surface binder and aggregate	Loss of aggregate (surface penetration 20 to 50mm)	Loss of material from joint (causing holes > 50mm deep)	Missing	
		.3	Sound	Minor tracking and flow of binder	Moderate tracking and flow of binder	Major tracking and flow of binder	Disintegrated	
	Nosing Defects	.4	Sound	Minor cracking along nosing	Moderate cracking along nosing, some break-up	Break-up of nosing material	Disintegrated	
	Elasto-meric and others	.5	Minor signs of wear	One bolt missing at cross section	Numerous bolts missing at cross section	Majority of bolts missing at a cross section	Failure due to missing bolts	
		.6	Strip sealant sound	Strip sealant loose/poor, compression seal dropped and/or worn	Sealant breached, strip sealant breached	Sealant missing, strip sealant missing/out	Failure	
		.7	Sound road surface adjacent to joint	Minor break up of road surface adjacent to joint	Moderate break up of road surface adjacent to joint, some debris in joint seal	Major break up of road surface adjacent to joint, significant debris in joint seal	Joint failure due to deteriorated condition of adjacent road surface	
		.8	Sound fixings	Bolt sealer missing	Fixings loose	Fixings missing, plates and angles loose	Failure due to missing fixtures	
		.9	Sound components	Initiation of cracking or tearing of components	Crack/tear < 20% of width of component	Crack/tear > 20% but <50% of width of component	Failure of expansion joint components	

(Left vertical label: Expansion Joints)

Table G.10 – Severity Descriptions (continued)

Continued

Table G.10 – Severity Descriptions (continued)

No	Item			Severity				
				1	2	3	4	5
10	Expansion Joints	Buried Joint	.10	Reasonably sound	Minor surfacing cracking	Moderate surfacing cracking	Major surfacing cracking	Failure
			.11	Sealant for induced crack is sound	Minor cracking or break up of sealant for induced crack	Moderate cracking or break up of sealant for induced crack	Major cracking or break up of sealant for induced crack	Disintegrated or missing sealant for induced crack
		Joint leakage	.12	No visible signs of leakage	Minor leakage through joint	Moderate leakage through joint	Major leakage through joint causing minor structural damage	Open joint causing major structural damage
11	Embank-ments		.1	Sound No deformation	Minor subsidence Minor deformation	Minor slip/ settlement causing slight cracking of carriageway	Major slip/ settlement causing major cracking of carriageway	Critical slip/ settlement
12	Bearings (also see 1: Metalwork)		.1	Negligible rusting, minor weathering	Minor rusting, moderate weathering	Moderate rusting	Major rusting	Failed or seized due to rusting
			.2	Correct position	Minor offset	Moderate offset/tilt	Dislodged	Off bearing/ missing
			.3	Sliding bearing in correct position	Sliding bearing in slightly skewed (off centre) position at normal temp	Sliding bearing at end of travel in normal temperatures	Designed extent of travel at normal temperatures	Sliding bearing failed
			.4	No crazing	External crazing	External breakdown	Major breakdown (PTFE, laminations, rubber, etc.)	Complete breakdown
			.5	Sliding plate sound	Minor deformation of sliding plate	Moderate deformation of sliding plate	Major deformation of sliding plate	Bearings seized by sliding plate deformations
			.6	Bearings sound	Minor cracks	Moderate cracks or loose	Splitting and deformation	Disintegrated

Continued

No	Item		Severity				
			1	**2**	**3**	**4**	**5**
13	Impact Damage	.1	No damage	Slight surface scoring, minor displacement of element, e.g. marking and chipping of beam faces, several bricks across arch barrel width, slight impact deformation of steelwork	Moderate displacement of element, e.g. beam slightly offset on bearings, significant number of bricks knocked out across arch barrel width, moderate impact deformation of steelwork	Severe displacement of element, e.g. beam dislodged off bearings, many bricks knocked out across arch barrel width, major impact deformation of steelwork	Knocked down, Broken, collapsing
14	Water-proofing (exclude leaks through joints)	.1	No visible sign of seepage	Minor seepage through deck/arch, etc. (slow dripping)	Moderate seepage through deck/arch, etc. (some resistance to seepage)	Major seepage (little resistance) through deck/arch, etc. causing structural damage	Non-functional Causing critical structural damage
		.2	No visible sign of seepage	Damp surface, slight water stains on soffit	Wet surface, drops of water falling and significant staining	Very wet surface and stalactites causing structural damage	Major structural damage caused by waterproofing not functioning properly
15	Stone slab bridges	.1	Sound, no defects or damage	Minor cracking	Moderate cracking but no visible displacement	Major cracking and/or displacement	Collapsed

Continued

Table G.10 – Severity Descriptions (continued)

No	Item	Severity						
				1	**2**	**3**	**4**	**5**
16	Timber	.1	No sign of damage	Minor signs of damage	Moderate signs of damage	Major signs of damage	Disintegrated through damage
		.2	No loss of section	Minor section loss (decay less than 5% of section)	Moderate section loss causing some reduction in functionality (decay 5 to 20% of section thickness)	Major section loss causing significant reduction in functionality (decay more than 20% of section thickness)	Collapsed or collapsing
		.3	No visible signs of open joints	Joints/shakes open slightly on surface or cracked coating at joints/shakes	Open joints/shakes < 50% width of beam, in areas of low flexure or < 25% in areas of high flexure	Open joints/shakes > 50% width of beam, in areas of low flexure or > 25% in areas of high flexure	Beam separated into multiple elements
		.4	No signs of rusting or damage to fixings	Non structural bolts loose, minor corrosion of fixings	Non structural bolts missing, moderate corrosion of fixings	Structural fixings missing	Failure of element due to missed/failed fixings

The column headed 'Def' on the bridge inspection pro-forma is for the Defect Type e.g. '3.2' for masonry pointing. If there are no defects then insert '0'.

Multiple Defects on an Element

G.5.6   When an element has more than one type of defect/damage, the guidelines contained in Table G.11 and Table G.12 should be used to assess its condition.

Table G.11 – Dominant Defect is Present

Severity	When the severity of one defect is adjudged to be at least one severity category higher (see examples in Table G.13) than any other defect on the element, the Severity for the element is defined based on this dominant defect, AND Other defects do not reduce the functionality of the element beyond that caused by the dominant defect.
Extent	The extent code in this case should correspond to the area affected by the dominant defect alone.

Table G.12 – Interacting Defects, or No Dominant Defect Present	
Severity	Where the cumulative effect of several defects is adjudged to be the same as, or worse than, the effect of the dominant defect then the severity code should be reported based on the cumulative effect of all the defects on the element, OR Where no dominant defect is evident, the severity should be based on the cumulative effect of the defects the inspector feels are relevant.
Extent	The extent code in this case should correspond to the area affected by all defects considered in assessing the severity.

G.5.7 The inspector should record the worst condition for the element at all times from either dominant or interacting defects and enter the severity/extent codes on the front page of the bridge inspection pro-forma.

G.5.8 The dominant and interacting defects are described in terms of the damage to a single element. The same guidelines also apply when assessing the condition of multiple elements. For example, if one primary beam, out of a total of 10, has a severity of 4 and all the others are 2 then the severity recorded is 4 and the extent recorded is C (i.e. 10% of elements), giving a condition of 4C. However, if all the beams were in condition 2 then the extent category would be E, giving a condition of 2E. Some examples of interacting defects are shown in Table G.13.

Table G.13 – Examples of Interacting Defects					
Element		**Individual Defects**	**S/Ex**	**Interacting Effect**	**S/Ex**
1	RC Abutment	10% of concrete spalled, general corrosion of steel	3C	Extent increases; Severity does not increase, abutment is generally in compression therefore anchorage of steel not critical	3D
		15% delaminated (signifies corrosion of underlying steel)	3C		
2	RC Beam	10% of concrete spalled, general corrosion of main tensile steel.	3C	Extent increases; Severity also increases because anchorage of the tensile steel is critical to the functionality of the element	4D
		15% of main tensile steel cover delaminated	3C		
		Cracking parallel to tensile reinforcement	3B		
3	Masonry Arch	Arch ring separation (<25mm)	3E	Extent already maximum of E; Severity increases because all defects interrupt the load path and together have a significant influence on functionality	4E
		10 to 25mm of pointing lost	3E		
		Pockets of bricks missing and loose	3C		
4	Masonry Retaining Wall	Few bricks missing at base of retaining wall	3B	Extent is low due to small area of wall damaged; Severity increased because stability of bulge is directly influenced by missing bricks	4B
		Moderate bulging above missing bricks	3B		
5	Metal beam	Slight corrosion of girder weld run between web and bottom flange at mid span	2B	Extent stays the same; Severity increases because the corrosion is concentrated at the critical section of the member	3B
		Minor section loss of flange and web cross section at mid span	2B		

G.5.9 If the inspector feels that one condition entry is not sufficient for assisting maintenance management for the structure, then they can provide additional severity/extent codes, for up to three defects per element, in the multiple defects section of the bridge inspection pro-forma:

- Enter an 'M' in the Defect column (Def) on the front page of the pro-forma to indicate that Multiple defects have been recorded for this element on the reverse of the pro-forma.

- The element number, from the front page of the pro-forma, is entered in the first column (Element No.) of the Multiple Defects section.

- The severity, extent and defect code for the most severe defect on the element are entered in the Defect 1 columns.

- The severity, extent and defect code for the defect with the next highest severity are entered in the Defect 2 columns.

- The severity, extent and defect code for the defect with the next highest severity are entered in the Defect 3 columns.

- Additional notes can be entered into the Comments column.

Defect Code

G.5.10 The Defect code helps in the identification of Work Required, Priority and Cost. This also provides valuable information about defect types, their frequency of occurrence and cost of repairs.

G.5.11 When the observed defect relates to a defect described in Table G.10 the appropriate reference should be recorded in the defect column of the pro-forma. The defect code is recorded as:

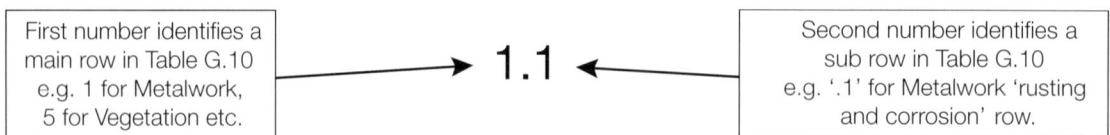

| First number identifies a main row in Table G.10 e.g. 1 for Metalwork, 5 for Vegetation etc. | → 1.1 ← | Second number identifies a sub row in Table G.10 e.g. '.1' for Metalwork 'rusting and corrosion' row. |

G.5.12 The severity code is not used in the defect code because it is reported for the elements in the severity column on the pro-forma. If the defect is not covered by the codes in Table G.10 then a description, consistent with the generic descriptions in Table G.8, should be entered in the comments box.

G.5.13 The inspector should record the most relevant or dominant defect. If other defects are also felt to be appropriate to work requirements (type, priority and cost) then their code/description should be entered in the comments column.

G.6   OTHER ENTRY FIELDS ON THE BRIDGE INSPECTION PRO-FORMA

Comments

G.6.1   Space for comments is provided on the front and back of the pro-forma. Comments should be used by the inspector to provide additional information that will be beneficial to the engineer and for the development of a computer database, e.g. clearly define if the bridge has several construction types.

G.6.2   Space is also provided for the engineer to add comments to the pro-forma. This may include an assessment of the overall condition of the bridge.

Work Required

G.6.3   Space is provided for identifying work required. The details of the information to be recorded in this area are not covered in this document and should be defined by individual authorities.

Signing Off

G.6.4   The inspector, engineer and data processing personnel must print their name, sign and date the pro-forma in the appropriate sections. The signing of the pro-forma is essential for future reference and traceability.

G.7 INSPECTION REPORTING FOR RETAINING WALLS

General

G.7.1 Retaining walls represent a significant proportion of the highway structure asset for many authorities. The guidance provided for retaining walls aims to promote a consistent approach and covers topics such as structural forms (paragraphs G.7.2-G.7.3), material types (paragraph G.7.4), inspection elements (paragraph G.7.5); and Retaining Wall inspection pro-forma (paragraph G.7.6).

Structural Forms

G.7.2 The structural forms for retaining walls are listed in Table G.14 along with the associated retaining wall code.

Table G.14 – Retaining Wall Structural Form Code	
Structural Form	**Code**
Gravity	R1
Cantilever on foundation	R2
Embedded	R3
Reinforced soil	R4
Gabions	R5
Cribwork	R6
Other	R7

G.7.3 Schematics of different retaining wall structural forms, along with typical elements, are contained in Volume 1: Part B: Section 3.3. The schematics do not provide comprehensive coverage of retaining wall arrangements; they should be used as a general guide along with local knowledge to ensure the appropriate elements are recorded for each wall.

Material Types

G.7.4 The material type code for a retaining wall is based on the material of the main structural element (element number 2 in Table G.16) and selected from Table G.15. When a retaining wall has a composite construction, e.g. soldier piles with lagging, then the primary structural form is used to define the element type, see Figure G.3.

Table G.15 – Retaining Wall Material Type Code	
Material Type	**Code**
Mass concrete	RA
Reinforced concrete	RB
Masonry	RC
Steel	RD
Timber	RE
FRP/Plastic	RF
Other	RO

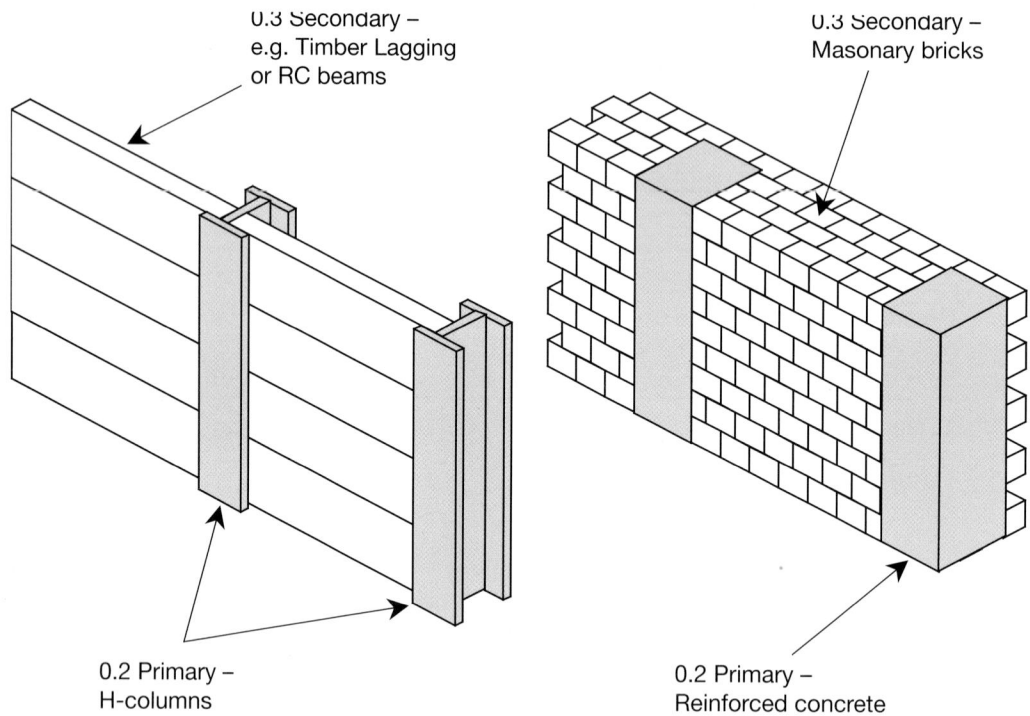

Figure G.3 – Retaining walls with primary and secondary elements

Inspection Elements

G.7.5   The inspection elements on a retaining wall are shown in Table G.16.

No	Element		Comment
	Table G.16 – Retaining Wall Inspection Elements		
No	**Element**		**Comment**
1	Foundations		Assessed by signs of distress on retaining wall
2	Retaining wall	Primary	See Figure G.3
3		Secondary	See Figure G.3
4	Parapet beam/plinth		Longitudinal beam/plinth on top of wall to support parapet/handrail
5	Drainage		Weep holes, back of wall drainage, drainage of supported material
6	Movement/Expansion joints		Normally non critical for retaining walls
7	Surface finishes: wall		e.g. painting, cladding, tiles
8	Surface finishes: handrail/parapet		e.g. painting, cladding, tiles
9	Handrail/Parapets/Safety fences		Along top of retaining wall (not foot of wall)
10	Carriageway	Top of wall	Defects may indicate movement or instability
11		Foot of wall	Defects may indicate movement or instability
12	Footway/Verge	Top of wall	Defects may indicate movement or instability
13		Foot of wall	Defects may indicate movement or instability
14	Embankment	Top of wall	Defects may indicate movement or instability
15		Foot of wall	Defects may indicate movement or instability
16	Invert/River bed		If watercourse is alongside the wall
17	Aprons		If watercourse is alongside the wall
18	Signs		Attached to the retaining wall
19	Lighting		Attached to the retaining wall
20	Services		Attached to the retaining wall

Retaining Wall Inspection Pro-Forma

G.7.6 The retaining wall inspection pro-forma retains the general layout of the bridge inspection pro-forma. The retaining wall inspection pro-forma is shown overleaf. The Retaining Wall length and retained height (average and maximum) are required.

Retaining Wall Inspection Pro Forma

Version: July 2004

☐ Superficial	☐ General	☐ Principal	☐ Special	**Form _____ of _____ for this wall**

Inspector:	Date:	Next Inspection Type/Date:

Wall Name:	Wall Ref/No:	Road Ref/No:

District:	Map Ref:	O.S.E	O.S.N

Panel of	Retained Height (m):	Max:	Ave:	Wall/Panel Length (m):	**Retaining Wall Code**

All above ground elements inspected: YES ☐ NO ☐	Photographs? YES ☐ NO ☐	Structural form Table G.14
Number of construction forms in wall/panel* length: 1 ☐ 2 ☐ 3 ☐ more ☐ (*delete as appropriate)		Material Table G.15

Set	No	Element Description		S	Ex	Def	W	P	Cost	Comments/Remarks
Main Elements	1	Foundations								
	2	Retaining Wall	Primary							
	3		Secondary							
	4	Parapet beam/plinth								
Durability Elements	5	Drainage								
	6	Movement/Expansion Joints								
	7	Surface finishes: wall								
	8	Surface finishes: handrail/parapet								
Safety Elements	9	Handrail/parapets/safety fences								
	10	Carriageway	Top of Wall							
	11		Foot of Wall							
	12	Footway/verge	Top of Wall							
	13		Foot of Wall							
Other Elements	14	Embankment	Top of Wall							
	15		Foot of Wall							
	16	Invert/river bed								
	17	Aprons								
Ancillary Elements	18	Signs								
	19	Lighting								
	20	Services								
	21									
	22									
	23									
	24									

Defect sketches:

S – severity, **Ex** – extent, **Def** – defect, **W** – work required, **P** – work priority, **Cost** – Cost of work

Element No.	Defect 1			Defect 2			Defect 3			Comments
	S	Ex	Def	S	Ex	Def	S	Ex	Def	

MULTIPLE DEFECTS

INSPECTOR'S COMMENTS

Name:	Signed:	Date:

ENGINEER'S COMMENTS

Name:	Signed:	Date:

WORK REQUIRED

Ref. No	Suggested Remedial Work	Priority	Estimated Cost	Action/Work Ordered?

Name:	Signed:	Date:

G.8 INSPECTION REPORTING FOR SIGN/SIGNAL GANTRIES

General

G.8.1 Sign/signal gantries do not represent a significant proportion of the highway structure asset; however they are sufficiently common and unique to merit separate inspection guidance from bridges and retaining walls. The guidance provided for sign/signal gantries aims to promote a consistent approach and covers topics such as structural forms (paragraph G.8.2-G.8.3), material types (paragraph G.8.4), inspection elements (paragraph G.8.5); and Signal Gantry inspection pro-forma (paragraph G.8.6).

Structural Forms

G.8.2 The structural forms for sign/signal gantries are show in Table G.17 along with the associated code.

Table G.17 – Sign/Signal Gantry Structural Form Code		
Structural Form		**Code**
Spanning carriageway	Truss	S1
	Beam	S2
Cantilever	Truss	S3
	Beam	S4
Other		S5

G.8.3 Schematics of the different sign/signal gantry structural forms are contained in Volume1: Part B: Section 3.4. The schematics do not provide comprehensive coverage of sign/signal gantry arrangements; they should be used as a general guide along with local knowledge to ensure the appropriate elements are recorded for each gantry.

Material Types

G.8.4 The material type code for a sign/signal gantry is based on the main structural elements (element number 2 from Table G.19) and selected from Table G.18.

Table G.18 – Sign/Signal Gantry Material Type Code	
Material Type	**Code**
Steel	SA
Aluminium	SB
Reinforced concrete	SC
Prestressed concrete	SD
FRP/Plastic	SE
Other	SF

Inspection Elements

G.8.5   The inspection elements on a sign/signal gantry are shown in Table G.19.

Table G.19 – Sign/Signal Gantry Inspection Elements		
No.	**Element**	**Comments**
1.	Foundations	Assessed by signs of distress on superstructure
2	Truss/Beams/Cantilever	See Volume 1:Part B: Section 3: Figure B.31 & B.32
3	Transverse members	See Volume 1:Part B: Section 3: Figure B.33
4	Columns/Supports/legs	See Volume 1:Part B: Section 3: Figure B.31, B.32 & B.33
5	Surface finishes: truss/beam/cantilever	e.g. painting, cladding, tiles
6	Surface finishes: column/support	e.g. painting, cladding, tiles
7	Surface finishes: other elements	For example elements 8, 9, 10 and 13
8	Access walkway/Deck	The elements that support personnel on the gantry
9	Access ladder	-
10	Handrails	Handrail on walkway
11	Base connections	Connection between the leg and the foundations, see Volume 1:Part B: Section 3: Figure B.31 & B.32
12	Support to longitudinal connection	The connection between the support and the longitudinal element, see Volume 1:Part B: Section 3: Figure B.31 & B.32
13	Sign and signal supports	The structural components that support the signs and signals, see Volume 1:Part B: Section 3: Figure B.33
14	Signs/Signals	Attached to the gantry
15	Lighting	Attached to the gantry
16	Services	Attached to the gantry

Sign/Signal Gantry Inspection Pro-Forma

G.8.6   The sign/signal gantry inspection pro-forma retains the general layout of the bridge inspection pro-forma. The sign/signal gantry inspection pro-forma is shown overleaf. Where length and height are defined as:

- Length = distance from centreline to centreline of supports (m).

- Height = minimum distance from road surface to underside of gantry (m).

Sign/Signal Gantry Inspection Pro Forma

Version: July 2004

☐ Superficial	☐ General	☐ Principal	☐ Special	**Form** _____ **of** _____ **for this gantry**

Inspector:	Date:	Next Inspection Type/Date:

Gantry Name:	Gantry Ref/No:	Road Ref/No:

District:	Map Ref:	O.S.E	O.S.N

Span _____ of _____	Height (m):	Length (m):	**Sign/Signal Gantry Code**

All above ground elements inspected: YES ☐ NO ☐	Photographs? YES ☐ NO ☐	Structural form _____ Table G.17

Access Information: Access Ladder/s YES ☐ NO ☐	Machine Aided Access YES ☐ NO ☐	Material _____ Table G.18

Set	No	Element Description	S	Ex	Def	W	P	Cost	Comments/Remarks
Load Bearing Elements	1	Foundations							
	2	Truss/beams/cantilever							
	3	Transverse members							
	4	Columns/supports/legs							
Durability Elements	5	Surface finishes: truss/beams/cant.							
	6	Surface finishes: columns/supports							
	7	Surface finishes: other elements							
Access	8	Access walkway/deck							
	9	Access ladder							
	10	Handrails							
Other	11	Base connections							
	12	Support to longitudinal connection							
	13	Sign and signal supports							
Ancillary	14	Signs/Signals							
	15	Lighting							
	16	Services							
	17								
	18								
	19								
	20								

Defect Sketches:

S – severity, **Ex** – extent, **Def** – defect, **W** – work required, **P** – work priority, **Cost** – Cost of work

Element No.	Defect 1			Defect 2			Defect 3			Comments
MULTIPLE DEFECTS										
	S	Ex	Def	S	Ex	Def	S	Ex	Def	

INSPECTOR'S COMMENTS

Name:	Signed:	Date:

ENGINEER'S COMMENTS

Name:	Signed:	Date:

WORK REQUIRED

Ref. No	Suggested Remedial Work	Priority	Estimated Cost	Action/Work Ordered?

Name:	Signed:	Date:

G.9 ALTERNATIVE BRIDGE INSPECTION PRO-FORMA

G.9.1 Some authorities may use bridge inspection pro-forma alternative to that described in paragraph G.2; one such pro-forma is illustrated below. Inspectors should always give due consideration to the needs and data recording requirements of each authority prior to any inspection being undertaken.

Trunk Road/Motorway Structure Inspection Report — BE11/07

Structure No. ☐☐☐☐☐☐☐☐☐☐   Grid Ref ☐☐☐☐☐☐ ☐☐☐☐☐

Agent Code ☐   Agent Name ☐

Structure Name ☐   From Span ☐☐   To Span ☐☐

Date of Inspection ☐☐☐☐☐☐☐☐☐ (e.g. 0 1 J U N 2 0 0 6)   Inspected By ☐

Type of Inspection* G ☐ P ☐ S ☐   Overall Assessment* G ☐ F ☐ P ☐

*Please tick

Defect Assessment	Estimated Cost (£)	Extent	Severity	Work	Priority	PD	Comments
Foundations							
2. Inverts and Aprons							
3. Fenders							
4. Piers and Columns							
5. Abutments							
6. Wing Walls							
7. Retaining Walls and Revetments							
8. Approach Embankments							
9. Bearings							
10. Main Beams / Tunnel Portals / Mast							
11. Transverse Beams / Catenary Cables							
12. Diaphragms or Bracings							
13. Concrete Slab							
14. Metal Deck Plates / Tunnel Linings							
15. Jack Arches							
16. Arch Ring / Corrugated Metal							
17. Spandrels							
18. Tie Rods							
19. Drainage Systems							
20. Waterproofing							
21. Surfacing							
22. Service Ducts							
23. Expansion Joints							
24. Parapets / Handrails							
25. Access Gantries or Walkways							
26. Machinery							
32. Dry Stone Walls							
33. Troughing							

Was the remedial work recommended at previous inspection satisfactorily completed? Please tick.   YES ☐   NO ☐

If 'NO' please comment and indicate any remedial work recommended and priority ☐

Reasons for Priority Allocation ☐

Signed ☐   Name ☐   Date ☐

Appendix H
Diving Report

HIGHWAYS AGENCY	**Trunk Road / Motorway Structure Diving Report**	DF/94

Structure Key `9 5 0 0 0`   Structure No. `_ _ / A 5 0 5 3 _ _ _ / _ / _ / 1 8 7 · 6 / _ / _ _`

Structure Name `DOGS LANE BURN BRIDGE`   Agent Name `P. WALTERS & PARTNERS`

Date of Inspection `2 5 J U N 1 9 9 2`
(eg 0 1 J U N 1 9 9 2)
Date of Previous Inspection `2 5 J U N 1 9 8 9`
Date of Next Inspection `2 5 J U N 1 9 9 4`

Diving Inspection	Notes	Sketch/Reference
1. Water depth	2.5m next Pier	
2. Scour holes (location/size)	3 No. (i) N. Abt. [W. Cnr.] (ii) C/R. PIER (W. Cnr.) (iii) S. Abt. (W. Cnr.) Under W. edges (N & S Abt.) and C/R PIER	(i) 4m x 2m x 2m. (ii) 4m x 2m x 2·5m (iii) 4m x 2m x 2m See Sketch No 2 : General Bed level reductions see sketch No. 1: L cal scour at W. ends, c/R. Pier abutment
3. Substructure (construction)	R.C. Spread footings (N & S Abts.) See original (& c/R PIER) Drgs.	See original Drgs. of plinths to c/R. pier
4. Substructure (condition) (G) Good (F) Fair (P) Poor	F - W. edges of N. & S Abts. roughened	
5. Masonry substructures (Joint sizes/condition)	Not Applicable	
6. Concrete (spalled/damaged)	W. Edges N. & S. Abts. damaged (possibly by debris in storm)	See sketch No. 2 (Section A - A)
7. Level grid m. c/s (10m upstream/downstream)	General Scour at Bridge has reduced bed level by appx. 1m	See section A-A (sketch No.2) Bed levels from soundings grid See sections B-B, C-C (sketch No.2) shown elsewhere

Please turn over

Diving Inspection (cont)		DF/94

	Notes	Sketch/Reference
8. Foundation (undercutting)	Undercutting to W. Corner N. & S. Abts.	Local scour holes - Plan: Sketch No.1
9. Bed material (debris present)	Coarse gravel / no debris	
10. Member sizes	c/R pier base : 3m wide x 1m deep Plinths to c/R pier : 1.5m wide x 2m deep	
11. Piles - length exposed (below water line)	Not applicable	

Signed *P Cousteau*

Name `P. Cousteau`
Date `25 JUNE 1992`

Inspection Width

Bore Hole No. 1

A B C

10.0m

North Abutment

Top of Bank

N

Scour Hole No. 1

Ç

Bridge

Scour Hole No. 2

Centre Pier

Stream Flow

Typical Grid Soundings – Referenced to Adjacent Structure

2m 2m

2m

2m

Scour Hole No. 3

45°

45°

South Abutment

Top of Bank

Bore Hole No. 2

A B C

10.0m

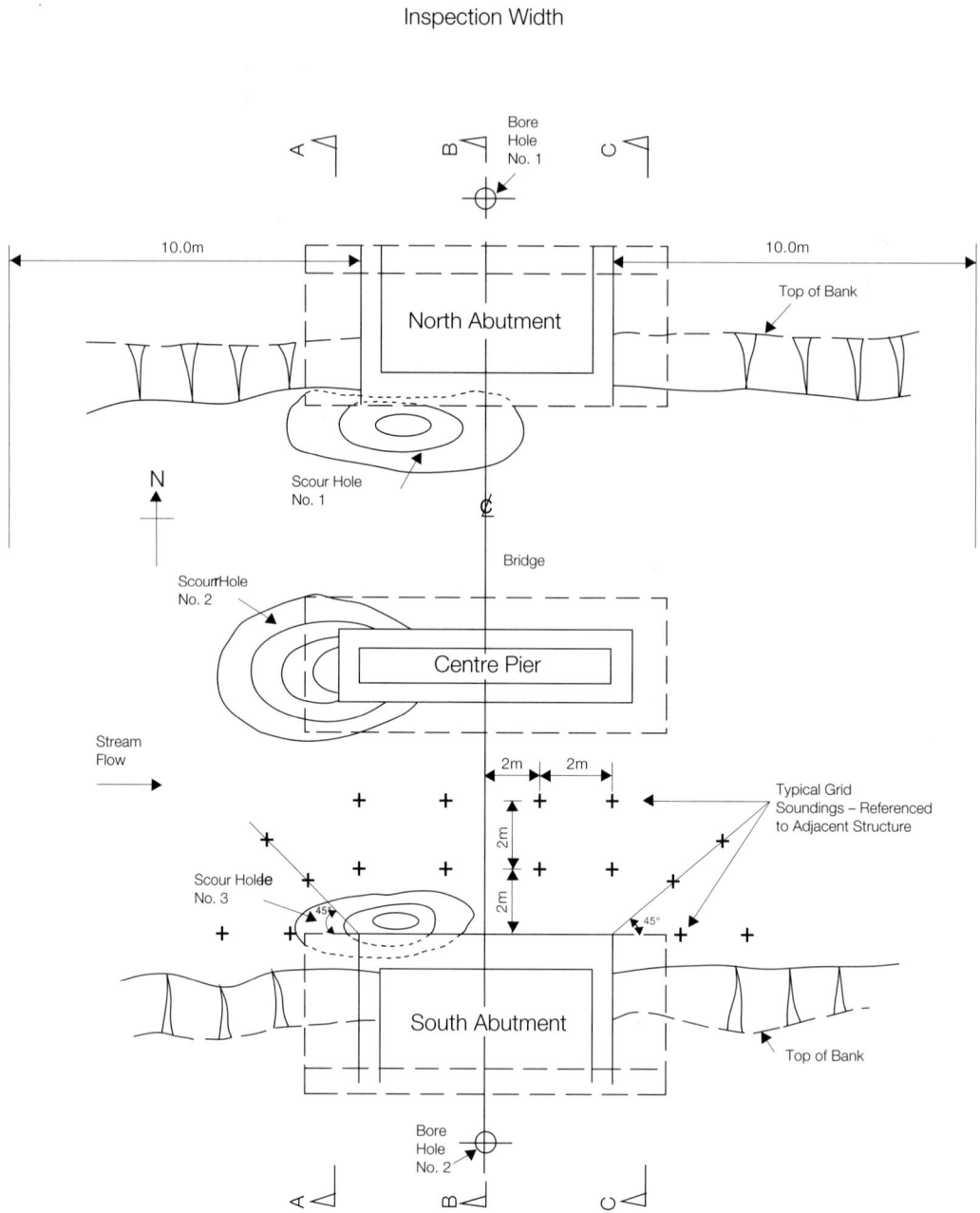

Plan – Dogs Lane Burn Bridge

Sketch No.1 Diving Inspection

Dogs Lane Burn Bridge

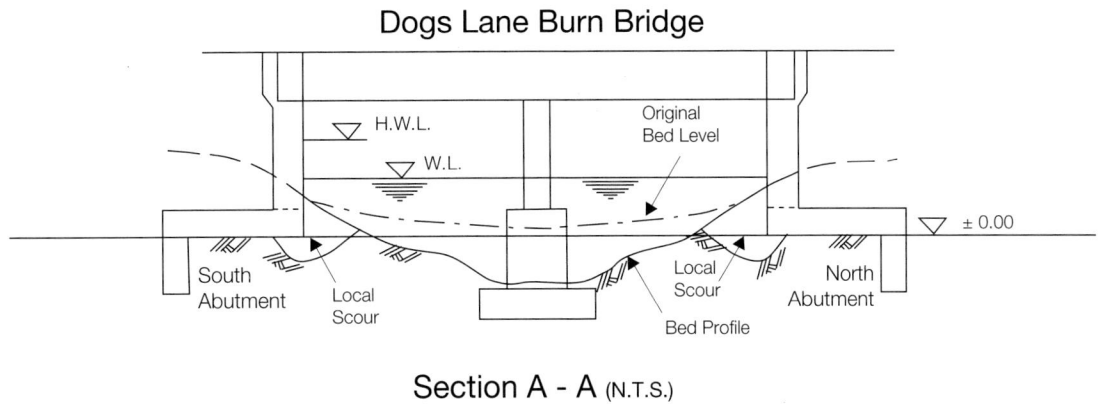

Section A - A (N.T.S.)

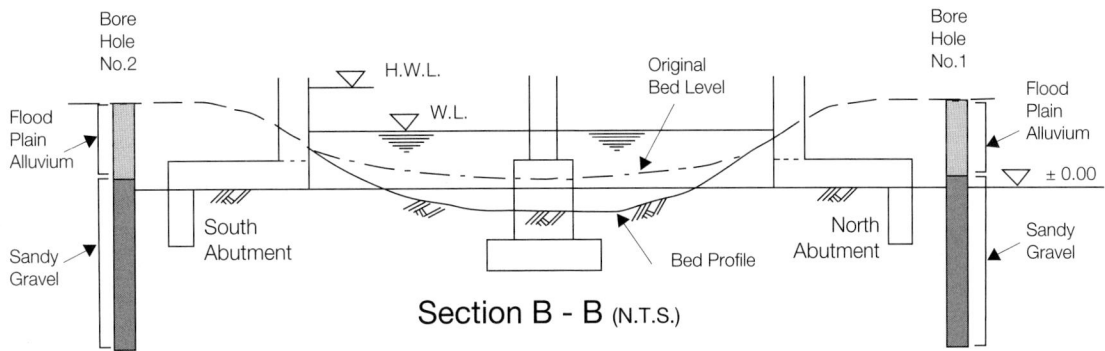

Section B - B (N.T.S.)

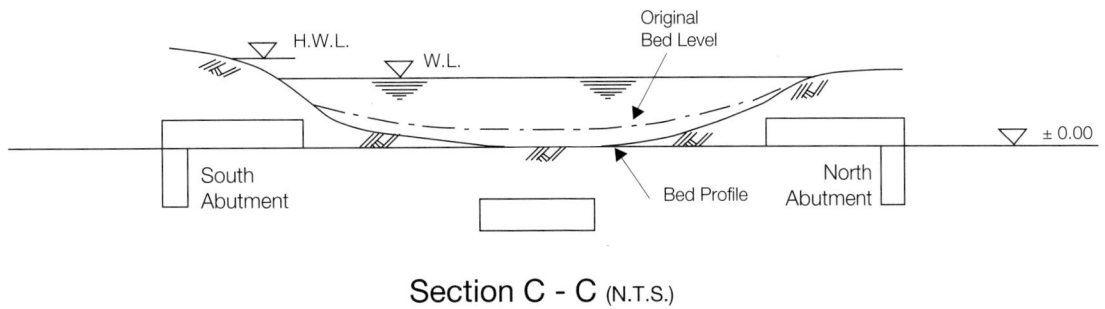

Section C - C (N.T.S.)

Sketch No.2 Diving Inspection

References for Part F

1. *Wildlife and Countryside Act 1981*, HMSO (Amended SI 2004, No.1487).

2. *The Protection of Badgers Act 1992*, HMSO.

3. *Wildlife (Northern Ireland) Order 1985 (SI 1985, No. 171 (NI 2))*, HMSO (Amended SI 1995, No. 761 (NI6)).

4. *The Conservation of Bats in Bridges Project: A Report on the Survey and Conservation of Bat Roosts in Bridges in Cumbria*, Billington GE & Norman GM, English Nature, Peterborough, 1997.

5. *Appraisal of Existing Iron and Steel Structures*, Bussell M, Steel Construction Institute, 1997.

6. *EC Directive 79/409/EEC: Council Directive of 2 April 1979 on the Conservation of Wild Birds*, Office for Official Publications of the European Communities, 2004.

7. *Bridge Condition Indicators: Volume 2: Bridge Inspection Reporting: Guidance Note on Evaluation of Bridge Condition Indicators*, CSS Bridges Group, 2002.

8. *Bridge Condition Indicators: Addendum to Volume 2: Bridge Inspection Reporting: Guidance Note on Evaluation of Bridge Condition Indicators*, CSS Bridges Group, 2004.

Acknowledgements

Project Sponsor

> Highways Agency

Technical Project Board Members

Brian Hill	Highways Agency
Brian Alison	Highways Agency
Martin Potts	Highways Agency
Alex Gardner	Transport Scotland
Martin Jackson	Herefordshire County Council
Richard McFarlane	Royal Borough of Kingston
Tudor Roberts	Transport Wales
Ronnie Wilson	Department for Regional Development Northern Ireland Roads Service

Project Team

Dr Vicky Vassou	Atkins
Dr Garry Sterritt	Atkins
Dr Roger Cole	Atkins
Iain Kennedy-Reid	Atkins
Dave Black	Atkins
Dr Navil Shetty	Atkins

Technical Advice and Assistance

Acknowledgement is due to the wide range of bridge managers, engineers, technicians and inspectors who have assisted in the development of the Manual through, previous work, attending workshops, reviewing drafts, providing material and photographs, and general support.

Disclaimer

The Highways Agency, the Technical Project Board, the Project Team and the Technical Advisors who produced this Document have endeavoured to ensure the accuracy of the contents. However, the guidance, recommendations and information given should always be reviewed by those using them in the light of the facts of their particular case and specialist advice be obtained as necessary. No liability for loss or damage that may be suffered by any person or organisation as a result of the use of any of the information contained here, or as the result of any errors or omissions in the information contained here, is accepted by the Highways Agency, the Technical Project Board, the Project Team, the Technical Advisors, and any agents or publishers working on their behalf.